INTRODUCTION TO THE HIGH-TEMPERATURE OXIDATION OF METALS

A straightfoward treatment describing the oxidation processes of metals and alloys at elevated temperatures. This new edition retains the fundamental theory but incorporates advances made in understanding degradation phenomena. Oxidation processes in complex systems are dicussed, from reactions in mixed environments to protective techniques, including coatings and atmosphere control. The authors provide a logical and expert treatment of the subject, producing a revised book that will be of use to students studying degradation of high-temperature materials and an essential guide to researchers requiring an understanding of this elementary process.

NEIL BIRKS was Professor Emeritus in the Department of Materials Science and Engineering at the University of Pittsburgh.

GERALD H. MEIER is William Kepler Whiteford Professor in the Department of Materials Science and Engineering at the University of Pittsburgh.

FRED S. PETTIT is Harry S. Tack Professor in the Department of Materials Science and Engineering at the University of Pittsburgh.

INTRODUCTION TO THE HIGH-TEMPERATURE OXIDATION OF METALS

2nd Edition

NEIL BIRKS
Formerly of University of Pittsburgh

GERALD H. MEIER
University of Pittsburgh

FRED S. PETTIT
University of Pittsburgh

CAMBRIDGE
UNIVERSITY PRESS

CAMBRIDGE UNIVERSITY PRESS
Cambridge, New York, Melbourne, Madrid, Cape Town, Singapore, São Paulo, Delhi

Cambridge University Press
The Edinburgh Building, Cambridge CB2 8RU, UK

Published in the United States of America by Cambridge University Press, New York

www.cambridge.org
Information on this title: www.cambridge.org/9780521485173

First edition © Edward Arnold Publishers, Ltd 1983
Second edition © Cambridge University Press 2006

First published 1983
Second edition published 2006
This digitally printed version 2009

A catalogue record for this publication is available from the British Library

ISBN 978-0-521-48042-0 hardback
ISBN 978-0-521-48517-3 paperback

Professor Neil Birks

This book is dedicated to one of its coauthors, Professor Neil Birks, who passed away during the preparation of the second edition. Neil was an accomplished researcher and educator in a number of fields including high-temperature oxidation, corrosion, erosion, and process metallurgy. He was also a good friend.

Neil's legacy to science and engineering is well established in his scholarly publications and the numerous students he mentored. It is our hope that this book will complete that legacy.

GHM
FSP

Contents

Acknowledgements

The authors gratefully acknowledge the scientific contributions of former and current students. Drs. J. M. Rakowski, M. J. Stiger, N. M. Yanar, and M. C. Maris-Sida are thanked for their assistance in preparing figures for this book.

The authors also greatly appreciate the helpful comments made by Professor J. L. Beuth (Carnegie Mellon University), Professor H. J. Grabke (Max-Planck Institüt für Eisenforschung), and Professor R. A. Rapp (Ohio State University) on parts of the manuscript.

Preface

Few metals, particularly those in common technological applications, are stable when exposed to the atmosphere at both high and low temperatures. Consequently, most metals in service today are subject to deterioration either by corrosion at room temperature or by oxidation at high temperature. The degree of corrosion varies widely. Some metals, such as iron, will rust and oxidize very rapidly whereas other metals, such as nickel and chromium, are attacked relatively slowly. It will be seen that the nature of the surface layers produced on the metal plays a major role in the behaviour of these materials in aggressive atmospheres.

The subject of high-temperature oxidation of metals is capable of extensive investigation and theoretical treatment. It is normally found to be a very satisfying subject to study. The theoretical treatment covers a wide range of metallurgical, chemical, and physical principles and can be approached by people of a wide range of disciplines who, therefore, complement each other's efforts.

Initially, the subject was studied with the broad aim of preventing the deterioration of metals in service, i.e., as a result of exposing the metal to high temperatures and oxidizing atmospheres. In recent years, a wealth of mechanistic data has become available. These data cover a broad range of phenomena, e.g., mass transport through oxide scales, evaporation of oxide or metallic species, the role of mechanical stress in oxidation, growth of scales in complex environments containing more than one oxidant, and the important relationships between alloy composition and microstructure and oxidation. Such information is obtained by applying virtually every physical and chemical investigative technique to the subject.

In this book the intention is to introduce the subject of high-temperature oxidation of metals to students and to professional engineers whose work demands familiarity with the subject. The emphasis of the book is placed firmly on supplying an understanding of the basic, or fundamental, processes involved in oxidation.

In order to keep to this objective, there has been no attempt to provide an exhaustive, or even extensive, review of the literature. In our opinion this would increase the factual content without necessarily improving the understanding of the subject and would, therefore, increase both the size and price of the book without enhancing its objective as an introduction to the subject. Extensive literature quotation is already available in books previously published on the subject and in review articles. Similarly the treatment of techniques of investigation has been restricted to a level that is sufficient for the reader to understand how the subject is studied without involving an overabundance of experimental details. Such details are available elsewhere as indicated.

After dealing with the classical situations involving the straightforward oxidation of metals and alloys in the first five chapters, the final chapters extend the discussion to reactions in mixed environments, i.e., containing more than one oxidant, to reactions involving a condensed phase as in hot corrosion, and the added complications caused by erosive particles. Finally, some typical coatings for high-temperature applications and the use of protective atmospheres during processing are described.

Pittsburgh   GHM

2005   FSP

Introduction

The primary purpose of this book is to present an introduction to the fundamental principles that govern the interaction of reactive gaseous environments (usually containing oxygen as a component) and solid materials, usually metals, at high temperatures. These principles are applicable to a variety of applications, which can include those where oxidation is desirable, such as forming a resistive silica layer on silicon-based semiconductors or removing surface defects from steel slabs during processing by rapid surface oxidation. However, most applications deal with situations where reaction of the component with the gaseous atmosphere is undesirable and one tries to minimize the rate at which such reactions occur.

The term 'high-temperature' requires definition. In contrast to aqueous corrosion, the temperatures considered in this book will always be high enough that water, when present in the systems, will be present as the vapour rather than the liquid. Moreover, when exposed to oxidizing conditions at temperatures between 100 and 500 °C, most metals and alloys form thin corrosion products that grow very slowly and require transmission electron microscopy for detailed characterization. While some principles discussed in this book may be applicable to thin films, 'high temperature' is considered to be 500 °C and above.

In designing alloys for use at elevated temperatures, the alloys must not only be as resistant as possible to the effects produced by reaction with oxygen, but resistance to attack by other oxidants in the environment is also necessary. In addition, the environment is not always only a gas since, in practice, the deposition of ash on the alloys is not uncommon. It is, therefore, more realistic in these cases to speak of the high-temperature corrosion resistance of materials rather than their oxidation resistance.

The rate at which the reactions occur is governed by the nature of the reaction product which forms. In the case of materials such as carbon the reaction product is gaseous (CO and CO_2), and does not provide a barrier to continued reaction. The materials that are designed for high-temperature use are protected by the formation of a solid reaction product (usually an oxide) which separates the component and atmosphere. The rate of further reaction is controlled by transport of reactants through this solid layer. The materials designed for use at the highest temperatures are ones which form the oxides with the slowest transport rates for reactants (usually α-Al_2O_3 or SiO_2), i.e., those with the slowest growth rates. However, other materials are often used at lower temperatures if their oxides have growth rates which are 'slow enough' because they may have better mechanical properties (strength, creep resistance), may be easier to fabricate into components (good formability/weldability), or are less expensive.

In some cases, the barriers necessary to develop the desired resistance to corrosion can be formed on structural alloys by appropriate composition modification. In many practical applications for structural alloys, however, the required compositional changes are not compatible with the required physical properties of the alloys. In such cases, the necessary compositional modifications are developed through the use of coatings on the surfaces of the structural alloys and the desired reaction-product barriers are developed on the surfaces of the coatings.

A rough hierarchy of common engineering alloys with respect to use temperature would include the following.

- Low-alloy steels, which form M_3O_4 (M = Fe, Cr) surface layers, are used to temperatures of about 500 °C.
- Titanium-base alloys, which form TiO_2, are used to about 600 °C.
- Ferritic stainless steels, which form Cr_2O_3 surface layers, are used to about 650 °C. This temperature limit is based on creep properties rather than oxidation rate.
- Austenitic Fe–Ni–Cr alloys, which form Cr_2O_3 surface layers and have higher creep strength than ferritic alloys, are used to about 850 °C.
- Austenitic Ni–Cr alloys, which form Cr_2O_3 surface layers, are used to about 950 °C, which is the upper limit for oxidation protection by chromia formation.
- Austenitic Ni–Cr–Al alloys, and aluminide and MCrAlY (M = Ni, Co, or Fe) coatings, which form Al_2O_3 surface layers, are used to about 1100 °C.
- Applications above 1100 °C require the use of ceramics or refractory metals. The latter alloys oxidize catastrophically and must be coated with a more oxidation-resistant material, which usually forms SiO_2.

The exercise of 'alloy selection' for a given application takes all of the above factors into account. While other properties are mentioned from time to time, the emphasis of this book is on oxidation and corrosion behaviour.

1

Methods of investigation

The investigation of high-temperature oxidation takes many forms. Usually one is interested in the oxidation kinetics. Additionally, one is also interested in the nature of the oxidation process, i.e., the oxidation mechanism. Figure 1.1 is a simple schematic of the cross-section of an oxide formed on the surface of a metal or alloy. Mechanistic studies generally require careful examination of the reaction products formed with regard to their composition and morphology and often require examination of the metal or alloy substrate as well. Subsequent sections of this chapter will deal with the common techniques for measuring oxidation kinetics and examining reaction-product morphologies.

In measuring the kinetics of degradation and characterizing the corresponding microstructures questions arise as to the conditions to be used. Test conditions should be the same as the application under consideration. Unfortunately, the application conditions are often not precisely known and, even when known, can be extremely difficult to establish as a controlled test. Moreover, true simulation testing is usually impractical because the desired performance period is generally much longer than the length of time for which laboratory testing is feasible. The answer to this is accelerated, simulation testing.

Accelerated, simulation testing requires knowledge of microstructure and morphological changes. All materials used in engineering applications exhibit a microstructural evolution, beginning during fabrication and ending upon termination of their useful lives. In an accelerated test one must select test conditions that cause the microstructures to develop that are representative of the application, but in a much shorter time period. In order to use this approach some knowledge of the degradation process is necessary.

Measurements of reaction kinetics

In the cases of laboratory studies, the experimental technique is basically simple. The specimen is placed in a furnace, controlled at the required temperature, and

1

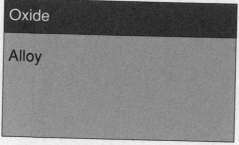

Figure 1.1 Schematic diagram of the cross-section of an oxide layer formed on the surface of a metal or alloy.

allowed to react for the appropriate time. The specimen is then removed, allowed to cool, and examined.

Although this procedure is simple, one drawback is that the start time for the reaction cannot be accurately established. Several starting procedures are commonly used.

(1) The specimen may simply be placed in the heated chamber containing the reactive atmosphere.
(2) The specimen may be placed in the cold chamber containing the atmosphere and then heated.
(3) The specimen may be placed in the cold chamber, which is then evacuated or flushed with inert gas, heated and then, at temperature, the reactive gas is admitted.

In all cases the start of the reaction is in doubt either because of the time required to heat the specimen or the inevitable formation of thin oxide layers, even in inert gases or under vacuum. This is true especially in the case of more reactive metals, so that when the reaction is started by admitting the reactive gas an oxide layer already exists.

Attempts have been made to overcome this by heating initially in hydrogen which is then flushed out by the reactive gas. This also takes a finite time and thus introduces uncertainty concerning the start of the reaction.

Thin specimens may be used to minimize the time required to heat the specimen. In this case care should be taken that they are not so thin and, therefore, of such low thermal mass that the heat of reaction, released rapidly during the initial oxidation period, causes severe specimen overheating.

The uncertainty concerning the start of the reactions usually only affects results for short exposure times up to about ten minutes and becomes less noticeable at longer times. However, in some cases such as selective oxidation of one element from an alloy, these effects can be quite long-lasting. In practice specimens and procedures must be designed with these factors in mind.

Many early investigations were simply concerned with oxidation rates and not with oxidation mechanisms. The rate of formation of an oxide in a metal according to the reaction (1.1),

$$2M + O_2 = 2MO, \tag{1.1}$$

can be investigated by several methods. The extent of the reaction may be measured by the following.

(1) The amount of metal consumed.
(2) The amount of oxygen consumed.
(3) The amount of oxide produced.

Of these only (2) can be assessed continuously and directly.

(1) *The amount of metal consumed*
   In practice this may be assessed by observing (a) the weight loss of the specimen or (b) the residual metal thickness. In both cases the specimen must be removed from the furnace, thus interrupting the process.
(2) *The amount of oxygen consumed*
   This may be assessed by observing either the weight gain or the amount of oxygen used. Both of these methods may be used on a continuous and automatic recording basis.
(3) *The amount of oxide produced*
   This may be assessed by observing the weight of oxide formed or by measuring the oxide thickness. Of course, in the latter case it is necessary to destroy the specimen, as it is with method (1).

Of the above methods, only those involving measurement of weight gain and oxygen consumption give the possibility of obtaining continuous results. The other methods require destruction of the specimen before the measurement can be achieved and this has the drawback that, in order to obtain a set of kinetic data, it is necessary to use several specimens. Where the specimen and the methods of investigation are such that continuous results can be obtained, one specimen will give the complete kinetic record of the reaction.

When representing oxidation kinetics, any of the variables mentioned above can be used, and can be measured as a function of time because, of course, they all result in an assessment of the extent of reaction. Nowadays it is most general to measure the change in mass of a specimen exposed to oxidizing conditions.

It is found experimentally that several rate laws can be identified. The principal laws are (1) linear law, (2) parabolic law, and (3) logarithmic law.

(1) *The linear law*, for which the rate of reaction is independent of time, is found to refer predominantly to reactions whose rate is controlled by a surface-reaction step or by diffusion through the gas phase.
(2) *The parabolic law*, for which the rate is inversely proportional to the square root of time, is found to be obeyed when diffusion through the scale is the rate-determining process.
(3) *The logarithmic law* is only observed for the formation of very thin films of oxide, i.e., between 2 and 4 nm, and is generally associated with low temperatures.

Under certain conditions some systems might even show composite kinetics, for instance niobium oxidizing in air at about 1000 °C initially conforms to the parabolic law but later becomes linear, i.e., the rate becomes constant at long times.

Discontinuous methods of assessment of reaction kinetics

In this case the specimen is weighed and measured and is then exposed to the conditions of high-temperature oxidation for a given time, removed, and reweighed. The oxide scale may also be stripped from the surface of the specimen, which is then weighed. Assessment of the extent of reaction may be carried out quite simply either by noting the mass gain of the oxidized specimen, which is the mass of oxygen taken into the scale, or the mass loss of the stripped specimen, which is equivalent to the amount of metal taken up in scale formation. Alternatively, the changes in specimen dimensions may be measured. As mentioned before these techniques yield only one point per specimen with the disadvantages that (a) many specimens are needed to plot fully the kinetics of the reaction, (b) the results from each specimen may not be equivalent because of experimental variations, and (c) the progress of the reaction between the points is not observed. On the other hand they have the obvious advantage that the techniques, and the apparatus required, are extremely simple. In addition, metallographic information is obtained for each data point.

Continuous methods of assessment

These methods fall into two types, those which monitor mass gain and those which monitor gas consumption.

Mass-gain methods

The simplest method of continuous monitoring is to use a spring balance. In this case the specimen is suspended from a sensitive spring whose extension is measured, using a cathetometer, as the specimen mass changes as a result of oxidation, thus

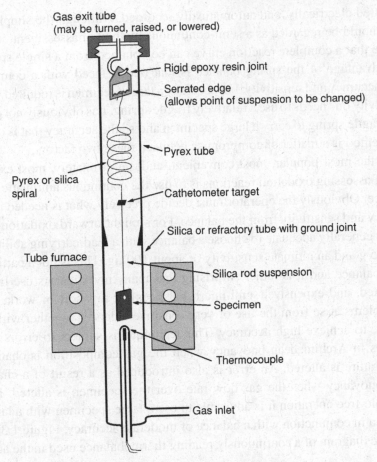

Figure 1.2 Features of a simple spring balance (for advanced design see ref. 1).

giving a semi-continuous monitoring of the reaction. Apparatus suitable for this is shown diagrammatically in Figure 1.2 which is self-explanatory.[1] An important feature is the design of the upper suspension point. In Figure 1.2 this is shown as a hollow glass tube which also acts as a gas outlet. The tube can be twisted, raised, or lowered to facilitate accurate placing of the specimen and alignment of the spring. A suspension piece is rigidly fastened to the glass tube and provides a serrated horizontal support for the spring whose suspension point may thus be adjusted in the horizontal plane. These refinements are required since alignment between the glass tube containing the spring and the furnace tube is never perfect and it is prudent to provide some means of adjustment of the spring position in order to ensure accurate placing of the specimen. It would be possible of course to equip a spring balance with a moving transformer, which would enable the mass gain to be measured electrically, and automatically recorded. Although the simple spring balance should be regarded as a semi-continuous method of assessment, it has the advantage that a complete reaction curve can be obtained from a single specimen. The disadvantage of the spring balance is that one is faced with a compromise between accuracy and sensitivity. For accuracy a large specimen is required whereas for sensitivity one needs to use a relatively fragile spring. It is obviously not possible to use a fragile spring to carry a large specimen and so the accuracy that is obtained by this method is a matter of compromise between these two factors.

By far the most popular, most convenient, and, unfortunately, most expensive method of assessing oxidation reactions is to use the continuous automatic recording balance. Obviously the operator must decide precisely what is needed in terms of accuracy and sensitivity from the balance. For straightforward oxidation experiments it is generally adequate to choose a balance with a load-carrying ability of up to about 25 g and an ultimate sensitivity of about 100 µg. This is not a particularly sensitive balance and it is rather surprising that many investigators use far more sophisticated, and expensive, semi-micro balances for this sort of work. In fact many problems arise from the use of very sensitive balances together with small specimens, to achieve high accuracy. This technique is subject to errors caused by changes in Archimedean buoyancy when the gas composition is changed or the temperature is altered. An error is also introduced as a result of a change in dynamic buoyancy when the gas flow rate over the specimen is altered. For the most trouble-free operation it is advisable to use a large specimen with a large flat surface area in conjunction with a balance of moderate accuracy. Figure 1.3 shows a schematic diagram of a continuously reading thermobalance used in the authors' laboratory.

Using the automatic recording balance it is possible to get a continuous record of the reaction kinetics and, in this way, many details come to light that are hidden by other methods. For instance spalling of small amounts of the oxide layer is immediately detected as the balance registers the mass loss; separation of the scale from the metal surface is recorded when the balance shows a slow reduction in rate well below rates that would be expected from the normal reaction laws; cracking of the scale is indicated by immediate small increases in the rate of mass gain. These features are shown in Figure 1.4. The correct interpretation of data yielded by the automatic recording balance requires skill and patience but, in almost every case, is well worthwhile.

Gas consumption

Continuous assessment based on measurement of the consumption of gas may be done in two ways. If the system is held at constant volume then the fall in pressure may be monitored, continuously or discontinuously. Unfortunately, allowing the

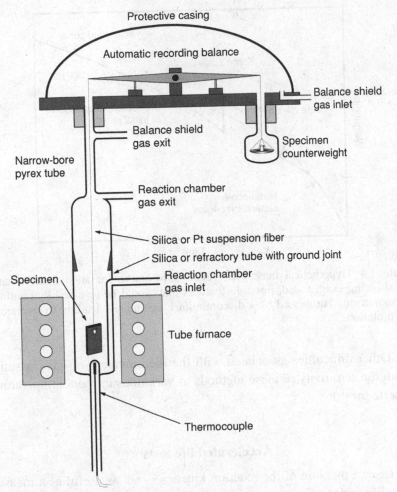

Figure 1.3 Typical experimental arrangement for measuring oxidation kinetics with an automatic recording balance.

pressure of the oxidant to vary appreciably during the reaction is almost certain to cause changes in the rate at which the reaction proceeds. The second method is to maintain the apparatus at constant pressure and to measure the fall in volume as oxidizing gas is consumed; this can be done automatically and the results plotted using a chart recorder. This technique is only useful when the atmosphere is a pure gas, for instance pure oxygen. In the case of a mixture of gases such as air, the fall in volume would represent the amount of oxygen consumed but, since nitrogen is not being consumed, the partial pressure of oxygen would fall, leading to possible changes in oxidation rate. This problem can, to some extent, be overcome by using a very large volume of gas compared with the amount of gas used during the reaction. In this case there remains the difficulty of measuring a small change in a large

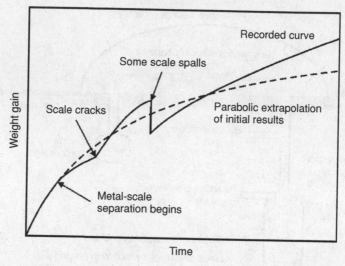

Figure 1.4 Hypothetical mass-gain versus time curve for an oxidation reaction showing exaggerated, possible features that would be revealed by continuous monitoring but missed by a discontinuous technique – leading to erroneous interpretation.

volume. Other difficulties associated with the measurement of gas consumption arise from the sensitivity of these methods to variations in room temperature and atmospheric pressure.

Accelerated life tests

In some cases a measure of the reaction kinetics is not as useful as a measure of the useful life under a given set of experimental conditions. Cyclic oxidation tests using weight-change versus time measurements and visual examination of exposed specimens can be useful to determine times beyond which protective scales, such as alumina, are no longer formed on alloy test coupons.

In the case of alloys used in electric heaters the useful life of an alloy is determined (American Society for Testing of Materials, ASTM, B76) by electrically heating a section of wire in air to a given temperature where it is held for two minutes followed by two minutes of cooling. This cycle is repeated until a 10% increase in electrical resistivity is noted and the time is reported as the useful life of the alloy.

Methods of specimen examination

The oxidation morphology may be examined by a wide range of techniques depending on the particular type of information being sought. The following procedure is recommended in most instances. Initially, the specimen should be examined by the naked eye and under a low-power binocular microscope, note being taken of whether the scale surface is flat, rippled, contains nodules, is cracked, or whether there is excessive attack on the edges or in the centers of the faces. This type of examination is important because subsequent examination generally involves microscopy of a section at a specific place and it is important to know whether this is typical or if there are variations over the specimen surface. It is often worthwhile examining the surface in the scanning electron microscope, before sectioning, and, as with all examination, if an area is seen that has significant features, photographs must be taken immediately. It is usually too late if one waits for a better opportunity. In addition, modern scanning electron microscopes allow the chemical composition of surface features to be determined. These techniques are described below.

Having examined the scale, it is generally useful to perform X-ray diffraction (XRD) on it to obtain information on the phases present. Studies using XRD can also provide auxiliary information such as the mechanical stress state in the scale, texture, etc.

Preparation of cross-section specimens

Once all the required information on the surface features has been obtained, it is generally useful to examine a cross-section of the specimen. This allows observation of the scale thickness and microstructure and any changes that have been produced in the underlying substrate.

Mounting

Metallographic examination involves polishing of a section and, under the polishing stresses, the scale is likely to fall away from the specimen. The specimen with its surrounding scale must, therefore, be mounted. By far the best type of mounting medium has been found to be the liquid epoxy resins. The process is very easy – a dish is greased, the specimen is positioned and the resin is poured in. At this stage the dish should be evacuated in a dessicator, left under vacuum for a few minutes, and then atmospheric pressure reapplied. This has the effect of removing air from crevices caused by cracks in the scale and on reapplying pressure the resin is forced into cavities left by evacuation. This sort of mounting procedure requires several hours for hardening, preferably being left in a warm place overnight, but the good support which is given to the fragile scales is well worth the extra time and effort. When emphasis must be placed on edge retention, it may be necessary to backup the specimen with metal shims prior to the mounting operation.

Polishing

It is important at this stage to realize that polishing is carried out for the oxides and not for the metal. A normal metallurgical polish will not reveal the full features of the oxide. Normally a metal is polished using successive grades of abrasive paper – the time spent on each paper being equivalent to the time required to remove the scratches left by the previous paper. If this procedure is carried out it will almost certainly produce scales which are highly porous. It is important to realize that, because of their friability, scales are damaged to a greater depth than that represented by the scratches on the metal, and, when going to a less abrasive paper, it is important to polish for correspondingly longer times in order to undercut the depth of damage caused by the previous abrasive paper. In other words, polishing should be carried out for much longer times on each paper than would be expected from examination of the metal surface. If this procedure is carried out, scales which are nicely compact and free from induced porosity will be revealed. The polished specimens may be examined using conventional optical metallography or scanning electron microscopy (SEM).

Etching

Finally, it may be found necessary to etch a specimen to reveal detail in either the metal or the oxide scale. This is done following standard metallographic preparation procedure with the exception that thorough and lengthy rinsing in a neutral solvent (alcohol) must be carried out. This is necessary because of existence of pores and cracks in the oxide, especially at the scale–metal interface, that retain the etchant by capillary action. Thus extensive soaking in rinse baths is advisable, followed by rapid drying under a hot air stream, if subsequent oozing and staining is to be avoided.

Description of specialized examination techniques

It is assumed that the reader is familiar with the techniques of optical microscopy. There are, however, a number of other specialized techniques, which are useful for examining various features of oxidation morphologies. These techniques mainly generate information from interactions between the specimen and an incident beam of electrons, photons, or ions. The basis for the various techniques will be described here. Examples of their application will be presented in subsequent chapters.

Figure 1.5 is a schematic diagram of some of the important interactions between a solid specimen and an incident beam of electrons. The incident beam, if sufficiently energetic, can knock electrons from inner shells in the atoms in the solid giving rise to characteristic X-rays from each of the elements present as higher-energy

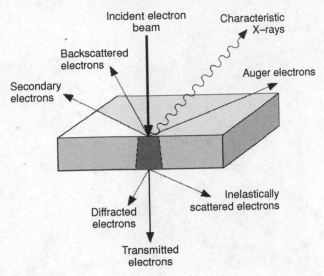

Figure 1.5 Schematic diagram of the interactions of an incident electron beam with a solid specimen.

electrons fall into the empty states. Some of the electrons will be elastically scattered in the backward direction with the number of backscattered electrons, which exit the specimen increasing as the average atomic mass of the solid increases and as the local inclination of the specimen, with respect to the incident beam, increases. Secondary electrons are the result of inelastic scattering of electrons in the conduction band of the solid, by incident or backscattered electrons, which gives them enough energy to exit the solid. Auger electrons are ejected from the solid by a process similar to that which generates characteristic X-rays, i.e., when a higher-energy electron falls into a vacated state, the energy released results in the ejection of an electron from an outer shell rather than creation of an X-ray. If the specimen is thin enough, some of the electrons will pass through the specimen. These will comprise a transmitted beam, which will contain electrons from the incident beam that have not undergone interactions or have lost energy as the result of inelastic scattering processes, and a diffracted beam of electrons, which have satisfied particular angular relations with lattice planes in the solid.

Figure 1.6 is a schematic diagram of some of the important interactions between a solid specimen and an incident X-ray beam. These will result in diffracted X-rays, which have satisfied particular angular relations with lattice planes in the solid and will, therefore, contain crystallographic information. The interactions will also result in the ejection of photoelectrons, which will have energies equal to the difference between that of the incident photon and the binding energy of the electron in the state from which it is ejected.

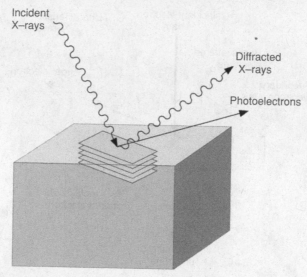

Incident
X–rays

Diffracted
X–rays

Photoelectrons

Figure 1.6 Schematic diagram of the interactions of an incident X-ray beam with a solid specimen.

The following is a brief description of the electron-optical and other specialized techniques, which may be used for examining oxidation morphologies, and the type of information which may be obtained from each technique. The reader is directed to the texts and reviews, referenced for each technique, for a more detailed description.

Scanning electron microscopy (SEM)

Scanning electron microscopy is currently the most widely used tool for characterising oxidation morphologies. In this technique[2,3] an incident electron beam is rastered over the specimen surface while the secondary electrons or backscattered electrons enter a detector creating an electrical signal which is used to modulate the intensity of a TV monitor, which is being rastered at the same rate as the electron beam. This produces an image which indicates the topography of the specimen, since the number of secondary or backscattered electrons which escape the specimen will depend on the local tilt of the specimen surface, and does so with great depth of focus. Also, information regarding the average atomic mass can be obtained if the image is formed using backscattered electrons. Modern scanning electron micrographs,[3] equipped with field-emission electron guns, can achieve magnifications as high as 100 000 × and resolutions of 1000 nm.

Since the incident electrons also produce characteristic X-rays from the elements in the solid, information can be obtained with regard to the elements present and, if suitable corrections are made, their concentration. The output from the X-ray detector can also be used to modulate the intensity of the monitor, which allows a map of the relative concentration of a given element to be produced. The ability to provide information with regard to microstructure and composition, simultaneously, makes SEM an extremely valuable tool in studying oxidation morphologies.

X-ray diffraction (XRD)

In this technique the specimen is bombarded with a focused beam of monochromatic X-rays.[4] The beam will be diffracted by the lattice planes, which satisfy Bragg's law, Equation (1.2), where λ is the wavelength of the X-rays, d_{hkl} is

$$n\lambda = 2d_{hkl} \sin \theta \tag{1.2}$$

the interplanar spacing of the lattice plane, θ is the angle between the incident beam and the lattice plane, and n is the order of the reflection.

Determination of the angles at which diffraction occurs allows determination of the spacing of the various lattice planes and, therefore, the crystal structure of the phase or phases present in the solid. Comparison of the d values with tabulated values for various substances can be used to identify the phases present in the specimen. This has resulted in XRD being used for many years to identify the phases present in oxide layers formed on metals and alloys. Modern X-ray diffractometers, with glancing-angle capabilities, can analyze oxide layers as thin as 1000 nm.

An additional useful feature of XRD is that the positions of the diffraction peaks shift if the solid is strained. Use of specialized diffraction techniques[5] allow these strains to be measured, which then allows determination of the state of stress in the oxide layer and, in some cases, the underlying metal substrate.

Transmission electron microscopy (TEM)

When higher spatial resolution examination of specimens is needed, transmission electron microscopy (TEM) is utilized.[3,6] Here a specimen thin enough to transmit the electron beam is required. This can be achieved by thinning the specimen directly or mounting it and thinning a cross-section.[7] If small precipitates are the feature of interest, they can sometimes be examined by preparing carbon extraction replicas from the specimen.

Figure 1.5 indicates that electrons, which have undergone several types of interaction with the specimen, exit the bottom side. The transmitted beam has passed through the specimen with essentially no interaction. The diffracted beams contain electrons, which have not lost energy (elastically scattered) but have satisfied angular criteria (similar to Bragg's law for X-rays) and have been scattered into a new path. The transmitted and diffracted beams can be used to form images of the specimen with very high resolution (on the order of 1 nm). Additionally the diffracted electrons can be used to form a selected area diffraction (SAD) pattern from a small region of the specimen, which yields crystallographic information in a manner analogous to XRD patterns.

Electrons which have lost energy by interaction with the specimen (inelastically scattered electrons) also exit the specimen. The amounts of energy lost by the electrons can be measured and form the basis for several types of electron energy-loss spectroscopies (EELS) which allow high-resolution chemical analysis to be performed.

Finally, characteristic X-rays are also produced in TEM, as they are in SEM, and can be analyzed to yield chemical-composition information. The X-rays are generated in a smaller specimen volume in TEM than SEM and, therefore, smaller features can be analyzed.

Surface analytical techniques

In many cases it is of interest to have information about compositions in very thin layers of materials, such as the first-formed oxide on an alloy. A group of techniques, which are highly surface sensitive, are available for such determinations.[8] Grabke et al.[9] have presented a good review of the application of these techniques in the analysis of oxidation problems.

The most widely used technique is Auger electron spectroscopy (AES). Since the Auger electrons are emitted as the result of transitions between electron levels, their energies are element specific and all elements, with the exception of H and He, can be detected.[9] Also, since the energies of Auger electrons are low (20 eV–2.5 keV) they are only emitted from shallow depths, of a few atomic layers, giving the technique its surface sensitivity. Auger spectra can be quantified using standards and published correction factors, and if the incident beam is scanned over the specimen surface, can be used to generate composition maps analogous to the X-ray maps from a scanning electron micrograph. The incident electron beam in modern instruments can be focussed to give analyses with lateral spatial resolution on the order of 50 nm.[9]

The photoelectrons produced by bombarding the specimens with X-rays can also be used to obtain chemical information in a technique called X-ray photoelectron spectroscopy (XPS). Photoelectrons have energies similar to Auger electrons so the specimen depth analyzed is similar to that in AES. However, since it is difficult to bring an X-ray beam to a fine focus the spatial resolution is poorer than that achievable in AES. An advantage of XPS is that, since the photoelectron energy is a function of the binding energy of the electron in the solid, it can give information regarding the ionization state of an element, i.e., whether it is present in elemental form, as an oxide, as a nitride, etc.

Secondary-ion mass spectroscopy (SIMS) is another useful technique, which involves sputtering away the surface of the specimen with an ion beam and analyzing the sputtered ions in a mass spectrometer. The SIMS technique can provide very precise chemical analysis of very thin surface layers and, as sputtering proceeds, a concentration–depth profile through the specimen.

The application of the above techniques is an important part of the study of oxidation, so much so that regular conferences are now held on this subject.[10–14] It is important to emphasize that proper investigation of oxidation mechanisms involves as many of the above techniques of observing kinetics and morphologies as feasible, and careful combination of the results.

References

1. S. Mrowec and A. J. Stoklosa, *J. Therm. Anal.*, **2** (1970), 73.
2. J. I. Goldstein, D. E. Newbury, P. Echlin, *et al.*, *Scanning Electron Microscopy and X-Ray Microanalysis*, New York, NY, Plenum Press, 1984.
3. M. H. Loretto, *Electron Beam Analysis of Materials*, 2nd edn., London, UK, Chapman and Hall, 1994.
4. B. D. Cullity, *Elements of X-Ray Diffraction*, 2nd edn, Reading, MA, Addison Wesley, 1978.
5. I. C. Noyan and J. B. Cohen, *Residual Stresses*, New York, NY, Springer-Verlag, 1987.
6. D. B. Williams and C. B. Carter, *Transmission Electron Microscopy*, New York, NY, Plenum Press, 1996.
7. M. Rühle, U. Salzberger and E. Schumann. High resolution transmission microscopy of metal/metal oxide interfaces. In *Microscopy of Oxidation 2*, eds. S. B. Newcomb and M. J. Bennett, London, UK, The Institute of Materials, 1993, p. 3.
8. D. P. Woodruff and T. A. Delchar, *Modern Techniques of Surface Science*, Cambridge, UK, Cambridge University Press, 1989.
9. H. J. Grabke, V. Leroy and H. Viefhaus, *ISIJ Int.*, **35** (1995), 95.
10. *Microscopy of Oxidation 1*, eds. M. J. Bennett and G. W. Lorimer, London, UK, The Institute of Metals, 1991.
11. *Microscopy of Oxidation 2*, eds. S. B. Newcomb and M. J. Bennett, London, UK, The Institute of Materials, 1993.
12. *Microscopy of Oxidation 3*, eds. S. B. Newcomb and J. A. Little, London, UK, The Institute of Materials, 1997.
13. *Microscopy of Oxidation 4*, eds. G. Tatlock and S. B. Newcomb, *Science Rev.*, **17** (2000), 1.
14. *Microscopy of Oxidation 5*, eds. G. Tatlock and S. B. Newcomb, *Science Rev.*, **20** (2003).

2

Thermodynamic fundamentals

Introduction

A sound understanding of high-temperature corrosion reactions requires the determination of whether or not a given component in a metal or alloy can react with a given component from the gas phase or another condensed phase, and to rationalize observed products of the reactions. In practice, the corrosion problems to be solved are often complex, involving the reaction of multicomponent alloys with gases containing two or more reactive components. The situation is often complicated by the presence of liquid or solid deposits, which form either by condensation from the vapour or impaction of particulate matter. An important tool in the analysis of such problems is, of course, equilibrium thermodynamics which, although not predictive, allows one to ascertain which reaction products are possible, whether or not significant evaporation or condensation of a given species is possible, the conditions under which a given reaction product can react with a condensed deposit, etc. The complexity of the corrosion phenomena usually dictates that the thermodynamic analysis be represented in graphical form.

The purpose of this chapter is to review the thermodynamic concepts pertinent to gas–metal reactions and then to describe the construction of the thermodynamic diagrams most often used in corrosion research, and to present illustrative examples of their application. The types of diagrams discussed are

(1) Gibbs free energy versus composition diagrams and activity versus composition diagrams, which are used for describing the thermodynamics of solutions.
(2) Standard free energy of formation versus temperature diagrams which allow the thermodynamic data for a given class of compounds, oxides, sulphides, carbides, etc., to be presented in a compact form.
(3) Vapour-species diagrams which allow the vapour pressures of compounds to be presented as a function of convenient variables such as partial pressure of a gaseous component.

16

(4) Two-dimensional, isothermal stability diagrams, which map the stable phases in systems involving one metallic and two reactive, non-metallic components.
(5) Two-dimensional, isothermal stability diagrams which map the stable phases in systems involving two metallic components and one reactive, non-metallic component.
(6) Three-dimensional, isothermal stability diagrams which map the stable phases in systems involving two metallic and two reactive, non-metallic components.

Basic thermodynamics

The question of whether or not a reaction can occur is answered by the second law of thermodynamics. Since the conditions most often encountered in high-temperature reactions are constant temperature and pressure, the second law is most conveniently written in terms of the Gibbs free energy (G') of a system, Equation (2.1),

$$G' = H' - TS',\tag{2.1}$$

where H' is the enthalpy and S' the entropy of the system. Under these conditions the second law states that the free-energy change of a process will have the following significance: $\Delta G' < 0$, spontaneous reaction expected; $\Delta G' = 0$, equilibrium; $\Delta G' > 0$, thermodynamically impossible process.

For a chemical reaction, e.g. Equation (2.2),

$$a\mathrm{A} + b\mathrm{B} = c\mathrm{C} + d\mathrm{D},\tag{2.2}$$

$\Delta G'$ is expressed as in Equation (2.3),

$$\Delta G' = \Delta G^{\circ} + RT \ln \left(\frac{a_{\mathrm{C}}^{c} a_{\mathrm{D}}^{d}}{a_{\mathrm{A}}^{a} a_{\mathrm{B}}^{b}} \right),\tag{2.3}$$

where ΔG° is the free-energy change when all species are present in their standard states; a is the thermodynamic activity, which describes the deviation from the standard state for a given species and may be expressed for a given species i as in Equation (2.4),

$$a_{\mathrm{i}} = \frac{p_{\mathrm{i}}}{p_{\mathrm{i}}^{\circ}}.\tag{2.4}$$

Here p_{i} is either the vapour pressure over a condensed species or the partial pressure of a gaseous species and p_{i}° is the same quantity corresponding to the standard state of i. Expressing a_{i} by Equation (2.4) requires the reasonable approximation of ideal gas behaviour at the high temperatures and relatively low pressures usually encountered. The standard free-energy change is expressed for a reaction such as Equation (2.2) by Equation (2.5),

$$\Delta G^{\circ} = c\Delta G_{\mathrm{C}}^{\circ} + d\Delta G_{\mathrm{D}}^{\circ} - a\Delta G_{\mathrm{A}}^{\circ} - b\Delta G_{\mathrm{B}}^{\circ},\tag{2.5}$$

where ΔG_C°, etc., are standard molar free energies of formation which may be obtained from tabulated values. (A selected list of references for thermodynamic data such as ΔG° is included at the end of this chapter.) For the special cases of equilibrium ($\Delta G' = 0$) Equation (2.3) reduces to Equation (2.6),

$$\Delta G^\circ = -RT \ln \left(\frac{a_C^c a_D^d}{a_A^a a_B^b} \right)_{eq}. \tag{2.6}$$

The term in parentheses is called the equilibrium constant (K) and is used to describe the equilibrium state of the reaction system.

Construction and use of thermodynamic diagrams

(a) Gibbs free energy versus composition diagrams and activity versus composition diagrams

The equilibrium state of a system at constant temperature and pressure is characterized by a minimum in the Gibbs free energy of the system. For a multicomponent, multiphase system, the minimum free energy corresponds to uniformity of the chemical potential (μ) of each component throughout the system. For a binary system, the molar free energy (G) and chemical potentials are related by Equation (2.7),

$$G = (1 - X)\mu_A + X\mu_B, \tag{2.7}$$

where μ_A and μ_B are the chemical potentials of components A and B, respectively, and $(1 - X)$ and X are the mole fractions of A and B. Figure 2.1(a) is a plot of G versus composition for a single phase which exhibits simple solution behaviour. It can readily be shown,[1,2] starting with Equation (2.7), that for a given composition μ_A and μ_B can be defined as in Equations (2.8) and (2.9),

$$\mu_A = G - X\frac{dG}{dX}; \tag{2.8}$$

$$\mu_B = G + (1 - X)\frac{dG}{dX}, \tag{2.9}$$

which, as indicated in Figure 2.1(a), means that a tangent drawn to the free-energy curve has intercepts on the ordinate at $X = 0$ and $X = 1$, corresponding to the chemical potentials of A and B, respectively. Alternatively, if the free energy of the unmixed components in their standard states, $(1 - X)\mu_A^\circ + X\mu_B^\circ$ is subtracted from each side of Equation (2.7), the result is the molar free energy of mixing, Equation (2.10):

$$\Delta G^M = (1 - X)(\mu_A - \mu_A^\circ) + X(\mu_B - \mu_B^\circ). \tag{2.10}$$

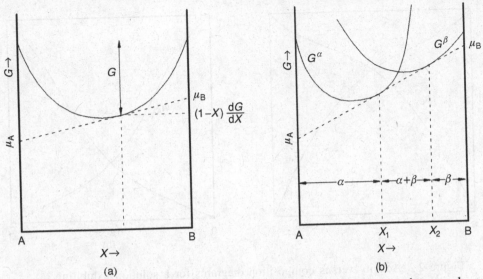

Figure 2.1 Free energy versus composition diagrams for (a) one phase, and (b) two coexisting phases.

A plot of ΔG^M versus X may be constructed in which case the tangent intercepts are $(\mu_A - \mu_A^\circ)$ and $(\mu_B - \mu_B^\circ)$, respectively. These quantities, the partial molar free energies of mixing, are directly related to the activities of the given component through Equations (2.11) and (2.12):

$$\mu_A - \mu_A^\circ = RT \ln a_A, \tag{2.11}$$

$$\mu_B - \mu_B^\circ = RT \ln a_B. \tag{2.12}$$

Figure 2.1(b) shows the free energy versus composition diagram for two phases. Application of the lever rule shows that the free energy for a mixture of the two phases is given by the point where the chord, drawn between the two individual free-energy curves, intersects the bulk composition. Clearly, for any bulk composition between X_1 and X_2 in Figure 2.1(b), a two-phase mixture will have a lower free energy than a single solution corresponding to either phase, and the lowest free energy of the system, i.e., the lowest chord, results when the chord is in fact a tangent to both free-energy curves. This 'common-tangent construction' indicates X_1 and X_2 to be the equilibrium compositions of the two coexisting phases and the intercepts of the tangent at $X = 0$ and $X = 1$ show the construction is equivalent to the equality of chemical potentials, i.e., Equations (2.13) and (2.14):

$$\mu_A^\alpha = \mu_A^\beta, \tag{2.13}$$

$$\mu_B^\alpha = \mu_B^\beta. \tag{2.14}$$

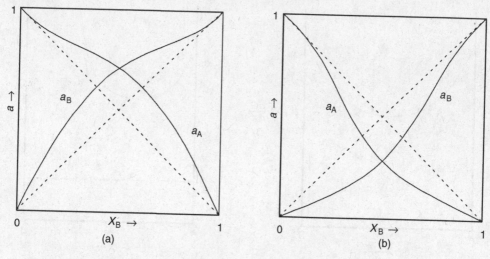

Figure 2.2 Activity versus composition diagrams for a solution exhibiting (a) positive- and (b) negative-deviation from Raoult's law.

Equations 2.11 and 2.12 indicate that similar information may be expressed in terms of activity. Figure 2.2 shows activity versus composition plots for a system with a free-energy curve such as that in Figure 2.1(a); with Figure 2.2(a) corresponding to a slightly positive deviation from Raoult's law ($a > X$) and Figure 2.2(b) corresponding to a slightly negative deviation ($a < X$). For cases where a two-phase mixture is stable, Figure 2.1(b), Equations (2.13) and (2.14) indicate the activities are constant across the two-phase field. Figure 2.3 illustrates the above points and their relationship to the conventional phase diagram for a hypothetical binary eutectic system. For a comprehensive treatment of free-energy diagrams and their applications to phase transformations the reader is referred to the excellent article by Hillert.[3]

Clearly, if the variation of G or ΔG^M with composition for a system can be established, either by experimental activity measurements or by calculations using solution models, the phase equilibria can be predicted for a given temperature and composition. Considerable effort has been spent in recent years on computer calculations of phase diagrams, particularly for refractory metal systems (see for example ref. 4). Furthermore, the concepts need not be limited to binary elemental systems. For example, ternary free-energy behaviour may be described on a three-dimensional plot, the base of which is a Gibbs triangle and the vertical axis represents G (or ΔG^M). In this case, the equilibrium state of the system is described using a 'common tangent plane', rather than a common tangent line, but the principles are unchanged. Systems with more than three components may be treated, but graphical interpretation becomes more difficult. Also, the above concepts may be

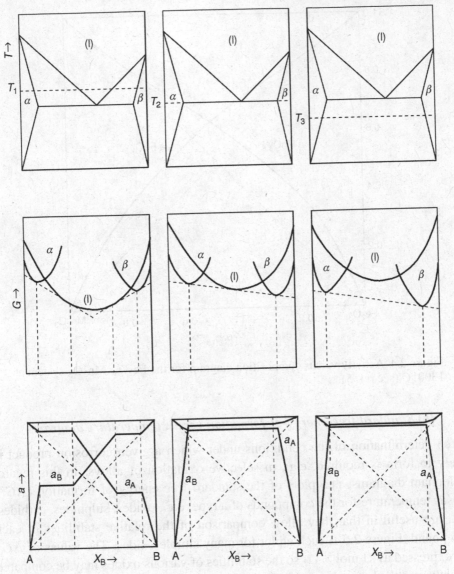

Figure 2.3 Free energy versus composition and activity diagrams for a hypothetical eutectic system.

applied to pseudo-binary systems where the components are taken as compounds of fixed stoichiometry. Figure 2.4 shows the activity versus composition plot[5] for such a system, Fe_2O_3–Mn_2O_3. This type of diagram may be constructed for any of the systems encountered in high-temperature corrosion, e.g., sulphides, oxides, sulphates, etc., and is useful in interpreting corrosion products and deposits formed on metallic or ceramic components.

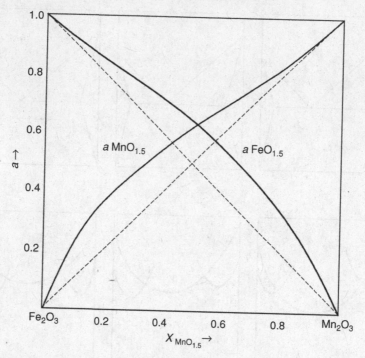

Figure 2.4 Activity versus composition diagram for the Fe$_2$O$_3$–Mn$_2$O$_3$ system at 1300 °C (see ref. 5).

(b) Standard free-energy-of-formation versus temperature diagrams

Often determination of the conditions under which a given corrosion product is likely to form is required, e.g., in selective oxidation of alloys. In this regard, Ellingham diagrams, i.e., plots of the standard free energy of formation (ΔG°) versus temperature for the compounds of a type, e.g., oxides, sulphides, carbides, etc., are useful in that they allow comparison of the relative stabilities of each compound. Figure 2.5 is such a plot for many simple oxides. The values of ΔG° are expressed as kJ mol^{-1} O$_2$ so the stabilities of various oxides may be compared directly, i.e., the lower the position of the line on the diagram the more stable is the oxide. Put another way, for a given reaction, Equation (2.15),

$$M + O_2 = MO_2, \tag{2.15}$$

if the activities of M and MO$_2$ are taken as unity, Equation (2.6) may be used to express the oxygen partial pressure at which the metal and oxide coexist, i.e., the dissociation pressure of the oxide, Equation (2.16):

$$p_{O_2}^{M/MO_2} = \exp \frac{\Delta G^\circ}{RT}. \tag{2.16}$$

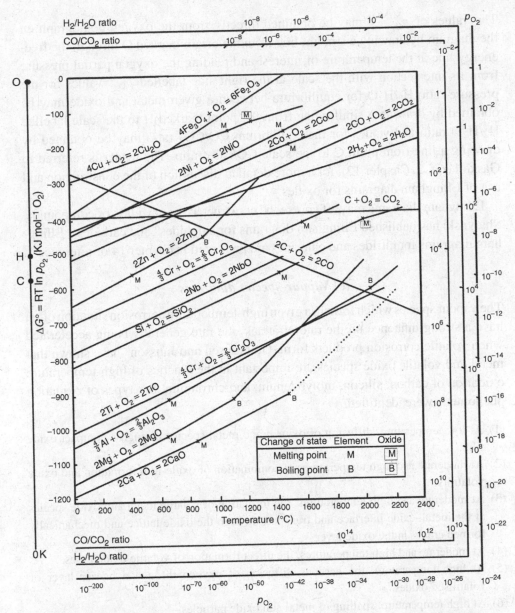

Figure 2.5 Standard free energy of formation of selected oxides as a function of temperature.

Naturally, in considering alloy oxidation, the activity of the metal and oxide must be taken into account, i.e., Equation (2.17):

$$p_{O_2}^{eq} = \frac{a_{MO_2}}{a_M} \exp \frac{\Delta G^{\circ}}{RT} = \frac{a_{MO_2}}{a_M} p_{O_2}^{M/MO_2}. \qquad (2.17)$$

The values of $p_{O_2}^{M/MO_2}$ may be obtained directly from the oxygen nomograph on the diagram by drawing a straight line from the origin marked O through the free-energy line at the temperature of interest and reading the oxygen partial pressure from its intersection with the scale at the right side labelled p_{O_2}. Values for the pressure ratio H_2/H_2O for equilibrium between a given metal and oxide may be obtained by drawing a similar line from the point marked H to the scale labelled H_2/H_2O ratio and values for the equilibrium CO/CO_2 ratio may be obtained by drawing a line from point C to the scale CO/CO_2 ratio. The reader is referred to Gaskell (ref. 2, Chapter 12), for a more detailed discussion of the construction and use of Ellingham diagrams for oxides.

Ellingham diagrams may, of course, be constructed for any class of compounds. Shatynski has published Ellingham diagrams for sulphides[6] and carbides.[7] Ellingham diagrams for nitrides and chlorides are presented in chapter 14 of reference 1.

(c) Vapour-species diagrams

The vapour species which form in a given high-temperature corrosion situation often have a strong influence on the rate of attack, the rate generally being accelerated when volatile corrosion products form. Gulbransen and Jansson[8] have shown that metal and volatile oxide species are important in the kinetics of high-temperature oxidation of carbon, silicon, molybdenum, and chromium. Six types of oxidation phenomena were identified.

(1) At low temperature, diffusion of oxygen and metal species through a compact oxide film.
(2) At moderate and high temperatures, a combination of oxide-film formation and oxide volatility.
(3) At moderate and high temperatures, the formation of volatile metal and oxide species at the metal–oxide interface and transport through the oxide lattice and mechanically formed cracks in the oxide layer.
(4) At moderate and high temperatures, the direct formation of volatile oxide gases.
(5) At high temperature, the gaseous diffusion of oxygen through a barrier layer of volatilised oxides.
(6) At high temperature, spalling of metal and oxide particles.

Sulphidation reactions follow a similar series of kinetic phenomena as has been observed for oxidation. Unfortunately, few studies have been made of the basic kinetic phenomena involved in sulphidation reactions at high temperature. Similarly, the volatile species in sulphate and carbonate systems are important in terms of evaporation/condensation phenomena involving these compounds on alloy or ceramic surfaces. Perhaps the best example of this behaviour is the rapid degradation of protective scales on many alloys, termed 'hot corrosion', which occurs when Na_2SO_4 or other salt condenses on the alloy.

Table 2.1

Species	Log K_p
Cr_2O_3 (s)	33.95
Cr (g)	−8.96
CrO (g)	−2.26
CrO_2 (g)	4.96
CrO_3 (g)	8.64

The diagrams most suited for presentation of vapour-pressure data in oxide systems are log $p_{M_xO_y}$, for a fixed T, versus log p_{O_2}; and log $p_{M_xO_y}$, for a fixed p_{O_2}, versus $1/T$ diagrams. The principles of the diagrams will be illustrated for the Cr–O system at 1250 K. Only one condensed oxide, Cr_2O_3, is formed under conditions of high-temperature oxidation. Thermochemical data are available for Cr_2O_3 (s),[9] Cr (g),[10] and the three gaseous oxide species CrO (g), CrO_2 (g), and CrO_3 (g).[11] These data may be expressed as standard free energies of formation ($\Delta G°$), or, more conveniently, as log K_p defined by Equation (2.1):

$$\log K_p = \frac{-\Delta G°}{2.303RT}.\tag{2.18}$$

The appropriate data at 1250 K are listed in Table 2.1. The vapour pressures of species in equilibrium with Cr metal must be determined for low oxygen pressures and those of species in equilibrium with Cr_2O_3 at high oxygen pressures. The boundary between these regions is the oxygen pressure for the Cr/Cr_2O_3 equilibrium, obtained from the equilibrium given in Equation (2.17),

$$2Cr\,(s) + \frac{3}{2}O_2\,(g) = Cr_2O_3\,(s),\tag{2.19}$$

for which log p_{O_2} has the value given in Equation (2.20):

$$\log p_{O_2} = -\frac{2}{3}\log K_p^{Cr_2O_3} = -22.6.\tag{2.20}$$

The boundary is represented by the vertical line in Figure 2.6. At lower oxygen pressures the pressure of Cr (g) is independent of p_{O_2}:

$$Cr\,(s) = Cr\,(g),\tag{2.21}$$

$$\log p_{Cr} = \log K_p^{Cr\,(g)} = -8.96.\tag{2.22}$$

For oxygen pressures greater than Cr/Cr_2O_3 equilibrium, the vapour pressure of Cr may be obtained from Equation (2.23),

$$Cr_2O_3\,(s) = 2\,Cr\,(g) + \frac{3}{2}O_2\,(g),\tag{2.23}$$

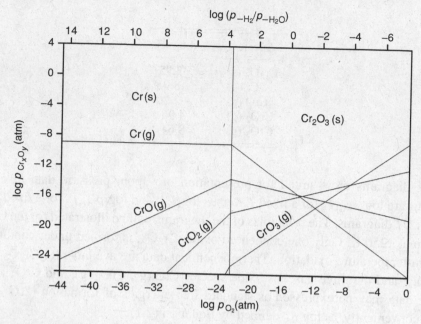

Figure 2.6 Volatile species in the Cr–O system at 1250 K.

with the equilibrium constant, K_{23}, determined from Equation (2.24):

$$\log K_{23} = 2 \log K_p^{\mathrm{Cr\,(g)}} - \log K_p^{\mathrm{Cr_2O_3}} = -51.9. \tag{2.24}$$

Therefore, Equations (2.25) or (2.26) are obtained:

$$2 \log p_{\mathrm{Cr}} + \frac{3}{2} \log p_{\mathrm{O_2}} = -51.9, \tag{2.25}$$

$$\log p_{\mathrm{Cr}} = -25.95 - \frac{3}{4} \log p_{\mathrm{O_2}}. \tag{2.26}$$

The vapour pressure of Cr is then expressed in Figure 2.6 by plotting Equations (2.22) and (2.26). The lines corresponding to the other vapour species are obtained in the same manner. For example, for CrO_3 at oxygen pressures below Cr/Cr_2O_3 equilibrium we have Equations (2.27) and (2.28):

$$\mathrm{Cr\,(s)} + \frac{3}{2}\mathrm{O_2\,(g)} = \mathrm{CrO_3\,(g)}, \tag{2.27}$$

$$\log K_{27} = \log K_p^{\mathrm{CrO_3\,(g)}} = 8.64, \tag{2.28}$$

from which one finds $\log p_{\mathrm{CrO_3}}$, Equation (2.29):

$$\log p_{\mathrm{CrO_3}} = 8.64 + \frac{3}{2} \log p_{\mathrm{O_2}}. \tag{2.29}$$

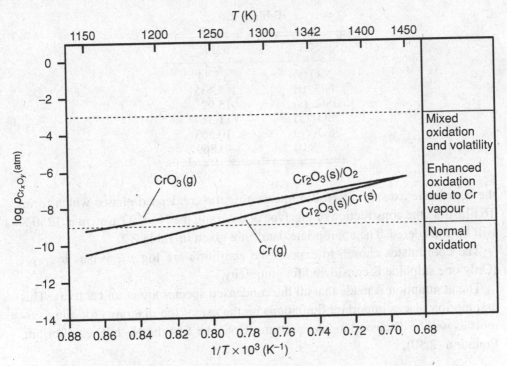

Figure 2.7 Cr–O system volatile species versus temperature. The line for $Cr(g)$ refers to the pressure at the $Cr–Cr_2O_3$ interface and the CrO_3 (g) line refers to the Cr_2O_3/gas interface for an oxygen partial pressure of 1 atm.

The other lines in Figure 2.6 are obtained by writing similar equilibrium equations. Figure 2.6 shows that significant vapour pressures of Cr are developed at low p_{O_2}, e.g., at the alloy–scale interface of a Cr_2O_3-forming alloy, and that very large pressures of CrO_3 are developed at high p_{O_2}. The latter are responsible for a phenomenon whereby Cr_2O_3 scales are thinned by vapour losses during oxidation at high p_{O_2}, particularly for high gas flow rates. Figure 2.7 indicates the temperature dependence of the various equilibria. Similar diagrams may be constructed for other oxides, sulphides, halides, etc.

(d) Two-dimensional, isothermal stability diagram: one metallic and two non-metallic components

When a metal reacts with a gas containing more than one oxidant, a number of different phases may form depending on both thermodynamic and kinetic considerations. Isothermal stability diagrams, usually constructed with the logarithmic values of the activities or partial pressures of the two non-metallic components as

Thermodynamic fundamentals

Table 2.2

Species	$\log K_p$
NiO (s)	5.34
NiS$_y$ (l)	3.435
NiSO$_4$ (s)	15.97
SO$_2$ (g)	11.314
SO$_3$ (g)	10.563
S (l)	−0.869

the coordinate axes, are useful in interpreting the condensed phases which form. To illustrate the construction of this type of diagram, the Ni–S–O system at 1250 K will be considered. The appropriate data[11] are given in Table 2.2.

The coordinates chosen to express the equilibria are $\log p_{S_2}$ versus $\log p_{O_2}$. (Only one sulphide is considered for simplicity.)

The assumption is made that all the condensed species are at unit activity. This assumption places important limitations on the use of the diagrams for alloy systems as will be discussed in later sections. Consider first the Ni–NiO equilibrium, Equation (2.30),

$$Ni\,(s) + \frac{1}{2}O_2(g) = NiO\,(s), \tag{2.30}$$

and $\log p_{O_2}$, defined in Equation (2.31),

$$\log p_{O_2} = -2\log K_p^{NiO} = -10.68, \tag{2.31}$$

resulting in the vertical phase boundary in Figure 2.8. Similarly, for NiS$_y$(l) ($y \approx 1$) we have Equations (2.32) and (2.33):

$$Ni\,(s) + \frac{y}{2}S_2\,(g) = NiS_y\,(l), \tag{2.32}$$

$$\log p_{S_2} = -6.87, \tag{2.33}$$

resulting in the horizontal boundary in Figure 2.8. Equations (2.31) and (2.33), therefore, define the gaseous conditions over which metallic Ni is stable. In a similar manner, the other phase boundaries are determined. For example, for the NiO–NiSO$_4$ equilibrium represented by Equations (2.34) and (2.35).

$$NiO\,(s) + \frac{1}{2}S_2\,(g) + \frac{3}{2}O_2\,(g) = NiSO_4\,(s), \tag{2.34}$$

$$\log K_{34} = \log K_p^{NiSO_4} - \log K_p^{NiO} = 10.63, \tag{2.35}$$

resulting in Equation (2.36) for the phase boundary:

$$\log p_{S_2} = -21.26 - 3\log p_{O_2}. \tag{2.36}$$

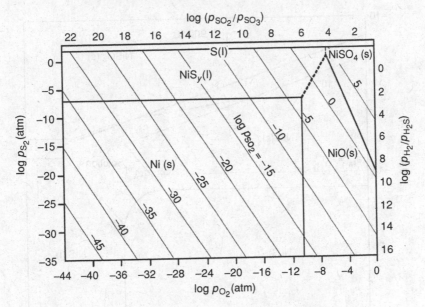

Figure 2.8 Condensed phase equilibria for the Ni–O–S system at 1250 K.

An additional feature of the diagrams is the set of superposed SO_2 isobars which define the S_2 and O_2 partial pressures which may result for a fixed SO_2 partial pressure. These isobars are obtained from the equilibrium shown in Equation (2.37),

$$\frac{1}{2}S_2\,(g) + O_2\,(g) = SO_2\,(g), \tag{2.37}$$

with $\log K_{37}$ given in Equation (2.38):

$$\log K_{37} = \log K_p^{SO_2} = 11.314; \tag{2.38}$$

these result in Equation (2.39):

$$\log p_{S_2} = -22.628 + 2\log p_{SO_2} - 2\log p_{O_2}. \tag{2.39}$$

The isobars appear as diagonal lines of slope -2. The horizontal line near the top of the diagram indicates the sulphur pressure at which S (1) will condense from the gas. This type of diagram will be used in later chapters to interpret the reactions involving metals in complex atmospheres and hot corrosion.

When significant intersolubility exists between neighbouring phases on stability diagrams, solution regions may be included as shown in Figure 2.9 taken from the work of Giggins and Pettit.[12] (Note the ordinate and abscissa are reversed from Figure 2.8 and the temperature is slightly lower.) The phase boundaries in Figure 2.9 were calculated from the equilibria, such as Equations (2.31), (2.33), and (2.36), but the boundaries of the solution regions were calculated taking the acitivtes of the essentially pure species as unity and of the minor species as some limiting value.

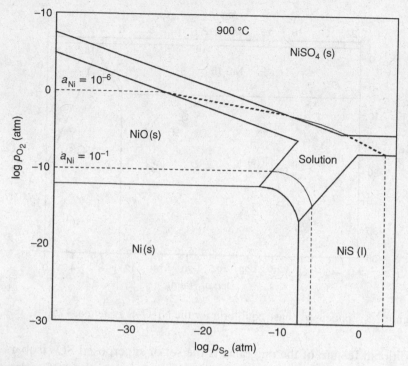

Figure 2.9 Stability diagram for the Ni–O–S system where regions of solution were assumed to exist between NiO, NiS, and NiSO₄ (from Giggins and Pettit[12]).

In constructing Figure 2.9, the activity of NiS was taken as 10^{-2} in equilibrium with essentially pure NiO and the activity of NiO was taken as 10^{-2} in equilibrium with essentially pure NiS. Similar approximations have been made for the other equilibria. Obviously experimental data regarding the extent of intersolubility of phases and activities in solution are essential to quantify such diagrams.

For some applications, different coordinate axes become more convenient for the construction of stability diagrams. For example, the phase equilibria for the Ni–S–O system in Figure 2.8 may be equally well represented on a plot of $\log p_{O_2}$ versus $\log p_{SO_3}$ since specification of these two variables fixes $\log p_{S_2}$ because of the equilibrium shown in Equation (2.40):

$$\frac{1}{2}S_2\,(g) + \frac{3}{2}O_2\,(g) = SO_3\,(g). \tag{2.40}$$

The equilibria of Figure 2.8 are replotted in Figure 2.10 using these coordinates. This choice of coordinates is particularly useful in analyzing hot-corrosion situations such as those that occur when a metal or alloy is oxidized while covered with a

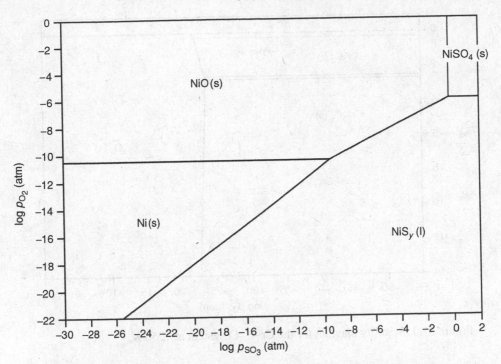

Figure 2.10 Stability data for the Ni–O–S system at 1250 K expressed as log p_{O_2} versus log p_{SO_3}.

deposit of sodium sulphate, since the components of Na_2SO_4 may be taken as Na_2O and SO_3. This phenomenon will be described in Chapter 8.

Stability diagrams of this type are not restricted to use with metal–sulphur–oxygen systems. As an example, Figure 2.11 is a stability diagram for the Cr–C–O system at 1250 K, where the carbon activity is an important variable. The coordinates are log a_C versus log p_{O_2} and the construction and use of the diagram is exactly analogous to those for the metal–sulphur–oxygen diagrams. Applications of stability diagrams involving a number of systems have been discussed by Jansson and Gulbransen[13,14] and compilations of diagrams are available for metal–sulphur–oxygen systems[11] and for a number of metals in combinations of oxygen, sulphur, carbon, and nitrogen.[15]

(e) Two-dimensional, isothermal stability diagrams: two metallic components and one non-metallic component

In considering the reaction of even binary alloys with gases containing mixed oxidants, the thermodynamic description of the condensed phase equilibria becomes

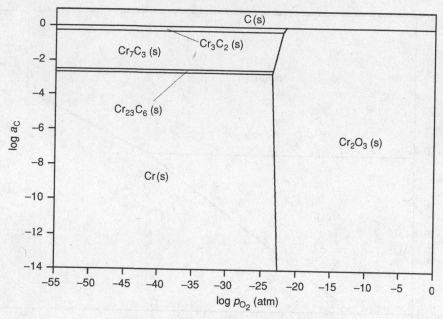

Figure 2.11 Stability diagram for the Cr–C–O system at 1250 K.

complicated by the addition of activity or composition of one metal component as a variable. However, before considering this problem, the condensed phase equilibria when only one oxidant is present must be considered. Figure 2.12, taken from the work of Sticher and Schmalzried,[5] is the stability diagram for the Fe–Cr–O system at 1300 °C. The construction of this type of diagram requires the establishment of the free-energy composition behaviour, as discussed in Section (a), for the system of interest and then the evaluation of the stability fields at a given temperature by minimising the free energy of the system, i.e., applying the common-tangent construction. The rules governing these types of diagrams are clearly explained by Pelton and Schmalzried.[16]

Few experimental data exist for the free-energy composition relationships in mixed-oxide or -sulphide solutions, for example $(Fe,Cr)_3O_4$ or $(Fe,Cr)_{1-x}S_x$, so it is necessary to use various solution models to estimate these data.[5,17] Furthermore, few data exist for the free energies of formation of compounds of fixed stoichiometry, such as the sulphide spinel $FeCr_2S_4$, and these must be estimated.[17]

The appearance of phases of variable composition and intermediate phases, such as spinels, as the mole fraction of a given metal is changed, is significant since superposition of diagrams for the individual pure metals will neglect these phases. For example, considering the corrosion of Fe–Cr alloys in S–O atmospheres by superposing the Fe–S–O and Cr–S–O diagrams will neglect some of the important phase equilibria and should only be done with adequate justification.

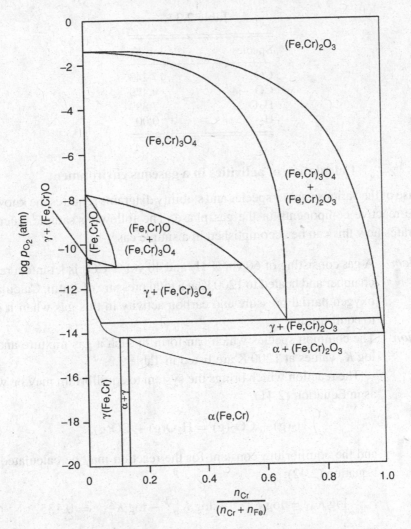

Figure 2.12 Stability diagram for the Fe–Cr–O system at 1300 °C.

(f) Three-dimensional, isothermal stability diagrams: two metallic and two non-metallic components

The corrosion of alloys in gases containing mixed oxidants is a problem of major interest. The proper thermodynamic treatment of this type of problem requires a combination of the concepts presented in Sections (d) and (e). This is a difficult task and a lack of experimental data for ternary compounds has prevented significant calculations of this type. One exception is the work of Giggins and Pettit[12] in which a simplified three-dimensional diagram (log p_{S_2} versus log p_{O_2} versus log a_{Cr}) has been constructed for the Ni–Cr–S–O system.

Table 2.3

Species	$\log K_p$
CO_2	17.243
CO	9.479
H_2O	7.899
H_2	0.00

Calculation of activities in a gaseous environment

The use of the various vapour species and stability diagrams requires the knowledge of the reactive components in the gas phase. The following sample calculation illustrates how this can be accomplished in a simple case.

Problem   A gas consisting of 60 vol % H_2 and 40 vol % CO_2 is let into a reaction chamber and heated to 1200 K at a total pressure of 1 atm. Calculate the oxygen partial pressure and carbon activity in this gas when it comes to equilibrium.

Solution   The common species which can form in such a gas mixture and their $\log K_p$ values at 1200 K are listed in Table 2.3.[10]

The reaction which brings the system to equilibrium may be written as in Equation (2.41),

$$H_2(g) + CO_2(g) = H_2O(g) + CO(g), \tag{2.41}$$

and the equilibrium constant for the reaction may be calculated from Equation (2.42):

$$\log K_{41} = \log K_p^{H_2O} + \log K_p^{CO} - \log K_p^{CO_2} = 0.135. \tag{2.42}$$

Therefore, K_{41} may be written as in Equation (2.43):

$$K_{41} = 1.365 = \frac{p_{H_2O} \, p_{CO}}{p_{H_2} \, p_{CO_2}}. \tag{2.43}$$

The partial pressure of each species may be related to the number of moles of that species n_i through the relationship shown in Equation (2.44):

$$p_i = \frac{n_i}{n_{tot}} P_{tot}, \tag{2.44}$$

which, on substitution into K_{41}, yields Equation (2.45):

$$K_{41} = 1.365 = \frac{n_{H_2O} n_{CO}}{n_{H_2} n_{CO_2}}. \tag{2.45}$$

Table 2.4

Species	Initial composition (moles)	Final composition (moles)
H_2	0.6	$0.6 - \lambda$
CO_2	0.4	$0.4 - \lambda$
CO	0	λ
H_2O	0	λ

Table 2.5

Species	Final composition (moles)	Partial pressure (atm)
H_2	0.3425	0.3425
CO_2	0.1425	0.1425
CO	0.2575	0.2575
H_2O	0.2575	0.2575

The remaining information required to calculate the equilibrium composition of the system can be obtained by applying a mass balance to the system. Take a basis of one mole of gas entering the reaction chamber and allow λ to represent the number of moles of CO formed as the system comes to equilibrium. Thus, the initial and final composition of the system will be that shown in Table 2.4.

Substitution of the final mole numbers into K_{41} results in Equation (2.46),

$$K_{41} = 1.365 = \frac{\lambda\lambda}{(0.6 - \lambda)(0.4 - \lambda)},$$ (2.46)

which yields a quadratic which may be solved for λ with the result:

$$\lambda = 0.2575 \text{ moles.}$$

This allows calculation of the numbers of moles of each species at equilibrium and, with the use of Equation (2.44), the partial pressures of each species, Table 2.5.

The equilibrium oxygen partial pressure may be calculated from either the CO/CO_2 or H_2/H_2O equilibria. For example, for the latter, Equation (2.47),

$$H_2\,(g) + 1/2O_2\,(g) = H_2O\,(g),$$ (2.47)

K_{47} can be written as in Equation (2.48),

$$K_{47} = 7.924 \times 10^7 = \frac{p_{H_2O}}{p_{H_2} p_{O_2}^{1/2}} = \frac{0.2575}{0.3425 p_{O_2}^{1/2}}, \qquad (2.48)$$

which yields

$$p_{O_2} = 9.0 \times 10^{-18} \text{ atm.}$$

The carbon activity may be calculated from the equilibrium shown in Equation (2.49),

$$2CO\,(g) = CO_2\,(g) + C\,(s), \qquad (2.49)$$

with $\log K_{49}$ and K_{49} given in Equations (2.50) and (2.51), respectively:

$$\log K_{49} = \log K_p^{CO_2} - 2 \log K_p^{CO} = -1.715, \qquad (2.50)$$

$$K_{49} = 0.193 = \frac{p_{CO_2} a_C}{p_{CO}^2} = \frac{0.1425 a_C}{(0.2575)^2}. \qquad (2.51)$$

Solution of Equation (2.51) yields $a_C = 0.009$.

The possibility of the formation of other species, such as CH_4, has been ignored in the above calculation because their concentrations would be extremely small. When the gas mixture contains more components, or if formation of additional species must be considered, the calculations become more complex, involving more equilibrium constants and more complicated mass balances. However, the calculations still involve the principles illustrated above. A number of computer programs are currently available to perform these, more complex, calculations.[18–22]

Summary

The basic types of thermodynamic diagrams useful in interpreting the results of high-temperature corrosion experiments have been discussed. While no attempt at a complete discussion of this subject has been made, it is hoped this chapter will facilitate a clearer understanding of the high-temperature corrosion phenomena to be described in the remainder of this book.

References

1. L. S. Darken and R. W. Gurry, *Physical Chemistry of Metals*, New York, McGraw-Hill, 1953.
2. D. R. Gaskell, *Introduction the Thermodynamics of Materials*, 3rd edn, Washington, DC, Taylor and Francis, 1995.

3. M. Hillert, The uses of Gibbs free energy–composition diagrams. In *Lectures on the Theory of Phase Transformations*, 2nd edn, ed. H. I. Aaronson, Warrendale, PA, The Minerals, Metals and Materials Society, 1999.
4. L. Kaufman and H. Bernstein, Computer calculations of refractory metal phase diagrams. In *Phase Diagrams, Materials Science and Technology*, ed. A. M. Alper, New York, Academic Press, 1970, vol. 1.
5. J. Sticher and H. Schmalzried, Zur geometrischen Darstellung thermodynamischer Zustandsgrossen in Mehrstoffsystemen auf Eisenbasis. Report, Clausthal Institute für Theoretische Huttenkunde und Angewandte Physikalische Chemie der Technischen Universität Clausthal, 1975.
6. S. R. Shatynski, *Oxid. Met.*, **11** (1977), 307.
7. S. R. Shatynski, *Oxid, Met.*, **13** (1979), 105.
8. E. A. Gulbransen and S. A. Jansson, In *Heterogeneous Kinetics at Elevated Temperatures*, eds. G. R. Belton and W. R. Worrell, New York, Plenum Press, 1970, p. 181.
9. C. E. Wicks and F. E. Block, Thermodynamic properties of 65 elements, their oxides, halides, carbides, and nitrides. Bureau of Mines, Bulletin 605, Washington, DC, US Government Printing Office, 1963, p. 408.
10. JANAF Thermochemical Tables, *J. Phys. Chemi. Ref. Data*, **4** (1975), 1–175.
11. E. A. Gulbransen and G. H. Meier, Themodynamic stability diagrams for condensed phases and volatility diagrams for volatile species over condensed phases in twenty metal–sulfur–oxygen systems between 1150 and 1450 K. University of Pittsburgh, DOE Report on Contract no. DE-AC01–79-ET-13547, May, 1979.
12. C. S. Giggins and F. S. Pettit, *Oxid. Met.*, **14** (1980), 363.
13. G. H. Meier, N. Birks, F. S. Pettit, and C. S. Giggins, Thermodynamic analyses of high temperature corrosion of alloys in gases containing more than one reactant. In *High Temperature Corrosion*, ed. R. A. Rapp, Houston, TX, NACE, 1983, p. 327.
14. S. A. Jansson and E. A. Gulbransen, Thermochemical considerations of high temperature gas-solid reactions. In *High Temperature Gas–Metal Reactions in Mixed Environments*, ed. S. A. Jansson and Z. A. Foroulis, New York, American Institute of Mining, Metallurgical, and Petroleum Engineerings, 1973, p. 2.
15. P. L. Hemmings and R. A. Perkins, Thermodynamic phase stability diagrams for the analysis of corrosion reactions in coal gasification/combusion atmospheres. Research Project 716–1, Interim Report, Palo Alto, CA, Lockheed Palo Alto Research Laboratories for Electric Power Research Insitute, December, 1977.
16. A. D. Pelton and H. Schmalzried, *Met. Trans.*, **4** (1973), 1395.
17. K. T. Jacob, D. B. Rao, and H. G. Nelson, *Oxid. Met.*, **13** (1979), 25.
18. G. Eriksson, *Chem. Scr.*, **8** (1975), 100.
19. G. Eriksson and K. Hack, *Met. Trans.*, **B21** (1990), 1013.
20. E. Königsberger and G. Eriksson, *CALPHAD*, **19** (1995), 207.
21. W. C. Reynolds, STANJAN, version 3, Stanford University, CA, Department of Mechanical Engneering, 1986.
22. B. Sundman, B. Jansson, and J.-O. Andersson. *CALPHAD*, **9** (1985), 153.

Selected thermodynamic data references

1. JANAF Thermochemical Data, including supplements, Midland, MI, Dow Chemical Co.; also National Standard Reference Data Series, National Bureau of Standards 37, Washington DC, US Government Printing Office, 1971; supplements 1974, 1975, and 1978.
2. *JANAF Thermochemical Tables*, 3rd edn, Midland, MI, Dow Chemical Co., 1986.

3. H. Schick, *Thermodynamics of Certain Refractory Compounds*, New York, Academic Press, 1966.
4. C. W. Wicks and F. E. Block, Thermodynamic properties of 65 elements, their oxides, halides, carbides, and nitrides, Bureau of Mines, Bulletin 605, Washington DC, US Government Printing Office, 1963.
5. R. Hultgren, R. L. Orr, P. D. Anderson, and K. K. Kelley, *Selected Values of Thermodynamic Properties of Metals and Alloys*, New York, Wiley, 1967.
6. O. Kubaschewski and C. B. Alcock, *Metallurgical Thermochemistry*, 5th edn, Oxford, Pergamon Press, 1979.
7. D. D. Wagman, W. H. Evans, V. B. Parker, *et al.*, Selected values of chemical thermodynamic properties, elements 1 through 34. NBS Technical Note 270–3, Washington, DC, US Government Printing Office, 1968; elements 35 through 53, NBS Technical Note 270–4, 1969.
8. K. H. Stern and E. A. Weise, High Temperature Properties and Decomposition of Inorganic Salts. Part I. Sulfates. National Standard Reference Data Series, National Bureau of Standards 7, Washington DC, US Government Printing Office, 1966. Part II. Carbonates, 1969.
9. D. R. Stull and G. C. Sinke, *Thermodynamic Properites of the Elements*. Advances in Chemistry, Monograph 18, Washington DC, American Chemical Society, 1956.
10. K. C. Mills, *Thermodynamic Data for Inorganic Sulphides, Selenides, and Tellurides*, London, Butterworth, 1974.
11. M. O. Dayhoff, E. R. Lippincott, R. V. Eck, and G. Nagarajan, Thermodynamic equilibrium in prebiological atmospheres of C, H, O, P, S, and Cl. NASA SP-3040, Washington, DC, 1964.
12. I. Barin, *Thermochemical Data of Pure Substances*, Weinheim, VCH Verlagsgesellschaft, 1993.
13. The NBS tables of chemical thermodynamic properties, Washington, DC, National Bureau of Standards, 1982.
14. J. B. Pankratz, Thermodynamic properties of elements and oxides, Bureau of Mines Bulletin 672, Washington, DC, US Government Printing Office, 1982.
15. J. B. Pankratz, Thermodynamic properties of halides, Bureau of Mines Bulletin 674, Washington, DC, US Government Printing Office, 1984.

3

Mechanisms of oxidation

Introduction

In this chapter the mechanisms by which ions and electrons transport through a growing oxide layer are described, using typical examples of both n-type and p-type semiconducting oxides. Using these concepts, the classical Wagner treatment of oxidation rates, controlled by ionic diffusion through the oxide layer, is presented showing how the parabolic rate constant is related to such fundamental properties as the partial ionic and electronic conductivities of the oxide and their dependence on the chemical potential of the metal or oxygen in the oxide. Finally, the mechanisms leading to observation of linear and logarithmic rate laws are discussed.

The discussion of defect structure has been simplified to that necessary to provide an understanding of oxidation mechanisms, however, an extensive review of defect theory has been provided by Kröger[1] and the defect structures of many oxides have been reviewed by Kofstad.[2] Smyth[3] has recently reviewed the literature on selected oxides. The reader is encouraged to consult these sources for more detailed treatment of specific oxides.

Mechanisms of oxidation

From consideration of the reaction

$$M(s) + \frac{1}{2}O_2(g) = MO(s)$$

it is obvious that the solid reaction product MO will separate the two reactants as shown below in Scheme 1.

$$\left| \begin{array}{c} M \\ Metal \end{array} \right| \begin{array}{c} MO \\ Oxide \end{array} \left| \begin{array}{c} O_2 \\ Gas \end{array} \right|$$

Scheme 1

In order for the reaction to proceed further, one or both reactants must penetrate the scale, i.e., either metal must be transported through the oxide to the oxide–gas

39

interface and react there, or oxygen must be transported to the oxide–metal interface and react there.

Immediately, therefore, the mechanisms by which the reactants may penetrate the oxide layer are seen to be an important part of the mechanism by which high-temperature oxidation occurs. Precisely the same applies to the formation and growth of sulphides and other similar reaction products.

Transport mechanisms

Since all metal oxides and sulphides are ionic in nature, it is not practicable to consider the transport of neutral metal, or non-metal, atoms through the reaction product. Several mechanisms are available to explain the transport of ions through ionic solids and these can be divided into mechanisms belonging to stoichiometric crystals and those belonging to non-stoichiometric crystals. In fact, although many compounds belong firmly to one or the other group, the correct view is probably that all of the possible defects are present to some extent or other in all compounds, but that in many cases certain types of defects predominate.

Highly stoichiometric ionic compounds

Predominant defects in these compounds, which represent one limiting condition, are the Schottky and Frenkel defects.

- *Schottky defects*. Ionic mobility is explained by the existence of ionic vacancies. In order to maintain electroneutrality, it is necessary to postulate an equivalent number, or concentration, of vacancies on both cationic and anionic sub-lattices. This type of defect occurs in the alkali halides and is shown for KCl in Figure 3.1. Since vacancies exist on both sub-lattices, it is to be expected that both anions and cations will be mobile.
- *Frenkel defects*. The case where only the cation is mobile can be explained by assuming that the anion lattice is perfect but that the cation lattice contains cation vacancies and interstitials in equivalent concentrations to maintain electroneutrality for the whole crystal. This type of defect is found in the silver halides and is shown for AgBr in Figure 3.2. The cation in this case is free to migrate over both vacancy and interstitial sites.

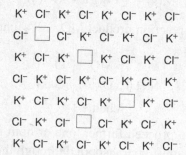

Figure 3.1 Schottky defects in potassium chloride.

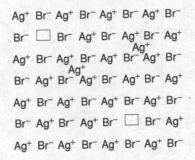

Ag^+ Br^- Ag^+ Br^- Ag^+ Br^- Ag^+ Br^-

Br^- ☐ Br^- Ag^+ Br^- Ag^+ Br^- Ag^+
Ag^+

Ag^+ Br^- Ag^+ Br^- Ag^+ Br^- Ag^+ Br^-
Ag^+

Br^- Ag^+ Br^- Ag^+ Br^- Ag^+ Br^- Ag^+

Ag^+ Br^- Ag^+ Br^- Ag^+ Br^- Ag^+ Br^-

Br^- Ag^+ Br^- Ag^+ Br^- ☐ Br^- Ag^+

Ag^+ Br^- Ag^+ Br^- Ag^+ Br^- Ag^+ Br^-

Figure 3.2 Frenkel defects in silver bromide.

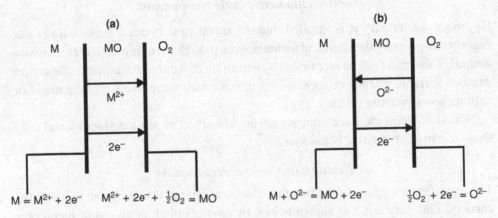

Figure 3.3 Interfacial reactions and transport processes for high-temperature oxidation mechanisms: (a) cation mobile and (b) anion mobile.

However, it is apparent that neither of these defects can be used to explain material transport during oxidation reactions because neither defect structure provides a mechanism by which electrons may migrate.

Considering a diagrammatic representation of the oxidation process shown in Figure 3.3, it is seen that either neutral atoms or ions and electrons must migrate in order for the reaction to proceed. In these cases, the transport step of the reaction mechanism links the two phase-boundary reactions as indicated. It will be noticed that there is an important distinction between scale growth by cation migration and scale growth by anion migration in that cation migration leads to scale formation at the scale–gas interface whereas anion migration leads to scale formation at the metal–scale interface.

In order to explain simultaneous migration of ions and electrons it is necessary to assume that the oxides, etc., that are formed during oxidation are non-stoichiometric compounds.

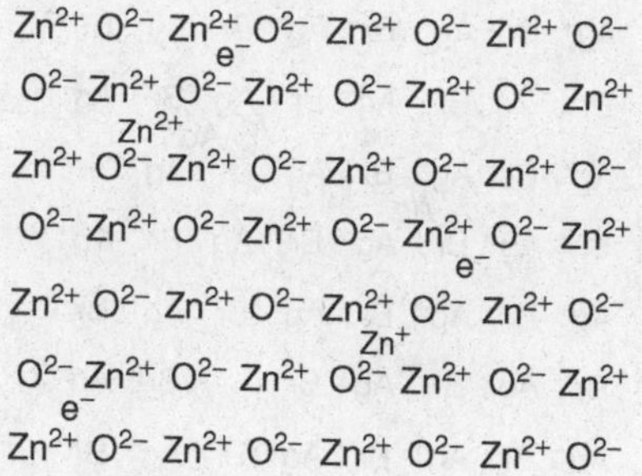

$$Zn^{2+}\ O^{2-}\ Zn^{2+}\underset{e^-}{O^{2-}}\ Zn^{2+}\ O^{2-}\ Zn^{2+}\ O^{2-}$$

$$O^{2-}\ Zn^{2+}\ O^{2-}\ Zn^{2+}\ O^{2-}\ Zn^{2+}\ O^{2-}\ Zn^{2+}$$

$$\underset{Zn^{2+}}{}$$

$$Zn^{2+}\ O^{2-}\ Zn^{2+}\ O^{2-}\ Zn^{2+}\ O^{2-}\ Zn^{2+}\ O^{2-}$$

$$O^{2-}\ Zn^{2+}\ O^{2-}\ Zn^{2+}\ O^{2-}\ Zn^{2+}\underset{e^-}{O^{2-}}\ Zn^{2+}$$

$$Zn^{2+}\ O^{2-}\ Zn^{2+}\ O^{2-}\ Zn^{2+}\ O^{2-}\ Zn^{2+}\ O^{2-}$$

$$\underset{Zn^+}{}$$

$$O^{2-}\ Zn^{2+}\ O^{2-}\ Zn^{2+}\ O^{2-}\ Zn^{2+}\ O^{2-}\ Zn^{2+}$$

$$\underset{e^-}{}$$

$$Zn^{2+}\ O^{2-}\ Zn^{2+}\ O^{2-}\ Zn^{2+}\ O^{2-}\ Zn^{2+}\ O^{2-}$$

Figure 3.4 Interstitial cations and excess electrons in ZnO – an n-type metal-excess semiconductor.

Non-stoichiometric ionic compounds

By non-stoichiometry it is implied that the metal to non-metal atom ratio is not exactly that given by the chemical formula, even though the compound is electrically neutral. These can only be reconciled by assuming that either the anion or the cation exhibits variable valency on its sub-lattice. It is much more likely that the metal or cation shows variable valency.

Non-stoichiometric ionic compounds are classified as semiconductors and may show negative or positive behaviour.

Negative (n-type) semiconductors

The classification refers to the fact that electrical charge is transferred by negative carriers. This may arise, as shown below, by having either an excess of metal or a deficit of non-metal.

Metal excess

The chemical formula is given as $M_{1+\delta}O$ and the best known example is zinc oxide (ZnO). In order to allow extra metal in the compound, it is necessary to postulate the existence of interstitial cations with an equivalent number of electrons in the conduction band. The structure may be represented as shown in Figure 3.4. Here, both Zn^+ and Zn^{2+} are represented as possible occupiers of interstitial sites. Cation conduction occurs over interstitial sites and electrical conductance occurs by virtue of having the 'excess' electrons excited into the conduction band. These, therefore, are called 'excess' or 'quasi-free' electrons.

The formation of this defect may be visualized, conveniently, as being formed from a perfect ZnO crystal by losing oxygen; the remaining unpartnered Zn^{2+} leaving the cation lattice and entering interstitial sites, and the two negative charges of the oxygen ion entering the conduction band. In this way one unit of ZnO crystal is destroyed and the formation of the defect may be represented as in Equation (3.1):

$$ZnO = Zn_i^{\cdot\cdot} + 2e' + \frac{1}{2}O_2. \tag{3.1}$$

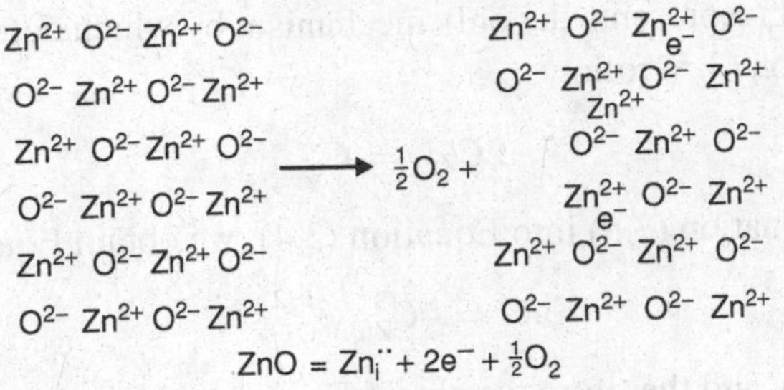

$$
\begin{array}{ll}
\text{Zn}^{2+}\ \text{O}^{2-}\ \text{Zn}^{2+}\ \text{O}^{2-} & \text{Zn}^{2+}\ \text{O}^{2-}\ \underset{\text{e}^-}{\text{Zn}^{2+}}\ \text{O}^{2-} \\
\text{O}^{2-}\ \text{Zn}^{2+}\ \text{O}^{2-}\ \text{Zn}^{2+} & \text{O}^{2-}\ \underset{\text{Zn}^{2+}}{\text{Zn}^{2+}}\ \text{O}^{2-}\ \text{Zn}^{2+} \\
\text{Zn}^{2+}\ \text{O}^{2-}\ \text{Zn}^{2+}\ \text{O}^{2-} & \text{O}^{2-}\ \text{Zn}^{2+}\ \text{O}^{2-} \\
\text{O}^{2-}\ \text{Zn}^{2+}\ \text{O}^{2-}\ \text{Zn}^{2+} & \underset{\text{e}^-}{\text{Zn}^{2+}}\ \text{O}^{2-}\ \text{Zn}^{2+} \\
\text{Zn}^{2+}\ \text{O}^{2-}\ \text{Zn}^{2+}\ \text{O}^{2-} & \text{Zn}^{2+}\ \text{O}^{2-}\ \text{Zn}^{2+}\ \text{O}^{2-} \\
\text{O}^{2-}\ \text{Zn}^{2+}\ \text{O}^{2-}\ \text{Zn}^{2+} & \text{O}^{2-}\ \text{Zn}^{2+}\ \text{O}^{2-}\ \text{Zn}^{2+}
\end{array}
$$

with arrow $\longrightarrow \frac{1}{2}\text{O}_2 +$ between the two blocks.

$$\text{ZnO} = \text{Zn}_i^{\cdot\cdot} + 2\text{e}^- + \tfrac{1}{2}\text{O}_2$$

Figure 3.5 Formation of metal-excess ZnO with excess electrons and interstitial Zn ions from 'perfect' ZnO.

for the formation of $\text{Zn}_i^{\cdot\cdot}$, doubly charged Zn interstitials, or in Equation (3.2),

$$\text{ZnO} = \text{Zn}_i^{\cdot} + \text{e}' + \frac{1}{2}\text{O}_2, \tag{3.2}$$

for the formation of $\text{Zn}_i^{\cdot}$, singly charged Zn interstitials. These processes are shown diagrammatically in Figure 3.5.

The nomenclature used here is that of Kröger[1].

M_M M atom on M site
X_X X atom on X site
M_i M atom on interstitial site
X_i X atom on interstitial site
N_M impurity N on M site
V_M vacancy on M site
V_X vacancy on X site
V_i vacant interstitial site
e' electron in conduction band
$\text{h}^{\cdot}$ – electron hole in valence band

The charges on the defects are measured relative to the normal site occupation and are indicated by the superscripts ($\cdot$) and ($'$) for positive and negative charges, respectively, e.g., $\text{Zn}_i^{\cdot\cdot}$ represents a divalent zinc ion on an interstitial site and carries a charge of $+2$ relative to the normal, unoccupied, site.

The two equilibria shown above will yield to thermodynamic treatment giving Equation (3.3) for the equilibrium in Equation (3.1),

$$K_1 = a_{\text{Zn}_i^{\cdot\cdot}}\, a_{\text{e}'}^2\, p_{\text{O}_2}^{1/2}, \tag{3.3}$$

or, since the defects are in very dilute solution, we may assume that they are in the range obeying Henry's law when the equilibrium may be written in terms of concentrations $C_{\text{Zn}_i^{\cdot\cdot}}$ and $C_{\text{e}'}$ as in Equation (3.4):

$$K_1' = C_{\text{Zn}_i^{\cdot\cdot}}\, C_{\text{e}'}^2\, p_{\text{O}_2}^{1/2}. \tag{3.4}$$

If Equation (3.1) represents the only mechanism by which defects are created in ZnO, then Equation (3.5) follows:

$$2C_{Zn_i^{..}} = C_{e'}. \tag{3.5}$$

Hence, putting Equation (3.5) into Equation (3.4), we obtain Equation (3.6a),

$$K_1' = 4C_{Zn_i^{..}}^3 \, p_{O_2}^{1/2}, \tag{3.6a}$$

or Equation (3.6b), and therefore we obtain Equation (3.7):

$$C_{Zn_i^{..}} = (K_1'/4)^{1/3} p_{O_2}^{-1/6} = const. \, p_{O_2}^{-1/6}, \tag{3.6b}$$

$$C_{e'} \propto p_{O_2}^{-1/6}. \tag{3.7}$$

Similarly, applying the same analysis to the reaction shown in Equation (3.2) the result shown in Equation (3.8) is obtained:

$$C_{Zn_i} = C_{e'} \propto p_{O_2}^{-1/4}. \tag{3.8}$$

Clearly each mechanism predicts a different dependence of defect concentration on oxygen partial pressure. Since the electrical conductivity depends on the concentration of conduction-band electrons, measurement of electrical conductivity as a function of oxygen partial pressure should serve to define the dependence law and indicate which defect is predominant. Initial experiments of this type carried out between 500 and 700 °C[4] indicated that the conductivity varied with oxygen partial pressure having exponents between 1/4.5 and 1/5. This indicates that neither defect mechanism predominates, and the actual structure could involve both singly and doubly charged interstitial cations.[2]

More recent work has confirmed that significant interstitial solution of zinc occurs in ZnO. The results of various studies have been interpreted in terms of neutral zinc interstitial atoms[5] and singly charged interstitial ions.[6,7] The latter are favoured presently, although it has also been suggested, on the basis of oxygen diffusion measurements, that oxygen vacancies are important above 1000 °C.[8] Kofstad[2] concludes that no overall consistent interpretation can be made regarding ZnO.

Inconsistencies between different investigations concerning the establishment of defect structures can frequently result from impurities in the sample, especially in the case of compounds which show only slight deviations from stoichiometry. As the purity of available materials improves, it would be prudent to repeat such measurements.

Non-metal deficit

Alternatively, n-type behaviour can be caused by non-metal deficit, as mentioned above. For oxides, this may be visualised as the discharge and subsequent

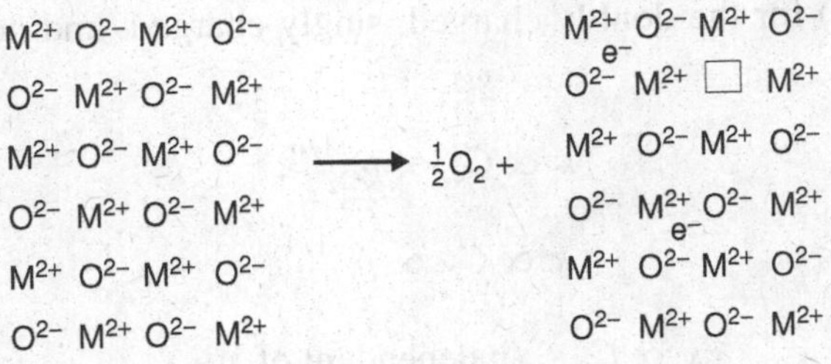

$$
\begin{array}{l}
M^{2+}\ O^{2-}\ M^{2+}\ O^{2-} \\
O^{2-}\ M^{2+}\ O^{2-}\ M^{2+} \\
M^{2+}\ O^{2-}\ M^{2+}\ O^{2-} \\
O^{2-}\ M^{2+}\ O^{2-}\ M^{2+} \\
M^{2+}\ O^{2-}\ M^{2+}\ O^{2-} \\
O^{2-}\ M^{2+}\ O^{2-}\ M^{2+}
\end{array}
\quad \longrightarrow \quad \tfrac{1}{2}O_2 + \quad
\begin{array}{l}
M^{2+}\ O^{2-}\ M^{2+}\ O^{2-} \\
\overset{e^-}{O^{2-}}\ M^{2+}\ \square\ M^{2+} \\
M^{2+}\ O^{2-}\ M^{2+}\ O^{2-} \\
O^{2-}\ \overset{e^-}{M^{2+}}\ O^{2-}\ M^{2+} \\
M^{2+}\ O^{2-}\ M^{2+}\ O^{2-} \\
O^{2-}\ M^{2+}\ O^{2-}\ M^{2+}
\end{array}
$$

$$ O_O = V_O^{\cdot\cdot} + 2e^- + \tfrac{1}{2}O_2 $$

Figure 3.6 Formation of oxygen-deficit MO with oxygen vacancies and excess electrons from 'perfect' MO.

evaporation of an oxygen ion; the electrons enter the conduction band and a vacancy is created on the anion lattice. The process is shown in Figure 3.6 and may be represented as in Equation (3.9):

$$ O_O = V_O^{\cdot\cdot} + 2e' + \frac{1}{2}O_2. \tag{3.9} $$

The vacant oxygen-ion site surrounded by positive ions represents a site of high positive charge to which a free electron may be attracted. It is therefore possible for the vacancy and the free electrons to become associated according to Equations (3.10) and (3.11):

$$ V_O^{\cdot\cdot} + e' = V_O^{\cdot}, \tag{3.10} $$

$$ V_O^{\cdot} + e' = V_O. \tag{3.11} $$

Therefore, we have the possibility that doubly and singly charged vacancies may exist as well as neutral vacancies.

The question of the charges on defects is not always obvious and must be considered from the aspect of the perfect lattice. If a site, which is normally occupied by an anion (O^{2-}), is vacant, then the lattice lacks two negative charges at that site compared with the normal lattice. The vacant anion site is, therefore, regarded as having two positive charges with respect to the normal lattice. Of course, such a site would represent an energy trough for electrons and so may capture conduction-band electrons, thus reducing the relative positive charge in the vacant site as indicated in Equations (3.10) and (3.11) above.

By noting the relationships between concentrations of anion vacancies and conduction-band electrons, the dependence of electrical conductivity κ on oxygen partial pressure can be derived as before. These are shown in Equations (3.12),

(3.13), and (3.14) for the doubly charged, singly charged, and neutral vacancies, respectively:

$$\kappa \propto C_{e'} \propto p_{O_2}^{-1/6}, \tag{3.12}$$

$$\kappa \propto C_{e'} \propto p_{O_2}^{-1/4} \tag{3.13}$$

$$\kappa \propto C_{e'} \quad \text{(independent of } p_{O_2}\text{)}. \tag{3.14}$$

Notice that in Equations (3.7), (3.8), (3.12), and (3.13) the power of the oxygen partial pressure is negative. This is the case for all n-type semiconductors.

Positive (p-type) semiconductors

In this case charge is transferred by positive carriers. This may arise from either a deficit of metal or an excess of non-metal.

Metal deficit

Positive (p-type) semiconduction arises from the formation of vacancies on the cation lattice together with electron holes, giving rise to conduction. The formula may be written $M_{1-\delta}O$. The value of δ can vary widely; from 0.05 in the case of $Fe_{0.95}O$ (wustite), 0.001 for NiO, to very small deviations in the case of Cr_2O_3 and Al_2O_3.

The possibility of forming electron holes lies in the ability of many metal ions, especially of the ions of transition metals, to exist in several valence states. The closer together the different valence states are in terms of energy of ionization, the more easily can cation vacancy formation be induced by the metal-deficit p-type mechanism.

The typical structure of this class of oxides is represented in Figure 3.7 taking NiO as an example. By virtue of the energetically close valence states of the cation it is relatively easy for an electron to transfer from a Ni^{2+} to a Ni^{3+} thus reversing the charges on the two ions. The site Ni^{3+} is thus seen to offer a low-energy position for an electron and is called an 'electron hole'.

The formation of this defect structure can easily be visualized if one considers the interaction of the NiO lattice with oxygen shown in Figure 3.8.

In step (b) the oxygen chemisorbs by attracting an electron from a Ni site thus forming a Ni^{3+} or hole. In step (c) the chemisorbed oxygen is fully ionized forming another hole and a Ni^{2+} ion enters the surface to partner the O^{2-}, thus forming a vacancy in the cation sub-lattice. Note that this process also forms an extra unit of NiO on the surface of the oxide, which should reflect in density changes if sufficiently sensitive measurements were made.

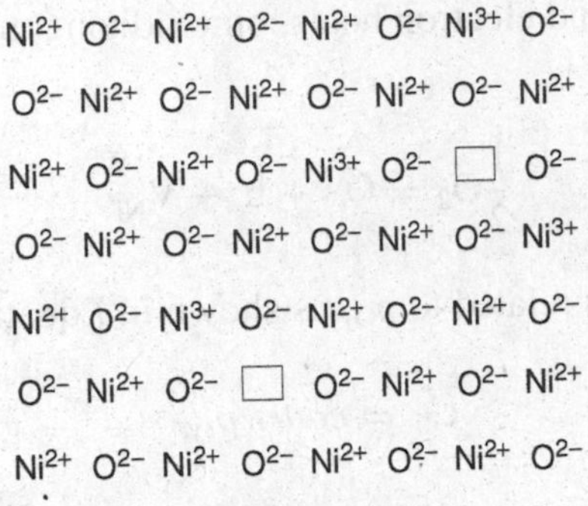

$$\text{Ni}^{2+} \ \text{O}^{2-} \ \text{Ni}^{2+} \ \text{O}^{2-} \ \text{Ni}^{2+} \ \text{O}^{2-} \ \text{Ni}^{3+} \ \text{O}^{2-}$$

$$\text{O}^{2-} \ \text{Ni}^{2+} \ \text{O}^{2-} \ \text{Ni}^{2+} \ \text{O}^{2-} \ \text{Ni}^{2+} \ \text{O}^{2-} \ \text{Ni}^{2+}$$

$$\text{Ni}^{2+} \ \text{O}^{2-} \ \text{Ni}^{2+} \ \text{O}^{2-} \ \text{Ni}^{3+} \ \text{O}^{2-} \ \square \ \text{O}^{2-}$$

$$\text{O}^{2-} \ \text{Ni}^{2+} \ \text{O}^{2-} \ \text{Ni}^{2+} \ \text{O}^{2-} \ \text{Ni}^{2+} \ \text{O}^{2-} \ \text{Ni}^{3+}$$

$$\text{Ni}^{2+} \ \text{O}^{2-} \ \text{Ni}^{3+} \ \text{O}^{2-} \ \text{Ni}^{2+} \ \text{O}^{2-} \ \text{Ni}^{2+} \ \text{O}^{2-}$$

$$\text{O}^{2-} \ \text{Ni}^{2+} \ \text{O}^{2-} \ \square \ \text{O}^{2-} \ \text{Ni}^{2+} \ \text{O}^{2-} \ \text{Ni}^{2+}$$

$$\text{Ni}^{2+} \ \text{O}^{2-} \ \text{Ni}^{2+} \ \text{O}^{2-} \ \text{Ni}^{2+} \ \text{O}^{2-} \ \text{Ni}^{2+} \ \text{O}^{2-}$$

Figure 3.7 Typical p-type metal-deficit semiconductor NiO with cation vacancies and positive holes.

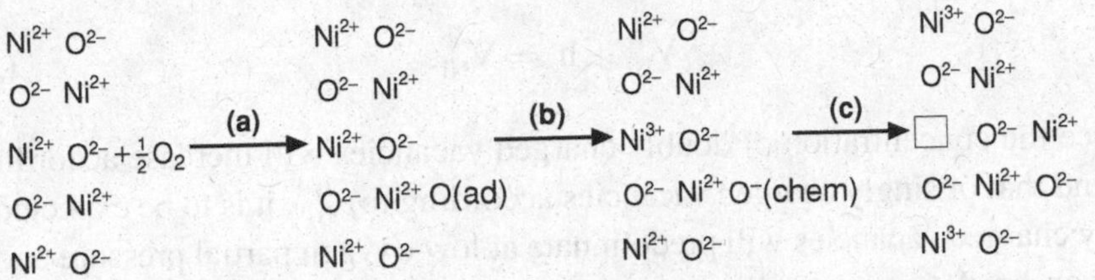

Figure 3.8 Formation of metal-deficit p-type semiconductor with cation vacancies and positive holes by incorporation of oxygen into the 'perfect' lattice. (a) Adsorption: $\frac{1}{2}\text{O}_2(\text{g}) = \text{O}(\text{ad})$; (b) chemisorption. $\text{O}(\text{ad}) = \text{O}^-(\text{chem}) + \text{h}^\cdot$; (c) ionization: $\text{O}^-(\text{chem}) = \text{O}_\text{O} + \text{V}_\text{N} + \text{h}^\cdot$. Overall reaction $\frac{1}{2}\text{O}_\text{O} + \text{V}_\text{N} + \text{Zh}^\cdot$.

The process can be represented by Equation (3.15),

$$\frac{1}{2}\text{O}_2 = \text{O}_\text{O} + 2\text{h}^\cdot + \text{V}_{\text{Ni}}'' \tag{3.15}$$

the equilibrium constant for which may be written as in Equation (3.16), assuming that Equation (3.15) represents the only mechanism by which the defects form and that these obey Henry's law:

$$C_{\text{h}^\cdot}^2 C_{\text{V}_{\text{Ni}}''} = K_{15} p_{\text{O}_2}^{1/2}. \tag{3.16}$$

By stoichiometry $C_{\text{h}^\cdot} = 2C_{\text{V}_{\text{Ni}}''}$ for electrical neutrality, and thus the relation in Equation (3.17) is obtained:

$$C_{\text{h}^\cdot} = const. \ p_{\text{O}_2}^{1/6}. \tag{3.17}$$

The electrical conductivity is expected to vary proportionally to the electron-hole concentration and so with the sixth root of the oxygen partial pressure. A further possibility, however, is that a cation vacancy may bond with an electron hole, i.e., in NiO a Ni^{3+} may be permanently attached to the Ni vacancy. In this case, the formation of vacancies and electron holes, on oxidizing, may be represented as in Equation (3.18):

$$\frac{1}{2}O_2 = O_O + h^\cdot + V'_{Ni}.$$
(3.18)

In this case we find $C_{h^\cdot}$ is related to p_{O_2} as shown in Equation (3.19).

$$C_{h^\cdot} = const.\, p_{O_2}^{1/4}$$
(3.19)

Experimental results on the variation of electrical conductivity with oxygen partial pressure have confirmed that both types of defect are found in NiO[2,4,9-17].

Equations (3.15) and (3.18) are clearly linked by Equation (3.20):

$$V''_{Ni} + h^\cdot = V'_{Ni}.$$
(3.20)

Since the concentration of doubly charged vacancies will increase according to $p_{O_2}^{1/6}$ and that of singly charged vacancies according to $p_{O_2}^{1/4}$, it is to be expected that doubly charged vacancies will predominate at low oxygen partial pressure and that singly charged vacancies will predominate at high oxygen partial pressures. Thus, the electrical conductivity is expected to vary with oxygen partial pressure according to a one-sixth power at low partial pressures and according to a one-fourth power at high partial pressures, corresponding to Equations (3.17) and (3.19), respectively. Some evidence exists for this behaviour[13,18,19].

It should be noticed that, in Equations (3.17) and (3.19), the power of the oxygen partial pressure is positive. This is so for all p-type semiconducting oxides.

Intrinsic semiconductors

There is another type of oxide which, although having quite close stoichiometry, shows relatively high electrical conductivity, which is independent of the oxygen partial pressure. Such behaviour is typical of so called 'intrinsic', or 'transitional' semiconduction when the concentration of electronic defects far exceeds that of ionic defects and the equilibrium of the electronic defects may be represented by the excitation of an electron from the valence band to the conduction band, producing a 'quasi-free' electron and an electron hole according to Equation (3.21):

$$Null = h^\cdot + e'.$$
(3.21)

The oxide CuO behaves in this way at $1000\,^\circ C$. Very few oxides belong to this class, which is not of great importance for an introductory study of the subject of high-temperature oxidation of metals.

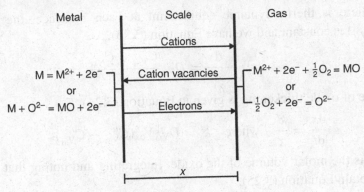

Figure 3.9 Simplified model for diffusion-controlled oxidation.

Rates of oxidation

It has already been seen that, for an oxidation process to proceed, under conditions where the two reactants are separated by the reaction product, it is necessary to postulate that ionic and electronic transport processes through the oxide are accompanied by ionizing phase-boundary reactions and formation of new oxide at a site whose position depends upon whether cations or anions are transported through the oxide layer.

Taking this very simple model and making still more simplifying assumptions, Wagner[20] was able to develop his celebrated theory of the high-temperature oxidation of metals. In fact, the theory describes the oxidation behaviour only for the case where diffusion of ions is rate determining and under highly idealized conditions. Before proceeding with the derivation of Wagner's expressions, a simplified treatment will be presented which may serve to emphasize the important features of diffusion-controlled oxidation.

Simplified treatment of diffusion-controlled oxidation

Assuming that cationic transport across the growing oxide layer controls the rate of scaling and that thermodynamic equilibrium is established at each interface, the process can be analyzed as follows. The outward cation flux, $j_{M^{2+}}$, is equal and opposite to the inward flux of cation defects (here taken to be vacancies). This model is shown in Figure 3.9.

Thus, $j_{M^{2+}}$ can be expressed as in Equation (3.22),

$$j_{M^{2+}} = -j_{V_M} = D_{V_M} \frac{C''_{V_M} - C'_{V_M}}{x}, \qquad (3.22)$$

where x is the oxide thickness, D_{V_M} is the diffusion coefficient for cation vacancies, and C'_{V_M} and C''_{V_M} are the vacancy concentrations at the scale–metal and scale–gas interfaces, respectively.

Since there is thermodynamic equilibrium at each interface, the value of $(C''_{V_M} - C'_{V_M})$ is constant and we have Equation (3.23),

$$j_{M^{2+}} = \frac{1}{V_{ox}}\frac{dx}{dt} = D_{V_M}\frac{C''_{V_M} - C'_{V_M}}{x}, \tag{3.23}$$

i.e., the rate of oxide thickening is given in Equation (3.24),

$$\frac{dx}{dt} = \frac{k'}{x} \quad \text{where} \quad k' = D_{V_M}V_{ox}(C''_{V_M} - C'_{V_M}), \tag{3.24}$$

where V_{ox} is the molar volume of the oxide. Integrating and noting that $x = 0$ at $t = 0$ we obtain Equation (3.25),

$$x^2 = 2k't, \tag{3.25}$$

which is the common parabolic rate law.

Furthermore, since it has been shown that the cation-vacancy concentration is related to the oxygen partial pressure by Equation (3.26),

$$C_{V_M} = const.p_{O_2}^{1/n}, $$

the variation of the parabolic rate constant with oxygen partial pressure can be predicted, i.e., Equation (3.26):

$$k' \propto \left[\left(p''_{O_2}\right)^{1/n} - \left(p'_{O_2}\right)^{1/n}\right]. \tag{3.26}$$

Since p'_{O_2} is usually negligible compared with p''_{O_2} we have Equation (3.27):

$$k' \propto \left(p''_{O_2}\right)^{1/n}. \tag{3.27}$$

Wagner theory of oxidation[20]

Figure 3.10 gives a summary of the conditions, later stated, for which the theory is valid. Assumptions are listed below.

(1) The oxide layer is a compact, perfectly adherent scale.
(2) Migration of ions or electrons across the scale is the rate-controlling process.
(3) Thermodynamic equilibrium is established at both the metal–scale and scale–gas interfaces.
(4) The oxide scale shows only small deviations from stoichiometry and, hence, the ionic fluxes are independent of position within the scale.
(5) Thermodynamic equilibrium is established locally throughout the scale.
(6) The scale is thick compared with the distances over which space charge effects (electrical double layer) occur.
(7) Oxygen solubility in the metal may be neglected.

Since thermodynamic equilibrium is assumed to be established at the metal–scale and scale–gas interfaces, it follows that activity gradients of both metal and

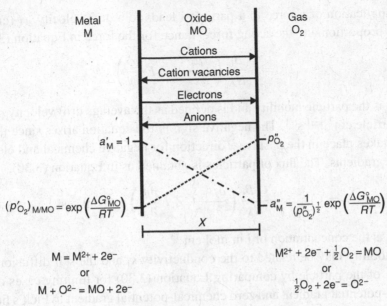

$M = M^{2+} + 2e^-$

or

$M + O^{2-} = MO + 2e^-$

$M^{2+} + 2e^- + \frac{1}{2}O_2 = MO$

or

$\frac{1}{2}O_2 + 2e^- = O^{2-}$

Overall reaction: $2M + O_2 = 2MO$; ΔG°_{MO}

Figure 3.10 Diagram of scale formation according to Wagner's model.

non-metal (oxygen, sulphur, etc.) are established across the scale. Consequently, metal ions and oxide ions will tend to migrate across the scale in opposite directions. Because the ions are charged, this migration will cause an electric field to be set up across the scale resulting in consequent transport of electrons across the scale from metal to atmosphere. The relative migration rates of cations, anions, and electrons are, therefore, balanced and no net charge transfer occurs across the oxide layer as a result of ion migration.

Being charged particles, ions will respond to both chemical and electrical-potential gradients, which together provide the net driving force for ion migration.

A particle, i, carrying a charge, Z_i, in a position where the chemical-potential gradient is $\dfrac{\partial \mu_i}{\partial x}$ and the electrical-potential gradient is $\dfrac{\partial \phi}{\partial x}$, is acted on by a force given by $\left(\dfrac{\partial \mu_i}{\partial x} + Z_i F \dfrac{\partial \phi}{\partial x} \right)$ J mol^{-1} cm^{-1} or by Equation (3.28):

$$\frac{1}{N_A} \left(\frac{\partial \mu_i}{\partial x} + Z_i F \frac{\partial \phi}{\partial x} \right) \text{ J particle}^{-1} \text{ cm}^{-1}, \qquad (3.28)$$

where N_A is Avogadro's number and F is the Faraday constant in coulombs equiv^{-1}. (Note that cm, rather than m, has been used as the unit of length since this will result in the various rate constants having units which are traditionally used.)

The application of a force to a particle i leads to a drift velocity, v_i (cm s^{-1}), which is proportional to the acting force, hence for the force in Equation (3.28),

$$v_i = -\frac{B_i}{N_A}\left(\frac{\partial \mu_i}{\partial x} + Z_i F \frac{\partial \phi}{\partial x}\right),$$

(3.29)

where B_i is the particle mobility and is defined as the average drift velocity per unit force, particle cm^2 J^{-1} s^{-1}. The negative sign in the equation arises since the drift velocity takes place in the positive x direction for negative chemical and electrical potential gradients. The flux of particles is obtained as in Equation (3.30),

$$j_i = C_i v_i = -\frac{C_i B_i}{N_A}\left(\frac{\partial \mu_i}{\partial x} + Z_i F \frac{\partial \phi}{\partial x}\right) \text{mol cm}^{-2}\,\text{s}^{-1}$$

(3.30)

where C_i is the concentration of i in mol cm^{-3}.

The mobility may be related to the conductivity κ_i and the self diffusion coefficient D_i of the particle by comparing Equation (3.30) for limiting cases of zero electrical-potential gradient and zero chemical-potential gradient to Fick's first law and Ohm's law, respectively. The relationships in Equation (3.31) are obtained,

$$k_B T B_i = D_i = \frac{RT\kappa_i}{C_i(Z_i F)^2}$$

(3.31)

where k_B is Boltzmann's constant. Upon using Equation (3.31) to replace the mobility B_i in Equation (3.30) with the partial conductivity one has Equation (3.32):

$$j_i = -\frac{\kappa_i}{Z_i^2 F^2}\left(\frac{\partial \mu_i}{\partial x} + Z_i F \frac{\partial \phi}{\partial x}\right)$$

(3.32)

Equation (3.32) can be used to describe the flux of cations, anions, or electrons through the oxide layer. Due to their different mobilities, different species would tend to move at different rates, however, this would set up electric fields tending to oppose this independence. In fact, the three species migrate at rates that are defined by the necessity of maintaining electroneutrality throughout the scale, i.e, such that there is no net charge across the oxide scale. This condition is usually achieved due to the very high mobility of electrons or electronic defects.

The original treatment by Wagner involved cations, anions, and electrons. However, the majority of oxides and sulphides show high electronic mobility, and the mobilities of the cation and anion species usually differ by several orders of magnitude. It is, therefore, possible to ignore the migration of the slower-moving ionic species and thus simplify the treatment somewhat.

The most frequently met case involves oxides and sulphides in which cations and electrons are the mobile species. Writing Z_c and Z_e for the charges on cations and electrons respectively, and using Equation (3.32), we have the respective fluxes given in Equations (3.33) and (3.34):

$$j_c = -\frac{\kappa_c}{Z_c^2 F^2}\left(\frac{\partial \mu_c}{\partial x} + Z_c F \frac{\partial \phi}{\partial x}\right), \tag{3.33}$$

$$j_e = -\frac{\kappa_e}{Z_e^2 F^2}\left(\frac{\partial \mu_e}{\partial x} + Z_e F \frac{\partial \phi}{\partial x}\right). \tag{3.34}$$

The condition for electrical neutrality is given by Equation (3.35),

$$Z_c j_c + Z_e j_e = 0, \tag{3.35}$$

and, using Equations (3.33)–(3.35), it is now possible to eliminate $\dfrac{\partial \phi}{\partial x}$ as Equation (3.36):

$$\frac{\partial \phi}{\partial x} = -\frac{1}{F(\kappa_c + \kappa_e)}\left[\frac{\kappa_c}{Z_c}\frac{\partial \mu_c}{\partial x} + \frac{\kappa_e}{Z_e}\frac{\partial \mu_e}{\partial x}\right]. \tag{3.36}$$

Substituting Equation (3.36) in Equation (3.33) we have for the flux of cations Equation (3.37).

$$j_c = -\frac{\kappa_c \kappa_e}{Z_c^2 F^2(\kappa_c + \kappa_e)}\left[\frac{\partial \mu_c}{\partial x} - \frac{Z_c}{Z_e}\frac{\partial \mu_e}{\partial x}\right]. \tag{3.37}$$

Since the electronic charge $Z_e = -1$, Equation (3.37) becomes Equation (3.38):

$$j_c = -\frac{\kappa_c \kappa_e}{Z_c^2 F^2(\kappa_c + \kappa_e)}\left[\frac{\partial \mu_c}{\partial x} + Z_c\frac{\partial \mu_e}{\partial x}\right]. \tag{3.38}$$

The ionization of a metal M is represented by $M = M^{Z_c+} + Z_c e$, and it follows that, at equilibrium, we have the relation given in Equation (3.39):

$$\mu_M = \mu_c + Z_c \mu_e. \tag{3.39}$$

Therefore, from Equations (3.38) and (3.39), Equation (3.40) is obtained:

$$j_c = -\frac{\kappa_c \kappa_e}{Z_c^2 F^2(\kappa_c + \kappa_e)}\frac{\partial \mu_M}{\partial x}. \tag{3.40}$$

Equation (3.40) is the expression for the cationic flux at any place in the scale where κ_c, κ_e, and $\dfrac{\partial \mu_M}{\partial x}$ are the instantaneous values at that place. Since all of these values may change with the position in the scale, it is necessary to integrate Equation (3.40) in order to define j_c in terms of the scale thickness and the measurable metal chemical potentials at the metal–scale, μ'_M, and scale–gas, μ''_M, interfaces. For phases with small deviations from stoichiometry, j_c can be assumed to be independent of x, hence defined by Equation (3.41a) or (3.41b),

$$j_c \int_0^x dx = -\frac{1}{Z_c^2 F^2} \int_{\mu'_M}^{\mu''_M} \frac{\kappa_c \kappa_e}{\kappa_c + \kappa_e} d\mu_M, \tag{3.41a}$$

$$j_c = -\frac{1}{Z_c^2 F^2 x} \int_{\mu'_M}^{\mu''_M} \frac{\kappa_c \kappa_e}{\kappa_c + \kappa_e} d\mu_M \text{ mol cm}^2 \text{ s}^{-1}. \tag{3.41b}$$

and, using Equation (3.35), Equation (3.42) is obtained:

$$j_e = -\frac{1}{Z_c F^2 x} \int_{\mu'_M}^{\mu''_M} \frac{\kappa_c \kappa_e}{\kappa_c + \kappa_e} d\mu_M \text{ mol cm}^{-2} \text{ s}^{-1}. \tag{3.42}$$

If the concentration of metal in the oxide scale is C_M mol cm^{-3} then the flux may also be expressed by Equation (3.43),

$$j_c = C_M \frac{dx}{dt}, \tag{3.43}$$

where x is the oxide-scale thickness.

The parabolic rate law is usually expressed by Equation (3.24), where k' is the parabolic rate constant in cm^2 s^{-1}.

Comparing Equations (3.41), (3.43), and (3.24), the parabolic rate constant is expressed by Equation (3.44):

$$k' = \frac{1}{Z_c^2 F^2 C_M} \int_{\mu''_M}^{\mu'_M} \frac{\kappa_c \kappa_e}{\kappa_c + \kappa_e} d\mu_M. \tag{3.44}$$

A similar treatment for the case where anions are more mobile than cations, i.e., the migration of cations may be neglected, yields Equation (3.45),

$$k' = \frac{1}{Z_a^2 F^2 C_X} \int_{\mu'_X}^{\mu''_X} \frac{\kappa_a \kappa_e}{\kappa_a + \kappa_e} d\mu_X, \tag{3.45}$$

where κ_a is the partial electrical conductivity of the anions and X is the non-metal, oxygen, or sulphur.

In general, it is found that the transport number of electrons, or electronic defects, is close to unity, compared with which the transport numbers of cations or anions are negligibly small. In these cases, Equations (3.44) and (3.45) reduce to Equations (3.46) and (3.47), respectively:

$$k' = \frac{1}{Z_c^2 F^2 C_M} \int_{\mu''_M}^{\mu'_M} \kappa_c d\mu_M \text{ cm}^2 \text{ s}^{-1}, \tag{3.46}$$

$$k' = \frac{1}{Z_a^2 F^2 C_X} \int_{\mu'_X}^{\mu''_X} \kappa_a d\mu_X \text{ cm}^2 \text{ s}^{-1}. \tag{3.47}$$

Using Equations (3.31), and (3.46), one has Equations (3.48) and (3.49),

$$k' = \frac{1}{RT} \int_{\mu''_M}^{\mu'_M} D_M d\mu_M \ \text{cm}^2 \, \text{s}^{-1}, \tag{3.48}$$

$$k' = \frac{1}{RT} \int_{\mu'_X}^{\mu''_X} D_X d\mu_X \ \text{cm}^2 \, \text{s}^{-1}, \tag{3.49}$$

for the parabolic rate constant, where D_M and D_X are the diffusion coefficients for metal, M, and non-metal, X, through the scale, respectively.

Equations (3.48) and (3.49) are written in terms of variables that can be measured relatively easily, although it is assumed that the diffusion coefficient is a function of the chemical potential of the species involved. Thus, in order to be able to calculate values of the parabolic rate constant, the relevant diffusion coefficient must be known as a function of the chemical potential of the mobile species. Such data are frequently not available or are incomplete. Furthermore, it is usually easier to measure the parabolic rate constant directly than to carry out experiments to measure the diffusion data. Thus, the real value of Wagner's analysis lies in providing a complete mechanistic understanding of the process of high-temperature oxidation under the conditions set out.

The predictions of Wagner's theory for n-type and p-type oxides in which cations are the mobile species should now be examined. The first class is represented by ZnO which is formed by the oxidation of zinc.

The defect structure of zinc oxide involves interstitial zinc ions and excess or quasi-free electrons according to Equations (3.1) and (3.2):

$$ZnO = Zn_i^{\cdot\cdot} + 2e' + \frac{1}{2}O_2, \tag{3.1}$$

$$ZnO = Zn_i^{\cdot} + e' + \frac{1}{2}O_2. \tag{3.2}$$

The partial conductivity of interstitial zinc ions is expected to vary with their concentration. As shown before, it is expected that this will vary according to the oxygen partial pressure, as shown in Equations (3.50) and (3.51),

$$\kappa_{Zn_i^{\cdot\cdot}} \propto C_{Zn_i^{\cdot\cdot}} \propto p_{O_2}^{-1/6}, \tag{3.50}$$

$$C_{Zn_i^{\cdot}} \propto p_{O_2}^{-1/4}, \tag{3.51}$$

i.e, $\kappa_{Zn_i^{\cdot\cdot}}$ and $\kappa_{Zn_i^{\cdot}}$ can be defined as in Equations (3.52) and (3.53):

$$\kappa_{Zn_i^{\cdot\cdot}} = const. \, p_{O_2}^{-1/6}, \tag{3.52}$$

$$\kappa_{Zn_i^{\cdot}} = const. \, p_{O_2}^{-1/4}. \tag{3.53}$$

Since ZnO shows only small deviations from stoichiometry we can write Equation (3.54):

$$\mu_{Zn} + \mu_O = \mu_{ZnO} \approx \text{constant};$$ (3.54)

also, μ_O may be expressed by Equation (3.55):

$$\mu_O = \frac{1}{2}\mu_{O_2}^o + \frac{1}{2}RT \ln p_{O_2}.$$ (3.55)

Therefore we obtain Equations (3.56a) (3.56b):

$$0 = d\mu_{Zn} + \frac{1}{2}RT d \ln p_{O_2},$$ (3.56a)

$$d\mu_{Zn} = -\frac{1}{2}RT d \ln p_{O_2}.$$ (3.56b)

Using Equations (3.46), (3.52), and (3.56b) we obtain Equation (3.57a):

$$k'_{ZnO} = -\text{const.} \int_{p''_{O_2}}^{p'_{O_2}} p_{O_2}^{-1/6} d \ln p_{O_2} = -\text{const.} \int_{p''_{O_2}}^{p'_{O_2}} p_{O_2}^{-7/6} dp_{O_2},$$ (3.57a)

which, upon integration, yields Equation (3.57b):

$$k'_{ZnO} = \text{const.''} \left[\left(\frac{1}{p'_{O_2}}\right)^{1/6} - \left(\frac{1}{p''_{O_2}}\right)^{1/6} \right].$$ (3.57b)

Alternatively, using Equations (3.46), (3.53), and (3.56) we obtain Equation (3.58):

$$k'_{ZnO} = \text{const.''} \left[\left(\frac{1}{p'_{O_2}}\right)^{1/4} - \left(\frac{1}{p''_{O_2}}\right)^{1/4} \right]$$ (3.58)

Generally p''_{O_2} is much greater than p'_{O_2} and so the parabolic rate constant for zinc is expected to be practically independent of the externally applied oxygen partial pressure, regardless of the relative abundances of $Zn_i^{..}$ and $Zn_i^{.}$.

A similar treatment can be applied to the oxidation of cobalt to CoO. Cobalt monoxide is a metal deficit p-type semiconductor forming cation vacancies and electron holes[21,22] according to Equation (3.59), where $K_{59} = C_{V'_{Co}} C_h \cdot p_{O_2}^{-1/2}$:

$$\frac{1}{2}O_2 = O_O + V'_{Co} + h^{.}.$$ (3.59)

Excluding defects caused by impurities and any contribution from intrinsic electronic defects we see from Equation (3.59) that, in order to satisfy both stoichiometry and electroneutrality, we have the relation shown in Equation (3.60):

$$C_{V'_{Co}} = C_h \cdot \propto p_{O_2}^{1/4},$$ (3.60)

and consequently the cationic partial conductivity is given by Equation (3.61):

$$\kappa_{Co} \propto p_{O_2}^{1/4}. \tag{3.61}$$

Thus, substituting Equation (3.61) into Equation (3.46), we obtain Equation (3.62) for the parabolic scaling constant (for Co being oxidized to CoO):

$$k'_{CoO} \propto \int_{p'_{O_2}}^{p''_{O_2}} p_{O_2}^{1/4} \mathrm{d} \ln p_{O_2}. \tag{3.62a}$$

Alternatively we can have Equation (3.62b),

$$k'_{CoO} \propto \int_{p'_{O_2}}^{p''_{O_2}} p_{O_2}^{-3/4} \mathrm{d} p_{O_2}, \tag{3.62b}$$

and integrating obtain Equation (3.63):

$$k'_{CoO} \propto \left[\left(p''_{O_2} \right)^{1/4} - \left(p'_{O_2} \right)^{1/4} \right]. \tag{3.63}$$

In Equation (3.63) the value of p'_{O_2} for the equilibrium between Co and CoO can be neglected compared with the value of p''_{O_2} in the atmosphere. This is valid because diffusion control of the oxidation reaction rate is achieved only under conditions where p''_{O_2} is sufficiently high to avoid control of the reaction rate by surface reactions or transport through the gas phase.

Under this assumption, Equation (3.63) simplifies to Equation (3.64).

$$k'_{CoO} \propto \left(p''_{O_2} \right)^{1/4} \tag{3.64}$$

The existence of doubly charged cation vacancies might also be postulated according to Equation (3.65),

$$\frac{1}{2} O_2 = O_O + V''_{Co} + 2h^{\cdot}, \tag{3.65}$$

for which K_{65} is expressed by Equation (3.66):

$$K_{65} = C_{V''_{Co}} C_h^2 p_{O_2}^{-1/2}. \tag{3.66}$$

If these defects were predominant then, following the same argument, the parabolic rate constant for oxidation of cobalt would be expected to vary with oxygen partial pressure as in Equation (3.67):

$$k'_{CoO} \propto \left(p''_{O_2} \right)^{1/6}. \tag{3.67}$$

Measurements of the electrical conductivity of CoO as a function of oxygen partial pressure by Fisher and Tannhäuser[21] show that the doubly charged vacancy model according to Equation (3.65) is obeyed up to values of oxygen partial pressure in the region of 10^{-5} atm. Above this value the singly charged model according to

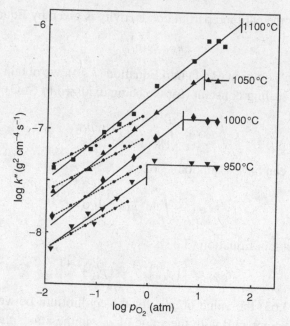

Figure 3.11 Variation of the parabolic rate constant with oxygen partial pressure and temperature for the oxidation of cobalt, showing the results of Bridges, Baur, and Fassell[24] (■▲◆▼) and Mrowec and Przybylski (●---●---●).[27]

Equation (3.59) is obeyed. Since the oxygen partial pressure at which the change occurs is so low, it may be expected that the CoO scale, growing on cobalt, would contain defects according to Equation (3.59) predominantly, except for a very thin layer immediately in contact with the metal.[22] The effect of this would be negligible.

Thus, if cobalt is oxidized in atmospheres where the oxygen partial pressure is substantially greater than 10^{-5} atm, the rate constant is expected to vary with $p_{O_2}^{1/4}$.

The oxidation of cobalt has been investigated[23,24] over a range of oxygen partial pressures and temperatures. A plot of $\log k''_{CoO}$ versus $\log p_{O_2}$ resulted in a series of lines of slope about 1/3 (varying between 1/2.6 and 1/3.6) compared with the expected. These results are shown in Figure 3.11 and can be interpreted by the presence, in the CoO, also of neutral cation vacancies. (Note that k'' is obtained from mass change measurements and is related to k' through conversions which will be found later in Table 3.2.) Neutral cation vacancies form according to Equation (3.68):

$$\frac{1}{2}O_2 = O_O + V_{Co}. \tag{3.68}$$

It is important to realize that, since their formation does not involve electrical defects, the presence of neutral vacancies will not be detected by electrical conductivity measurements made as a function of oxygen partial pressure. However, neutral vacancies can, and do, affect the diffusion of cations and, if Equation (3.68) represents the majority of defect formation in CoO, it would be expected that the diffusion coefficient of cobalt and the parabolic rate constant would vary according to Equations (3.69a) and (3.69b):

$$D_{Co} \propto p_{O_2}^{1/2}, \tag{3.69a}$$

$$k_{Co} \propto p_{O_2}^{1/2}. \tag{3.69b}$$

The results shown in Figure (3.11), which give the exponent of about 1/3, confirm that in CoO singly charged and neutral vacancies are the important species.

Also to be seen from Figure (3.11) is the fact that, as soon as the oxygen potential of the atmosphere exceeds that at which Co_3O_4 can form, the parabolic rate constant becomes insensitive to the external oxygen partial pressure. This is the result of the oxygen partial pressure being fixed at the CoO/ Co_3O_4 interface. The Co_3O_4 layer is more slowly growing than the CoO layer and does not contribute appreciably to the overall kinetics.

As a further example of how the presence of several defect types can influence the results, let us consider the oxidation of copper to Cu_2O. Cuprous oxide is a metal-deficit p-type semiconductor forming cation vacancies and electron holes. The formation of these defects may be represented by Equation (3.70):

$$\frac{1}{2}O_2 = O_O + 2V'_{Cu} + 2h^{.}. \tag{3.70}$$

However, it is also possible for the vacancies and electron holes to associate producing uncharged copper vacancies, according to Equation (3.71):

$$V'_{Cu} + h^{.} = V_{Cu}. \tag{3.71}$$

Alternatively, the formation of neutral cation vacancies may be represented by combining Equations (3.70) and (3.71) as in Equation (3.72):

$$\frac{1}{2}O_2 = O_O + 2V_{Cu}. \tag{3.72}$$

Following the same derivation as in the case of CoO it is seen that, if Equation (3.70) represents the defect structure of Cu_2O, then the parabolic rate constant for the oxidation of copper to Cu_2O should follow the relationship shown in Equation (3.73):

$$k_{Cu_2O} \propto p_{O_2}^{1/8}. \tag{3.73}$$

Alternatively, if the defect structure is represented by Equation (3.72), i.e., neutral cation vacancies predominate, then the relationship for the parabolic rate constant

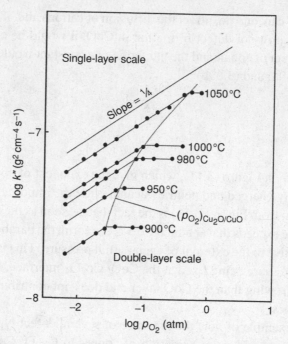

Figure 3.12 Variation of the parabolic rate constant with oxygen partial pressure and temperature for the oxidation of copper, following Mrowec and Stoklosa[25] and Mrowec et al.[26]

should be that shown in Equation (3.74):

$$k_{Cu_2O} \propto p_{O_2}^{1/4}. \tag{3.74}$$

Mrowec and Stoklosa[25] and Mrowec et al.[26] have investigated the oxidation of Cu to Cu_2O over a range of temperatures and oxygen partial pressures and their results are shown in Figure 3.12. From this it is clear that the exponent is very close to 1/4, indicating that neutral cation vacancies are the predominant defect species in Cu_2O.

A more complete test of the Wagner mechanism and treatment has been carried out by Fueki and Wagner[13] and by Mrowec and Przybylski[27,28] who derived values for the diffusion coefficient of nickel in NiO and cobalt in CoO from measurements of the parabolic oxidation rate constant. Since the parabolic rate constant can be expressed in the form of Equation (3.48),

$$k' = \frac{1}{RT} \int_{\mu_M''}^{\mu_M'} D_M d\mu_M, \tag{3.48}$$

it should be possible to calculate k' from knowledge of the diffusion coefficient D_M as a function of the chemical potential of M, μ_M. It is, however, easier to measure the oxygen partial pressure than the metal chemical potential and Equation (3.48) can be rearranged accordingly. For example, if the metal oxide is formed according to the general Equation (3.75),

$$2M + \frac{Z_c}{2}O_2 = M_2O_{Z_c},\qquad(3.75)$$

and assuming that deviations from stoichiometry are small, such that the chemical potential of $M_2O_{Z_c}$ is constant, we have Equation (3.76),

$$2\mu_M + Z_c\mu_O = \mu_{M_2O_{Z_c}} = const.,\qquad(3.76)$$

from which, since $\mu_O = \frac{1}{2}(\mu_{O_2}^0 + RT\ln p_{O_2})$, we obtain Equation (3.77):

$$d\mu_M = -\frac{Z_c}{4}RT\,d\ln p_{O_2}.\qquad(3.77)$$

Substituting Equation (3.77) in Equation (3.48) the result is Equation (3.78):

$$k' = \frac{Z_c}{4}\int_{p'_{O_2}}^{p''_{O_2}} D_M\,d\ln p_{O_2}.\qquad(3.78)$$

The difficulty in using Equation (3.78) to calculate values of k' lies in having to know how the value of D_M varies with defect concentration and, therefore, with oxygen potential.

Fueki and Wagner[13] have overcome this problem by using a differentiation technique introduced by C. Wagner and used earlier[29] to obtain transference numbers for Sb_2S_3. This technique involves differentiating Equation (3.78) with respect to $\ln p''_{O_2}$ for constant p'_{O_2}, i.e., Equation (3.79):

$$\frac{dk'}{d\ln p''_{O_2}} = \frac{Z_c}{4}D_M.\qquad(3.79)$$

If, therefore, the value of k' is measured as a function of the external oxygen partial pressure, p''_{O_2}, at constant temperature, then D_M can be obtained from the slope of the plot of k' versus $\ln p''_{O_2}$. Fueki and Wagner[13] determined the diffusion coefficient of nickel in NiO as a function of oxygen partial pressure in this way. The results indicated that D_{Ni} was proportional to $p_{O_2}^{1/6}$ at 1000 °C, to $p_{O_2}^{1/4}$ at 1300 °C, and to $p_{O_2}^{1/3.5}$ at 1400 °C, indicating a predominance of doubly charged vacancies at 1000 °C and singly charged vacancies at 1300 °C, with the existence of neutral vacancies at higher temperatures.

Mrowec and Przybylski have applied this method to the cobalt–oxygen system, since the defect structure is well known and the self-diffusion coefficient of cobalt has been measured accurately using tracer techniques by Carter and Richardson[30] and also Chen *et al.*[31] Discs of cobalt were oxidized between 900 and 1300 °C and between 10^{-6} and 1 atm of oxygen. The resulting plot of k″ against log p_{O_2} was curved showing, according to Equation (3.79), that k″ varied with oxygen partial

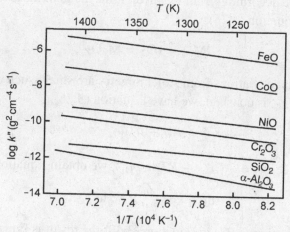

Figure 3.13 Order-of-magnitude parabolic rate constants for the growth of several oxides.

pressure. The slope of the curve, $\dfrac{dk''}{d \log p_{O_2}}$, was used to obtain values of D_{Co}, which give a convincing demonstration of the soundness of Wagner's analysis of parabolic oxidation. Tracer results: $D_{Co} = 0.0052 \exp(-19\,200/T)$ cm^2 s^{-1} (ref. 30); $D_{Co} = 0.0050 \exp(-19\,400/T)$ cm^2 s^{-1} (ref. 31). Whereas the oxidation measurement yielded $D_{Co} = 0.0050 \exp(-19\,100/T)$ cm^2 s^{-1} (ref. 28).

Thus, in principle, insight into the transport mechanisms by which oxidation reactions proceed is obtained by studying the variation of the oxidation rates with oxygen pressure and temperature. The absolute rates themselves are useful in comparing the oxidation kinetics of several metals. Some of these results are plotted in Figure 3.13 from which it can be seen that the highest oxidation rates are, by and large, obtained in systems whose oxides have the highest defect concentrations. However, the rate constants in Figure 3.13 are generally several orders of magnitude larger than those which one would calculate from lattice diffusion data from Equations (3.48) or (3.49). This indicates that some form of 'short-circuit transport' is contributing to growth of the oxide film.

Short-circuit diffusion

A detailed knowledge of transport paths during the growth of product films from the reaction of a gas and metal is usually not available. However, for some systems qualitative results are available. Atkinson *et al.*[32] have reported that the oxidation of Ni is controlled by the outward diffusion of Ni ions along grain boundaries in the NiO film at temperatures below about 1100 °C. Figure 3.14 indicates diffusion results collected for NiO.[33] The boundary diffusivities were extracted by assuming

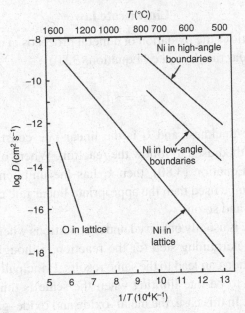

Figure 3.14 Summary of diffusion parameters for NiO, after Atkinson *et al.*[33] The Ni data were obtained at unit activity of oxygen and the O data at an oxygen partial pressure of 0.2.

a 'grain-boundary width' of 1 nm. The lattice-diffusion coefficient for Ni ions is much larger than that for O ions, as would be expected from the defect structure of NiO. However, the diffusion coefficients for Ni ions in low-angle and high-angle grain-boundaries are larger still. The relative contributions to scale growth of bulk and grain-boundary diffusion will depend on the temperature and the oxide grain size. It is generally observed that oxide scales growing on metals have rather fine grain sizes (on the order of a micrometer) so boundary diffusion can predominate to quite high temperatures. It is reported for NiO that boundary diffusion has a similar p_{O_2}-dependence to bulk diffusion[33] which suggests that similar point defects control both processes.

Measurements of transport in growing Cr_2O_3 scales, using $^{16}O/^{18}O$-exchange measurements, indicate that growth is primarily by Cr ions moving outward along grain boundaries.[34] Conversely, extensive studies of Al_2O_3, reviewed in ref. 35, indicate that alumina grows primarily by O ions moving inward along oxide grain boundaries, although there is some component of outward growth as well.

Irrespective of details of the transport processes, in the development of oxidation-resistant alloys, the aim is to evolve a composition that develops one of the slow-growing oxides as a scale. This is usually done to encourage formation of Al_2O_3, SiO_2, or Cr_2O_3 and will be discussed in Chapter 5.

Linear rate law

Under certain conditions the oxidation of a metal proceeds at a constant rate and is said to obey the 'linear rate law', i.e., Equation (3.80),

$$x = k_1 t, \tag{3.80}$$

where x is the scale thickness and k_1 is the linear rate constant. The units of k_1 depend upon the method used to follow the reaction. Where oxide scale thickness is measured, as in Equation (3.80), then k_1 has the units $cm\ s^{-1}$; if mass-gain measurement were to be used then the appropriate linear rate constant would have units of $g\ cm^{-2}\ s^{-1}$, and so on.

The linear rate law is usually observed under conditions where a phase-boundary process is the rate-determining step for the reaction, although other steps in the mechanism of oxidation can lead to the same results. Principally, diffusion through the scale is unlikely to be rate limiting when the scale is thin, i.e., in the initial stages of the process. In this case, the metal–oxide and oxide–gas interfaces cannot be assumed to be in thermodynamic equilibrium, although there are few known cases where the reaction at the metal–scale interface has been found to play a rate-determining role. It can, therefore, be assumed that the reactions occurring at the metal–scale interface, i.e., the ionization of metal in the case of cation conductors or ionization of metal and oxide formation in the case of anion conducting scales, are fast and, in general, attention can be turned to the process occurring at the scale–gas interface for an interpretation of the linear rate law.

At the scale–gas interface, the processes can be broken down into several steps. Reactant gas molecules must approach the scale surface and become adsorbed there, the adsorbed molecules then split to form adsorbed oxygen, which eventually attracts electrons from the oxide lattice to become initially chemisorbed and finally incorporated into the lattice. The removal of the electrons from the scale causes a change in the concentration of electron defects in the oxide at the scale–gas interface. These processes can be represented as in Equation (3.81):

$$\frac{1}{2}O_2(g) \underset{(1)}{\rightarrow} \frac{1}{2}O_2(ad) \underset{(2)}{\rightarrow} O(ad) \underset{(3)}{\overset{e^-}{\rightarrow}} O^-(chem) \underset{(4)}{\overset{e^-}{\rightarrow}} O^{2-}(latt). \tag{3.81}$$

which is written in terms of the oxygen molecule being the active oxidizing species in the gas phase. Except for cases of very low oxygen partial pressure, reactions carried out in such atmospheres do not lead to constant rate kinetics. It can, therefore, be assumed that the steps 1–4, which occur on the oxide surface, are fast reactions.

If, however, the oxidizing medium is a CO–CO_2 mixture then constant rate kinetics are readily observed.[36,37] In this case the surface reactions may be represented as in Equataion (3.82):

$$CO_2(g) \xrightarrow[(1)]{} CO_2(ad) \xrightarrow[(2)]{} CO(ad) + O(ad) \xrightarrow[(3)]{e^-}$$

$$O^-(chem) \xrightarrow[(4)]{e^-} O^{2-}(latt). \tag{3.82}$$

Primarily, CO_2 adsorbs onto the oxide surface and there dissociates to adsorbed CO and O species. The adsorbed oxygen then goes through the ionization stages already given in Equation (3.81). Two further processes are the desorption of adsorbed carbon monoxide and the possibility of the removal of adsorbed oxygen by reaction with carbon monoxide molecules from the gas phase. Since steps 3 and 4 can be assumed from above to be fast and since the rate of impingement of CO_2 on the scale surface, from atmospheres containing high concentrations of CO_2, is high, the most likely rate-determining step is step 2 (dissociation of adsorbed carbon dioxide to form adsorbed carbon monoxide molecules and oxygen atoms on the surface of the oxide scale).

Hauffe and Pfeiffer[36] investigated the oxidation of iron in $CO-CO_2$ mixtures between 900 and 1000 °C and found that, under a total gas pressure of 1 atm, the rate was constant and proportional to $(p_{CO_2}/p_{CO})^{2/3}$, i.e, to $p_{O_2}^{1/3}$. They suggested that this may indicate that chemisorption of oxygen is the rate-controlling step.

The same reaction was investigated in more detail by Pettit *et al.*[37] who assumed that the rate-determining step was the dissociation of CO_2 to give CO and adsorbed oxygen according to Equation (3.83):

$$CO_2(g) = CO(g) + O(ad). \tag{3.83}$$

If this assumption is correct, then the rate of the reaction per unit area $\dot{n}/A$, where $\dot{n}$ is the equivalents of oxide per unit time, would be given by Equation (3.84):

$$\frac{\dot{n}}{A} = k' p_{CO_2} - k'' p_{CO}. \tag{3.84}$$

where k' and k'' are constants specifying the rates of the forward and reverse reactions, respectively. During the initial stages of reaction, at least, the scale is largely in equilibrium with the iron substrate, thus substantial changes in oxide-defect concentrations at the scale–gas interface do not occur. Consequently, this factor plays no part in Equation (3.84).

If the gas had a composition in equilibrium with iron and wustite, then the reaction rate would be zero, i.e., $\dot{n}/A = 0$. Consequently we would have Equation (3.85),

$$\frac{k''}{k'} = \frac{p_{CO_2}}{p_{CO}} = K, \tag{3.85}$$

Mechanisms of oxidation

where K is the equilibrium constant for the reaction $Fe + CO_2 = FeO + CO$, and, therefore, from Equations (3.84) and (3.85), the rate under CO_2/CO ratios exceeding the equilibrium condition is given by Equation (3.86):

$$\frac{\dot{n}}{A} = k'(p_{CO_2} - Kp_{CO}).$$

(3.86)

The component partial pressures can be written in terms of the relevant mole fractions N_{CO_2} and N_{CO} and the total pressure P:

$$p_{CO_2} = PN_{CO_2},$$

(3.87)

$$p_{CO} = PN_{CO} = P(1 - N_{CO_2}).$$

(3.88)

Substituting Equations (3.87) and (3.88) in Equation (3.86) gives (Equation 3.89a),

$$\frac{\dot{n}}{A} = k'P\left[(1 + K)N_{CO_2} - K\right],$$

(3.89a)

which can be rearranged to Equation (3.89b),

$$\frac{\dot{n}}{A} = k'P(1 + K)\left[N_{CO_2} - N_{CO_2}(eq)\right],$$

(3.89b)

where $N_{CO_2}(eq)$ refers to the mole fraction of CO_2 in the CO_2/CO mixture in equilibrium with iron and wustite. According to Equation (3.89b) the oxidation rate at a given temperature should be proportional to both the mole fraction of CO_2 in the gas and the total pressure of the gas.

The experimental results for 925 °C shown in Figure 3.15 substantiate that both of the above conditions are met and, therefore, the rate-determining step in the reaction is the decomposition of carbon dioxide to carbon monoxide and adsorbed oxygen.

Rarefied atmospheres

Constant reaction rates can also be observed during reactions of a metal under rarified atmospheres and with reactive gases diluted with an inert, non-reactive gas.

A good example of this case is the heat treatment of metals under partial vacuum when the atmosphere consists of nitrogen and oxygen molecules at a low pressure, where the rate-determining step is likely to be the rate of impingement of oxygen molecules on the specimen surface. This rate can be calculated by using the Hertz–Knudsen–Langmuir equation derived from the kinetic theory of gases, Equation (3.90),

$$j_{O_2} = \alpha \frac{p_{O_2}}{\left(2\pi M_{O_2}RT\right)^{1/2}},$$

(3.90)

Table 3.1

Flux, j_{O_2}	Pressure, p_{O_2}	R	M_{O_2}
mol cm^{-2} s^{-1}	atm	cm^3 atm mol^{-1} K^{-1}	g mol^{-1}
mol cm^{-2} s^{-1}	dyn cm^{-2}	erg mol^{-1} K^{-1}	g mol^{-1}
mol cm^{-2} s^{-1}	N m^{-2}	N m mol^{-1} K^{-1}	kg mol^{-1}

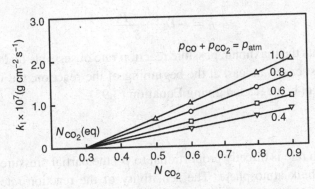

Figure 3.15 Variation of oxidation rate of iron in CO–CO$_2$ with CO$_2$ mole fraction, N. After Pettit *et al.*[37]

where p_{O_2} is the oxygen partial pressure in the partial vacuum, M_{O_2} is the molecular mass of oxygen, R is the gas constant, T is the absolute temperature, and α is the adherence coefficient; α has a maximum value of unity where all of the impinging molecules adsorb and enter into reaction. Thus the maximum possible reaction rate is given when $\alpha = 1$ in Equation (3.90). When using Equation (3.90) it is necessary to use compatible units and these are tabulated in Table 3.1. Evaluating Equation (3.90) for reaction of a metal with oxygen existing at a partial pressure of 10^{-6} atm in a partial vacuum at 1000 K, the maximum possible reaction rate is $j_{O_2} = 2.45 \times 10^{-7}$ mol cm^{-2} s^{-1} or 7.8×10^{-6} g cm^{-2} s^{-1}.

According to Equation (3.90), the reaction rate will vary with temperature for a given oxygen partial pressure according to Equation (3.91):

$$j_{O_2} = const. \; T^{-1/2}. \tag{3.91}$$

Dilute gases

When a metal is reacted with an atmosphere consisting of an active species diluted by an inactive or inert species, the active molecules are rapidly depleted in the gas layers immediately adjacent to the specimen surface. Subsequent reaction can then only proceed as long as molecules of the active species can diffuse through the denuded layer to the metal surface. When the concentration of the active species is low in the atmosphere this step can become rate controlling. The denuded gas layer next to the metal surface may be regarded as a boundary layer of thickness δ. If the concentrations of the active species at the metal surface and in the bulk gas are expressed, respectively, by the partial pressure p_i'' and p_i' then the rate at which active species, i, diffuse across the boundary layer may be expressed as in Equation (3.92),

$$j_i = -D_i \frac{p_i' - p_i''}{\delta},$$

(3.92)

which represents the maximum possible reaction rate observable under the circumstances. In most cases, at least at the beginning of the reaction, the value of p_i'' is very low and can be neglected giving Equation (3.93),

$$j_i \approx -D_i \frac{p_i'}{\delta},$$

(3.93)

i.e., the reaction rate is directly proportional to p_i', the partial pressure of the active species in the bulk atmosphere. The sensitivity of the reaction rate to temperature arises solely from the way in which D_i and, to a lesser extent, δ varies with temperature.

Reaction-rate control by this step can be identified most securely by the sensitivity of the reaction rate to the gas flow rate. If the gas flow rate is increased the thickness of the boundary layer is decreased and, correspondingly, the reaction rate increases according to Equation (3.92).

It is important to note that the concentration of adsorbed species and, therefore, the rates of surface reactions involving adsorbed species also vary with the partial pressure of the active species in the gas phase. Thus, in order to identify boundary-layer transport as the process controlling a constant rate, the sensitivity of the rate to both active-species partial pressure and gas flow rate must be established.

Transition from linear law to parabolic law

It was stated earlier that constant-rate kinetics are often observed with thin scales. This is obviously a relative term and denotes the initial stages of a reaction which may continue for a long, or short, time depending on the conditions under which the reaction occurs.

Initially, when a scale is formed and is very thin, diffusion through the scale will be rapid, establishing virtual equilibrium with the metal at the scale–gas interface. In other words, the metal activity at this interface will be maintained at a high value, initially close to unity, by rapid diffusion within the scale. Under these conditions, the rate of the reaction is likely to be controlled by one of the steps discussed above.

As the reaction proceeds at a constant rate the scale layer thickens, at the same time the flux of ions through the scale must be equivalent to the surface reaction rate. To maintain this constant flux, the activity of the metal at the scale–gas interface must fall as the scale thickens, eventually approaching the value in equilibrium with the atmosphere. Since the metal activity cannot fall below this value, further increase in scale thickness must result in a reduction in the metal activity gradient across the scale and, consequently, to a reduction in ionic flux and the reaction rate. At this point the transport of ions across the scale becomes the rate-controlling process and the rate falls with time according to a parabolic rate law. The transition from linear to parabolic kinetics has been documented and described for the oxidation of iron in CO_2–CO gas mixtures.[38]

Logarithmic rate law

When metals are oxidized under certain conditions, typically at low temperatures of up to about 400 °C, the initial oxide formation, up to the 1000 Å range, is characterized by an initial rapid reaction that quickly reduces to a very low rate of reaction. Such behaviour has been found to conform to rate laws described by logarithmic functions such as in Equations (3.94) and (3.95),

$$x = k_{log} \log (t + t_o) + A \text{ (direct log law)}, \quad (3.94)$$

$$1/x = B - k_{il} \log t \text{ (inverse log law)}, \quad (3.95)$$

where A, B, t_o, k_{log}, and k_{il} are constants at constant temperature.

Several interpretations of this type of behaviour have been provided. For a detailed description and summary, Kofstad[39] and Hauffe[40] should be consulted. Since the phenomenon is principally one that occurs at low temperatures, a strictly qualitative description of the various interpretations will be given here, since this phase of the oxidation reaction generally occurs during the heating period of an investigation into the formation of thick scales at high temperatures.

Interpretations of the logarithmic laws have been based on the adsorption of reactive species, the effects of electric fields developed across oxide layers, quantum-mechanical tunnelling of electrons through the thin scales, progressive blocking of low-resistance diffusion paths, non-isothermal conditions in the oxide layer, and nucleation and growth processes. A concise summary of these theories has been given by Kofstad.[39]

Two things should be borne in mind. Firstly, it is difficult experimentally to obtain accurate and reliable measurements of reaction rates relating to the formation of the initial oxide layers, although, lately, modern high-vacuum techniques may allow an accurate start to the reaction to be assumed by achieving thermal equilibrium before admitting the reactant gas. Secondly, in any thermodynamic treatment carried out to assess defect concentrations or concentration gradients, it may be necessary to include surface, or interfacial energies when considering such thin films. Thus, the actual deviations from stoichiometry in the film may be unknown. Similarly the physical properties of the film may vary significantly from those of the bulk material. Due to difficulties such as these, the interpretation of logarithmic rate behaviour is the least-understood area of metal oxidation.

Adsorption has been assumed to be the rate-determining process during early oxide-film formation. When a clean surface is exposed to an oxidizing gas, each molecule impinging on the surface may either rebound or adsorb. The fraction, α, that remains adsorbed on the metal surface should be constant for a constant temperature and oxygen partial pressure. Therefore, under these conditions a constant reaction rate is expected. However, the value of α is markedly lower on those parts of the surface covered with a monolayer or oxide nuclei. Thus, as adsorption or oxide nucleation proceeds, the reaction rate is expected to decrease accordingly until complete coverage of the surface by oxide has been achieved, when a much lower rate is expected to be observed.

These initial processes of adsorption and nucleation have been investigated principally by Benard and coworkers,[41] who have demonstrated convincingly that the initial nucleation of oxide occurs at discrete sites on the metal surface. The oxide islands proceed to grow rapidly over the metal surface until complete coverage is eventually achieved. Furthermore, it has been demonstrated that the step of adsorption begins at oxygen partial pressures significantly below the decomposition pressure of the oxide.

When the oxidizing gas chemisorbs onto the oxide surface growing on the metal, it captures, or withdraws, electrons from the oxide lattice according to Equation (3.81):

$$\frac{1}{2}O_2(g) \longrightarrow O(ad) \xrightarrow{e^-} O^-(chem).\qquad(3.81)$$

The result is to create additional electron holes in the surface of a p-type oxide or reduce the number of excess electrons in the surface of an n-type oxide. The electrons are withdrawn from a depth of about 100 Å over which, consequently, a strong electric field exists, known as a space charge. This concept has been used by Cabrera and Mott[42] to explain the inverse logarithmic law for the case of the formation of thin oxide films of very tight stoichiometry and, consequently, low ionic migration rates at low temperatures due to the low defect concentrations (the electronic mobility is assumed to be greater than that of the ions). The electric field set up by chemisorption of oxygen, aided by quantum-mechanical tunneling of electrons from metal through the oxide to the adsorbed oxygen, acts on the mobile ionic species and increases the migration rate. Clearly, since the potential difference between the metal and adsorbed oxygen species is assumed to be constant, the thinner the oxide layer the stronger the field and the faster the ionic diffusion. As the film thickens, the field strength is reduced and the reaction rate falls. When the oxide layer thickness exceeds about 100 Å, tunneling of electrons from the metal through the scale is not feasible and the full potential difference is no longer exerted. Thus, eventually, very low reaction rates are observed.

Inverse logarithmic rates have been reported by Young *et al.*[43] for copper at 70 °C, Hart[44] for aluminium at 20 °C and Roberts[45] for iron at temperatures up to 120 °C.

Mott[46] proposed an interpretation of the direct log law, later revised by Hauffe and Ilschner,[47] by explaining that the initial rapid reaction rates were due to quantum-mechanical tunneling of electrons through the thin oxide layer. Although electrons are not generally mobile in oxides at low temperatures, the electrical conductivity of thin oxide layers on metals is, surprisingly, much higher than would be expected from bulk electrical properties. Thus, if the transport of electrons were the rate-determining step, the reaction rate would be surprisingly high in the initial stages when the film is very thin. As the film thickens, the maximum thickness for which tunneling of electrons is valid is exceeded and the observed reaction rate falls away markedly.

Other interpretations of log laws have rejected the assumption of uniform transport properties across the oxide layers and assume instead that paths for rapid ionic transport exist, e.g., along grain boundaries, or dislocation pipes, or over the surface of pores. These paths are progressively blocked as growth proceeds, either by recrystallization and grain growth, or by blocking of pores by growth stresses in the oxide, or by a combination of all three processes. It is also possible that pores may act as vacancy sinks in the oxide and grow into cavities large enough to interfere effectively with ionic transport by reducing the cross-section of compact oxide through which transport may occur. The formation of second-phase alloy oxides of low ionic conductivity may also give the same progressive resistance to ionic migration. These theories have been considered primarily by Evans and co-workers.[48]

In view of the multiple interpretations available, it is clearly wise in practice to investigate thoroughly and judge each case on its merits rather than pass verdict from a simple examination of kinetics alone.

The presentation of Wagner's treatment of high-temperature oxidation of metals given here follows the original very closely. One of the assumptions made in this treatment is that the oxide scale under consideration is thick compared with the distance over which space-charge effects, discussed above, occur. Fromhold[49-51] has considered the effect of the presence of space charges at the scale–gas and metal–scale interfaces. His conclusions show that their presence would cause the rate constant to vary with scale thickness in the range of 5000–10 000 Å. This analysis is not considered further here since the oxide scales normally encountered are very much thicker than this.

Relationships between parabolic rate constants

In the above treatment the parabolic rate constant is derived for the units relating to measurement of the oxide thickness as the reaction parameter. As mentioned before, there are several methods of following the reaction depending on choice of reaction parameters, each of which produces its own particular parabolic rate constant, as shown below using Wagner's notation for the various parabolic rate constants.

(a) *Measurement of scale thickness (x):*

$$\frac{dx}{dt} = \frac{k'}{x}, \text{ i.e., } x^2 = 2k't.$$

(3.25)

In Equation (3.25) k' is the 'practical tarnishing constant' or 'scaling constant' and has units of $cm^2\ s^{-1}$.

(b) *Measurement of the mass increase of the specimen (m)*

The parabolic rate constant k'' is defined by Equation (3.96),

$$\left(\frac{m}{A}\right)^2 = k''t,$$

(3.96)

where A is the area over which reaction occurs; k'' is also referred to as the practical tarnishing constant or 'scaling constant' and has units of $g^2\ cm^{-4}\ s^{-1}$.

(c) *Measurement of metal surface recession (l)*

Measuring the thickness of metal consumed leads to the relationship defining k_c, Equation (3.97); k_c is called the 'corrosion constant' and has units of $cm^2\ s^{-1}$:

$$l^2 = 2k_ct.$$

(3.97)

(d) *Rate of growth of scale of unit thickness*

The rational rate constant is defined as the rate of growth over unit area, in equivalents per second, of a scale of unit thickness, i.e., Equation (3.98),

$$k_r = \frac{x}{A}\frac{d\tilde{n}}{dt},$$

(3.98)

where $\tilde{n}$ is the number of equivalents in the oxide layer of thickness x. k_r is sometimes called the theoretical tarnishing constant and has units of equivalents $cm^{-1}\ s^{-1}$.

It is easy to see that having so many different ways of expressing parabolic rate constants, and so many different symbols, produces a situation ripe for confusion.

Table 3.2 *Relationships between variously defined parabolic oxidation rate constants (for the scale stoichiometry expressed as $M_\nu X$).*

B	A	k' $cm^2\,s^{-1}$	k'' $g^2\,cm^{-4}\,s^{-1}$	k_c $cm^2\,s^{-1}$	k_r equiv. $cm^{-1}\,s^{-1}$
k' ($cm^2\,s^{-1}$)		1	$2[M_X/\overline{V}Z_X]^2$	$[\overline{V}_M/\overline{V}]^2$	$1/\overline{V}$
k'' ($g^2\,cm^{-4}\,s^{-1}$)		$\frac{1}{2}[\overline{V}Z_X]^2/M_X^2$	1	$\frac{1}{2}[\overline{V}_M Z_X/M_X]^2$	$\overline{V}/2[Z_X/M_X]^2$
k_c ($cm^2\,s^{-1}$)		$[\overline{V}/\overline{V}_M]^2$	$2[M_X/\overline{V}Z_X]^2$	1	$\overline{V}/\overline{V}_M^2$
k_r equiv. $cm^{-1}\,s^{-1}$		$\overline{V}$	$2/\overline{V}[M_X/Z_X]^2$	$\overline{V}_M^2/\overline{V}$	1

The relating factor F is given according to $A = FB$; A is listed horizontally and B vertically. The symbols have the following meaning: $\overline{V}$ = equivalent volume of scale (cm^3 equiv.$^{-1}$); $\overline{V}_M$ = equivalent volume of metal (cm^3 equiv.$^{-1}$); M_X = atomic mass of non-metal X (oxygen, sulphur, etc.); Z_X = valency of X (equiv. g-atom X^{-1}).

Consequently, it is necessary to check the definition of a rate constant very carefully when evaluating quantitative data.

It is easy to calculate the value of any rate constant from any other since they all represent the same process. The relationships are given in Table 3.2.

References

1. F. O. Kröger, *The Chemistry of Imperfect Crystals*, Amsterdam, North Holland Publishing Co., 1964.
2. P. Kofstad, *Nonstoichiometry, Diffusion, and Electrical Conductivity in Binary Metal Oxides*, New York, Wiley, 1972.
3. D. M. Smyth, *The Defect Chemistry of Metal Oxides*, Oxford, Oxford University Press, 2000.
4. H. H. von Baumbach and C. Z. Wagner, *Phys. Chem.*, **22** (1933), 199.
5. G. P. Mohanty and L. V. Azaroff, *J. Chem. Phys.*, **35** (1961), 1268.
6. D. G. Thomas, *J. Phys. Chem. Solids*, **3** (1957), 229.
7. E. Scharowsky, *Z. Phys.*, **135** (1953), 318.
8. J. W. Hoffmann and I. Lauder, *Trans. Faraday Soc.*, **66** (1970), 2346.
9. I. Bransky and N. M. Tallan, *J. Chem. Phys.*, **49** (1968), 1243.
10. N. G. Eror and J. B. Wagner, *J. Phys. Stat. Sol.*, **35** (1969), 641.
11. S. P. Mitoff, *J. Phys. Chem.*, **35** (1961), 882.
12. G. H. Meier and R. A. Rapp, *Z. Phys. Chem. NF*, **74** (1971), 168.
13. K. Fueki and J. B. Wagner, *J. Electrochem. Soc.*, **112** (1965), 384.
14. C. M. Osburn and R. W. Vest, *J. Phys. Solids*, **32** (1971), 1331.
15. J. E. Stroud, I. Bransky, and N. M. Tallan, *J. Chem. Phys.*, **56** (1973), 1263.
16. R. Farhi and G. Petot-Ervas, *J. Phys. Chem. Solids*, **39** (1978), 1169.
17. R. Farhi and G. Petot-Ervas, *J. Phys. Chem. Solids*, **39** (1978), 1175.

18. G. J. Koel and P. J. Gellings, *Oxid. Met.*, **5** (1972), 185.
19. M. C. Pope and N. Birks, *Corr. Sci.*, **17** (1977), 747.
20. C. Wagner, *Z. Phys. Chem.*, **21** (1933), 25.
21. B. Fisher and D. S. Tannhäuser, *J. Chem. Phys.*, **44** (1966), 1663.
22. N. G. Eror and J. B. Wagner, *J. Phys. Chem. Solids*, **29** (1968), 1597.
23. R. E. Carter and F. D. Richardson, *TAIME*, **203** (1955), 336.
24. D. W. Bridges, J. P. Baur, and W. M. Fassell, *J. Electrochem. Soc.*, **103** (1956), 619.
25. S. Mrowec and A. Stoklosa, *Oxid. Met.*, **3** (1971), 291.
26. S. Mrowec, A. Stoklosa, and K. Godlewski, *Cryst. Latt. Def.*, **5** (1974), 239.
27. S. Mrowec and K. Przybylski, *Oxid. Met.*, **11** (1977), 365.
28. S. Mrowec and K. Przybylski, *Oxid. Met.*, **11** (1977), 383.
29. F. S. Pettit, *J. Phys. Chem.*, **68** (1964), 9.
30. R. E. Carter and F. D. Richardson,, *J. Met.*, **6** (1954), 1244.
31. W. K. Chen, N. L. Peterson, and W. T., Reeves, *Phys. Rev.*, **186** (1969), 887.
32. A. Atkinson, R. I. Taylor, and A. E. Hughes, *Phil. Mag.*, **A45** (1982), 823.
33. A. Atkinson, D. P. Moon, D. W. Smart, and R. I. Taylor, *J. Mater. Sci.*, **21** (1986), 1747.
34. R. J. Hussey and M. J. Graham, *Oxid. Met.*, **45** (1996), 349.
35. R. Prescott and M. J. Graham, *Oxid. Met.*, **38** (1992), 233.
36. K. Hauffe and H. Pfeiffer, *Z. Elektrochem.*, **56** (1952), 390.
37. F. S. Pettit, R. Yinger, and J. B. Wagner, *Acta Met.*, **8** (1960), 617.
38. F. S. Pettit and J. B. Wagner, *Acta Met.*, **12** (1964), 35.
39. P. Kofstad, *High Temperature Corrosion*, New York, Elsevier Applied Science Publishers, Ltd, 1988.
40. K. Hauffe, *Oxidation of Metals*, New York, Plenum Press, 1965.
41. J. Benard, *Oxydation des Méteaux*, Paris, Gautier-Villars, 1962.
42. N. Cabrera and N. F. Mott, *Rept. Prog. Phys.*, **12** (1948), 163.
43. F. W. Young, J. V. Cathcart, and A. T. Gwathmey, *Acta Met.*, **4** (1956), 145.
44. R. K. Hart, *Proc. Roy. Soc.*, **236A** (1956), 68.
45. M. W. Roberts, *Trans. Faraday Soc.*, **57** (1961), 99.
46. N. F. Mott, *Trans. Faraday Soc.*, **36** (1940), 472.
47. K. Hauffe and B. Z. Ilschner, *Elektrochem.*, **58** (1954), 382.
48. U. R. Evans, *The Corrosion and Oxidation of Metals*, London, Edward Arnold, 1960.
49. A. T. Fromhold, *J. Phys. Chem. Solids*, **33** (1972), 95.
50. A. T. Fromhold and E. L. Cook, *J. Phys. Chem. Solids*, **33** (1972), 95.
51. A. T. Fromhold, *J. Phys. Soc. J.*, **48** (1980), 2022.

4

Oxidation of pure metals

Introduction

The following discussion will necessarily be idealized since it is difficult to find systems that conform completely to a single mode of rate control. This makes it correspondingly difficult to choose examples to illustrate completely any single rate-controlling process. Even such factors as specimen shape can influence the detailed mechanisms operating over the life of the specimen. For example, when a metal is exposed to an oxidizing atmosphere at high temperature, the initial reaction is expected to be very rapid since the oxide layer formed is very thin. However, if the parabolic rate law is extrapolated to zero scale thickness, an infinite rate is predicted. Clearly this is not so and the initial stage of oxidation must be controlled by some process other than ionic transport through thin oxide scales, e.g., the processing of gas molecules on the oxide surface according to Equation (4.1):

$$O_2(g) \rightarrow O_2(ad) \rightarrow 2O(ad) = 2O^-(chem) + 2h^{\cdot} \rightarrow 2O^{2-}(latt) + 4h^{\cdot}. \quad (4.1)$$

This describes the adsorption, dissociation, chemisorption, and ionization of oxygen and would lead to a constant reaction rate, if rate controlling. However, these processes are so rapid that the oxidation period over which they control the reaction rate is rarely observed. In fact, observation of this early period is extremely difficult since, under most conditions, the oxide scale formed during the period of heating the sample to temperature is sufficiently thick for ionic diffusion through the scale to be rate controlling by the time reaction monitoring is begun under isothermal conditions.

The initial, surface-process-controlled stage can, however, be observed under special conditions. This has been illustrated by Pettit, Yinger, and Wagner[1] (described previously) who chose a gas system with a very slow surface reaction. Alternatively, it may be possible to heat the specimens of some metals in an inert atmosphere or vacuum and suddenly introduce the oxidizing species when isothermal conditions have been reached. This has been shown to be feasible for Cu, and the low-temperature nucleation of Cu_2O has been studied *in situ* by transmission electron microscopy (TEM).[2] Experiments of this type for systems with more stable oxides are difficult because of the formation of oxides during heating, even at low oxygen partial pressures.

Once the oxidation of the metal reaches the stage where ionic diffusion is rate controlling, a parabolic rate law is found to hold for a period whose duration depends upon factors such as specimen geometry and scale mechanical properties.

Specimen geometry is important because, as the reaction proceeds, the metal core gets smaller and so the area of metal–scale interface gets smaller. Thus, when the reaction rate is expressed in mass gain per unit area, taking the initial surface area of the metal specimen to be constant can lead to small but significant errors. This aspect has been thoroughly treated by Romanski,[3] Bruckmann,[4] and Mrowec and Stoklosa[5] who showed that, unless corrections allowing for the reduction of the scale–metal interface area are made, the rate of the reaction will apparently fall below that predicted on the basis of the parabolic rate law.

The above consideration applies so long as the scale and metal maintain contact at the scale–metal interface. As the oxidation reaction proceeds, however, stresses are generated within the oxide, and these are called growth stresses. These stresses will be discussed in detail in the next chapter but a brief discussion is needed at this point. In systems where cations are mobile, growth stresses arise in the scale since the scale must relax to maintain contact with the metal as the metal atoms cross the scale–metal interface to diffuse outwards as cations and electrons. If the scale does not relax, voids will form at the scale–metal interface, separating the scale from the metal. On a planar interface there are no forces restraining such relaxation, but at edges and corners it is not possible for the scale to relax in two or three directions. The geometry of the scale in these regions is stabilized and resists such relaxation in much the same way as the sides of a cardboard box cannot be flexed in the corner regions. In these geometrically stabilized regions, the scale must creep to maintain contact with the metal at a rate that is determined by the oxidation rate of the metal, i.e., the rate at which metal is being removed. The adhesion between scale and metal is the maximum force that can be exerted to cause the scale to creep and maintain contact. Thus, unless the scaling rate is low or the scale is quite plastic, contact between scale and metal will be lost as the reaction proceeds. As the scale thickens, contact will eventually be lost in any real situation involving corners. With cylindrical specimens a similar situation arises.

As oxidation proceeds and contact between scale and metal is lost at edges and corners, spreading over the faces with time, the area across which cations can be supplied is, therefore, decreased. This means that, to supply the scale–gas interface over the separated areas, the cations have a longer diffusion distance (see Figure 4.1)

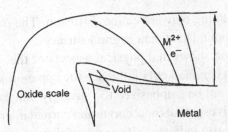

Figure 4.1 Schematic diagram showing longer diffusion distance for cations to supply scale–gas interface when scale–metal separation occurs at edges and corners.

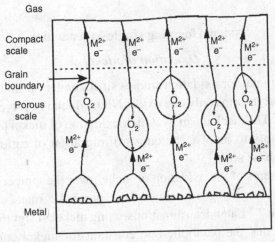

Figure 4.2 Mechanism of formation of porous zone resulting from scale–metal separation. Note that the scale decomposes more readily at the grain boundaries in the scale than at the grain surfaces.

and hence the reaction rate falls, unless vaporization of the metal across the void can keep up with the oxide growth. This leads to specific reaction rates that are lower than expected when the original metal surface area is used in the calculation.

Frequently this loss of contact causes the formation of a porous zone of scale between the outer compact layer and the metal. The mechanism of formation is quite straightforward, as outlined below and shown in Figure 4.2. When scale separation occurs, the metal activity at the inner surface of the scale is high and so cations still continue to migrate outwards. However, this causes the metal activity to fall and the oxygen activity to rise correspondingly. As the oxygen activity rises, the oxygen partial pressure in local equilibrium with the scale inner surface rises and oxygen 'evaporates' into the pore, diffuses across the pore, and forms oxide on the metal surface. In this way a porous layer can be formed next to the scale–metal interface while maintaining outward cation diffusion. The oxide scale dissociates more rapidly at grain boundaries than at grain surfaces.

The above discussion shows that physical aspects of the scale and scale–metal interface configuration can markedly influence the apparent kinetic behaviour and rate appearance. It cannot be emphasized too strongly that careful metallographic and other examination is essential for accurate and correct interpretation of observed kinetics. These factors also indicate that specimen size may influence the course of the kinetics. Thus, extrapolating from results on small specimens to predict results on large pieces or pieces of different geometry may not be easy or feasible. The above and other aspects will be brought out in the following discussion of the oxidation of selected pure metals.

Systems forming single-layer scales

Oxidation of nickel

Nickel is an ideal metal for oxidation studies since, under normal temperature and pressure conditions, it forms only one oxide, NiO, which is a p-type semiconductor with cation deficit. The mechanism by which oxidation of nickel proceeds is, therefore, expected simply to involve the outward migration of cations and electrons forming a single-phase scale.

Early measurements of the oxidation of nickel over the temperature range 700–1300 °C showed a surprising variation of the parabolic rate constant over four orders of magnitude.[6-8] Later determinations, using nickel of significantly improved purity,[9,10] showed that the parabolic rate constant for nickel containing 0.002% impurity was reliably reproducible and was lower than that for the less-pure nickel samples used earlier. Since the majority of impurity elements found in nickel are either divalent or trivalent, they will, when dissolved in the oxide, either have no effect on or increase the mobility of cations in NiO. It is, therefore, understandable that the impure nickel specimens were found in general to oxidize more rapidly than pure nickel. The activation energy of the process remained fairly constant regardless of impurity level indicating that the rate-determining step remained unchanged. The scale morphology has also been found to vary significantly with the purity of the metal. This is particularly well demonstrated by results obtained at 1000 °C.[11]

Using nickel of high purity a compact, adherent, single-layer scale of NiO is formed. Platinum markers, placed on the metal surface before reaction, are found afterwards at the scale–metal interface. This is very strong evidence that the scale forms entirely by the outward migration of cations and electrons.

Earlier work using nickel that contained low concentrations of impurities (0.1%) showed that a fundamentally different scale was produced. In this case the scale consisted of two layers, both of NiO, the outer compact and the inner porous.

Furthermore, platinum markers were found to lie at the interface between the 'outer compact' and inner-porous NiO layer after oxidation.[12–14]

The presence of the markers at the boundary between the compact and porous layers of the scale implies that the outer layer grows by outward cation migration and that the inner porous layer grows by inward migration of oxygen. The mechanism proposed to explain the formation of such a scale[12] assumes that the oxide is separated from the metal early in the oxidation process due to insufficient oxide plasticity, a situation which is aggravated by higher scaling rates of the impure nickel. Once the scale loses contact with the metal, the oxygen activity at the inner surface rises and the scale dissociates at a corresponding rate. The oxygen migrates across the pore or gap and begins to form new nickel oxide on the metal surface. Birks and Rickert[12] showed that the oxygen partial pressures likely to be involved in this mechanism were sufficient to account for the observed reaction rate. A diagrammatic representation of this process is the same as that given in Figure 4.2.

The oxidation of cold-worked Ni has been observed to be more rapid than well-annealed Ni.[15,16] The activation energy for the process was also lower for the cold-worked Ni. These results have been interpreted in terms of oxide grain boundaries providing easy diffusion paths through the fine-grained oxide formed on the cold-worked Ni. Such an interpretation is consistent with the diffusion data presented in Figure 3.14. In addition, Mrowec and Werber[17] propose that, because oxide disso-ciation is faster at grain boundaries than over grain faces, the outer compact layer is penetrated by microcracks in the grain boundaries and that, consequently, inward transport of oxygen molecules from the atmosphere also plays a role. Atkinson and Smart[18] have also argued that ^{18}O penetration through NiO scales was too rapid to be explained completely by lattice or grain-boundary diffusion. However, the oxygen-pressure dependence of Ni oxidation is what one would predict for bulk diffusion. Therefore, it is clear that point defects are important contributors to scale growth.

As discussed earlier, NiO is a p-type cation-deficit semiconductor and, therefore, the cations will migrate with electrons from the scale–metal interface to the scale–gas interface during oxidation. Correspondingly, there will be a flow of defects, cation vacancies and electron holes, in the opposite direction. Consequently, the driving force for the reaction will be reflected by the concentration gradient of cation vacancies across the scale. The nickel vacancies are formed according to Equation (4.2),

$$\frac{1}{2}O_2 = O_O + 2h^{\cdot} + V_{Ni}'', \qquad (4.2)$$

from which we obtain Equation (4.3):

$$C_{V_{Ni}''} = const. \, p_{O_2}^{1/6}. \qquad (4.3)$$

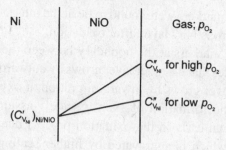

Figure 4.3 Variation of cation-vacancy concentration across a NiO scale for high and low oxygen partial pressures.

If associated vacancies V'_{Ni} are important, the reaction would be that shown in Equation (4.4),

$$\frac{1}{2}O_2 = O_O + h^{\cdot} + V'_{Ni},\qquad(4.4)$$

from which we obtain $C_{V'_{Ni}}$:

$$C_{V'_{Ni}} = const. \, p_{O_2}^{1/4}.\qquad(4.5)$$

The concentration gradient of cation vacancies across the scale is very sensitive to the oxygen partial pressure of the atmosphere, as shown diagrammatically in Figure 4.3. Correspondingly, the oxidation rate constant is expected to increase as the oxygen partial pressure of the gas is increased.

Fueki and Wagner[19] oxidized nickel over a range of oxygen partial pressures from the dissociation pressure of NiO to 1 atm at temperatures between 900 and 1400 °C. Their results showed that the observed parabolic rate constant did, in fact, vary with the oxygen partial pressure of the atmosphere to the $1/n$ power: the value of n varying from lower than 6 at 1000 °C to 3.5 at 1400 °C. They also determined the self-diffusion coefficient of nickel ions in nickel oxide, from their results, as a function of oxygen partial pressure. The values of the self-diffusion coefficients were independent of oxygen partial pressure in the low oxygen-pressure range and proportional to $p_{O_2}^{1/6-1/3.5}$ in the intermediate oxygen-pressure range.

They suggested that the values of n of 6 and 3.5 are consistent with the formation of doubly and singly charged vacancies according to Equations (4.1) and (4.3) above. However, considering the results described later for the oxidation of cobalt, it is possible that intrinsic defects may also be involved.

Oxidation of zinc

Zinc also forms one oxide ZnO and, therefore, a single phase, single-layered scale is expected when pure zinc is oxidized. However, ZnO is an n-type cation-excess semiconductor, i.e., having interstitial Zn ions and electrons within the conduction band.

According to the equilibria established for this defect structure, the concentration gradient of interstitial zinc ions across the scale will depend also upon the oxygen partial pressure of the atmosphere, i.e., for divalent interstitials formed according to Equation (4.6),

$$ZnO = Zn_i^{\cdot\cdot} + 2e' + \frac{1}{2}O_2, \tag{4.6}$$

with $C_{Zn_i^{\cdot\cdot}}$ given in Equation (4.7),

$$C_{Zn_i^{\cdot\cdot}} = const. \, p_{O_2}^{-1/6}; \tag{4.7}$$

and for monovalent interstitials formed according to Equation (4.8),

$$ZnO = Zn_i^{\cdot} + e' + \frac{1}{2}O_2, \tag{4.8}$$

with $C_{Zn_i^{\cdot}}$ given in Equation (4.9):

$$C_{Zn_i^{\cdot}} = const. \, p_{O_2}^{-1/4}. \tag{4.9}$$

In both cases, increasing the oxygen partial pressure reduces the concentration of defects. Thus, the difference in concentration of interstitial zinc ions can be represented as in Equation (4.10),

$$C_{Zn_i}^o - C_{Zn_i} = const. \left[\left(p_{O_2}^o \right)^{-1/n} - \left(p_{O_2} \right)^{-1/n} \right] \tag{4.10}$$

$$= const'. \left[1 - \left(\frac{p_{O_2}}{p_{O_2}^o} \right)^{-1/n} \right],$$

where $C_{Zn_i}^o$ represents the concentration at $p_{O_2}^o$, the oxygen partial pressure in equilibrium with Zn and ZnO, i.e., at the scale–metal interface, C_{Zn_i} and p_{O_2} represent conditions at the scale–gas interface, and n is either 6 or 4. For practical situations $p_{O_2} \gg p_{O_2}^o$, the value $(p_{O_2}/p_{O_2}^o)^{-1/n}$ is very small compared with unity and, therefore, the gradient of concentration of interstitial zinc ions across the scale is insensitive to the oxygen partial pressure in the atmosphere. Consequently, the scaling rate constant for zinc will be insensitive to the external oxygen partial pressure so long as this is high compared with the value for Zn/ZnO equilibrium. The situation is shown in Figure 4.4.

The independence of the parabolic rate constant for zinc in oxygen at 390 °C was demonstrated by Wagner and Grünewald,[20] who obtained values of 7.2×10^{-9} and 7.5×10^{-9} g^2 cm^{-4} h^{-1} for oxygen partial pressures of 1 and 0.022 atm, respectively. These results adequately confirm the conclusions drawn from a consideration of the defect structure of ZnO.

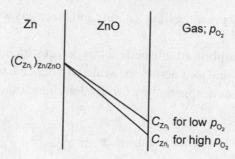

Figure 4.4 Variation of concentration of singly or doubly charged interstitial zinc ions across a ZnO scale for high and low oxygen partial pressures.

Oxidation of aluminium

The thermodynamically stable oxide of aluminium is α-Al_2O_3. This oxide has a rhombohedral structure consisting of a hexagonal packing of oxide anions with cations occupying two-thirds of the octahedral interstitial sites. Alpha is a slowly-growing oxide and is the protective film formed on many high-temperature alloys and coatings. This will be discussed in the next chapter. Alumina also exists in a number of metastable crystal structures.[21] These include γ-Al_2O_3, which is a cubic spinel, δ-Al_2O_3, which is tetragonal, and θ-Al_2O_3, which is monoclinic. Another form, κ-Al_2O_3, which is similar to α but has a shift in the stacking sequence on the anion sub-lattice, has also been identified.[22] Some of the metastable aluminas sometimes form prior to the stable α on high-temperature alloys and predominate for the oxidation of aluminium at temperatures below its melting point of 660 °C.

At room temperature Al is always covered with an 'air-formed film' 2–3 nm in thickness,[23] which consists of amorphous alumina. Oxidation at temperatures less than 350 °C results in growth of the amorphous film following inverse logarithmic kinetics.[24] At temperatures between 350 and 425 °C the amorphous film grows with parabolic kinetics.[24] At temperatures above 425 °C the kinetics are complex. Investigations by TEM[23,24] and secondary-ion mass spectroscopy (SIMS)[25] suggest the following sequence. The pre-existing film grows initially and, following an incubation period, γ-Al_2O_3 crystals nucleate at the amorphous oxide–Al interface. The γ nucleates heterogeneously on the surface at ridges resulting from specimen preparation and grows by oxygen transport through local channels in the amorphous film and anion attachment at the peripheries of the growing γ islands. The origin of the local channels have been proposed to be cracks in the amorphous oxide over the surface ridges.[23] Oxidation of a [111]-oriented Al single crystal which was sputter-cleaned and oxidized at 550 °C resulted in direct γ nucleation without the development of an amorphous layer.[25]

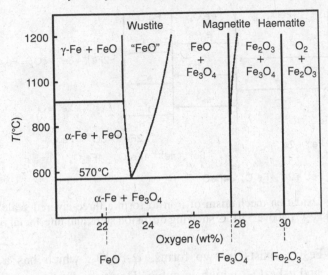

Figure 4.5 The iron–oxygen phase diagram.

Systems forming multiple-layer scales

Oxidation of iron

When iron oxidizes in air at high temperatures, it grows a scale consisting of layers of FeO, Fe_3O_4, and Fe_2O_3 and, thus, provides a good example of the formation of multi-layered scales. The importance of the oxidation of iron in society has led to it being extensively investigated and, consequently, relatively well understood with a very wide literature.

From the phase diagram of the iron–oxygen system, shown in Figure 4.5, it is clear that the phase wustite, FeO, does not form below 570 °C. Thus, Fe oxidized below this temperature would be expected to form a two-layer scale of Fe_3O_4 and Fe_2O_3 with the Fe_3O_4 next to the metal. Above 570 °C the oxide layer sequence in the scale would be FeO, Fe_3O_4, Fe_2O_3, with the FeO next to the metal.

The wustite phase, FeO, is a p-type metal-deficit semiconductor which can exist over a wide range of stoichiometry, from $Fe_{0.95}O$ to $Fe_{0.88}O$ at 1000 °C according to Engell.[26,27] With such high cation-vacancy concentrations, the mobilities of cations and electrons (via vacancies and electron holes) are extremely high.

The phase magnetite, Fe_3O_4, is an inverse spinel, which, therefore, has divalent ions, Fe^{2+}, occupying octahedral sites and half of the trivalent ions, Fe^{3+}, occupying tetrahedral sites. Defects occur on both sites and, consequently, iron ions may diffuse over both tetrahedral and octahedral sites. A certain degree of intrinsic semiconduction is shown and, thus, electrons may diffuse outward over electron holes and, as excess electrons, in the conduction band. Only slight variation of stoichiometry is found, except at high temperatures.

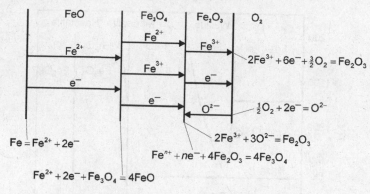

Figure 4.6 Oxidation mechanism of iron to form a three-layered scale of FeO, Fe_3O_4, and Fe_2O_3 above $570\,^\circ C$ showing diffusion steps and interfacial reactions.

Haematite, Fe_2O_3, exists in two forms, α-Fe_2O_3, which has a rhombohedral structure, and γ-Fe_2O_3, which is cubic. However, Fe_3O_4 oxidizes to form α-Fe_2O_3 above $400\,^\circ C$ and only this structure need be considered.[28] In the rhombohedral crystal, the oxygen ions exist in a close-packed hexagonal arrangement with iron ions in interstices. It has been reported that α-Fe_2O_3 shows disorder only on the anion sub-lattice[29] from which only the oxygen ions are expected to be mobile. Other studies have suggested that Fe_2O_3 growth occurs by outward cation migration.[30] Schwenk and Rahmel[31] consider both species to be mobile with the corresponding defects being $V_O^{\cdot\cdot}$ and $Fe_i^{\cdot\cdot\cdot}$ compensated electrically by excess electrons. Thus, there is still some question regarding the details of the defect structure of α-Fe_2O_3.

On the basis of the above knowledge of the structure and diffusion properties of the iron oxides, a relatively simple mechanism can be proposed to represent the oxidation of iron as shown in Figure 4.6. At the iron–wustite interface, iron ionizes according to Equation (4.11):

$$Fe = Fe^{2+} + 2e^-. \tag{4.11}$$

The iron ions and electrons migrate outward through the FeO layer over iron vacancies and electron holes, respectively. At the wustite–magnetite interface magnetite is reduced by iron ions and electrons according to Equation (4.12):

$$Fe^{2+} + 2e^- + Fe_3O_4 = 4FeO. \tag{4.12}$$

Iron ions and electrons surplus to this reaction proceed outward through the magnetite layer, over iron-ion vacancies on the tetrahedral and octahedral sites and over electron holes and excess electrons, respectively. At the magnetite–haematite interface, magnetite is formed according to Equation (4.13),

$$Fe^{n+} + ne^- + 4Fe_2O_3 = 3Fe_3O_4, \tag{4.13}$$

the value of n being 2 or 3 for Fe^{2+} or Fe^{3+} ions, respectively.

If iron ions are mobile in the haematite they will migrate through this phase over iron-ion vacancies V_{Fe}''' together with electrons and new haematite will form at the Fe_2O_3–gas interface according to Equation (4.14):

$$2Fe^{3+} + 6e^- + \frac{3}{2}O_2 = Fe_2O_3. \tag{4.14}$$

At this interface also, oxygen ionizes according to Equation (4.15):

$$\frac{1}{2}O_2 + 2e^- = O^{2-}. \tag{4.15}$$

If oxygen ions are mobile in the haematite layer, the iron ions and electrons, in excess of requirements for reduction of haematite to magnetite, will react with oxygen ions diffusing inwards through the Fe_2O_3 layer over oxygen vacancies forming new Fe_2O_3 according to Equation (4.16):

$$2Fe^{3+} + 3O^{2-} = Fe_2O_3. \tag{4.16}$$

The corresponding electrons then migrate outwards through the Fe_2O_3 to take part in the ionization of oxygen at the Fe_2O_3–gas interface.

The much greater mobility of defects in wustite causes this layer to be very thick compared with the magnetite and haematite layers. In fact the relative thicknesses of $FeO : Fe_3O_4 : Fe_2O_3$ are in the ratio of roughly $95 : 4 : 1$ at $1000\,°C$.[31]

At temperatures below $570\,°C$, the wustite phase does not form and only the magnetite and haematite layers are seen in the scale. Oxidation of etched Fe surfaces at 300 and $400\,°C$ in pure oxygen results in α-Fe_2O_3 forming first on the α-Fe with subsequent nucleation of Fe_3O_4 at the α-Fe_2O_3–Fe interface.[32] The rate of scaling is correspondingly low in the absence of wustite.

The rapid rate of reaction of iron above $570\,°C$ causes thick scales to develop quickly and, in spite of the relatively high plasticity of the FeO layer, scale–metal adhesion is lost and a porous inner layer of FeO is formed, next to the metal, by the mechanism described earlier. The stresses associated with rapidly growing scale undoubtedly induce physical defects in the outer scale and the penetration of gas molecules, especially those belonging to the CO–CO_2 and H_2–H_2O redox systems, will play a role in scale formation.

Since the scale formed on iron above $570\,°C$ is predominantly wustite, growth of this layer controls the overall rate of oxidation. However, since the defect concentrations in wustite at the iron–wustite and wustite–magnetite interfaces are fixed by the equilibria achieved there, for any given temperature, the parabolic rate constant will be relatively unaffected by the external oxygen partial pressure. Increasing the oxygen partial pressure in the gas phase should, theoretically, lead to an increase in the relative thickness of the haematite layer. However, since this layer only accounts for about 1% of the metal–scale thickness, any variation in rate constant with oxygen partial pressure will be difficult to detect.

A similar argument applies to temperatures below 570 °C in atmospheres of low oxygen partial pressure; low defect concentrations at the magnetite–iron and magnetite–haematite interfaces are fixed by the equilibria achieved there.

If it were possible to oxidize iron above 570 °C in atmospheres of low oxygen partial pressure, within the wustite existence range, under conditions leading to equilibrium being achieved at the wustite–gas interface, then a variation of parabolic rate constants with oxygen partial pressure should be observed. Unfortunately, the oxygen partial pressures required to demonstrate this are so low (10^{-12} atm at 1000 °C) that they can only be achieved using redox gas systems. Pettit, Yinger and Wagner[1] used the CO–CO$_2$ system to investigate the oxidation of iron under such conditions and found that the rate was controlled by the decomposition of adsorbed CO$_2$ on the scale surface leading to a constant reaction rate as described earlier. For this reason, it is not possible to observe the variation in parabolic rate constant with oxygen partial pressure when iron is oxidized to give a scale composed of wustite only.

Oxidation of cobalt

Cobalt forms two oxides, CoO and Co$_3$O$_4$, of NaCl and spinel structures, respectively; CoO is a p-type cation-deficit semiconductor through which cations and electrons migrate over cation vacancies and electron holes. In addition to the usual extrinsic defects, due to deviations from stoichiometry above 1050 °C, intrinsic Frenkel-type defects are also present.[33] The variations of oxidation-rate constant with oxygen partial pressure and with temperature are, therefore, expected to be relatively complex. Consequently, it is important to ensure that very accurate data are obtained for the oxidation reactions, over a wide range of oxygen pressure and temperature.

The early data obtained up to 1966 are summarized in the available texts on high-temperature oxidation of metals particularly those by Kofstad[34] and Mrowec and Werber.[17]

In a very careful study, Mrowec and Przybylski[35] investigated the oxidation of cobalt between 940 and 1300 °C at oxygen pressure between 6.58×10^{-4} and 0.658 atm. They introduced refinements to the measurements such as:

(a) taking into account the thermal expansion of the metal, which can increase the area at temperature by about 10% over the area measured at room temperature;

(b) using flat, thin ($19 \times 15 \times 0.5$ mm) specimens so that surface-area changes due to metal consumption during the reaction were held to about 3%.

Platinum markers, applied to the cobalt samples before oxidation, were found at the metal–scale interface after reaction over the whole range of oxygen pressures and temperatures investigated. This confirms that the compact CoO layer is formed by outward diffusion of the metal.

Since the oxygen pressures used were orders of magnitude higher than the decomposition pressure of CoO the variation of the parabolic rate constant k_p'' with oxygen partial pressures, Equation (4.17),

$$k_p'' = const. \left[p_{O_2}^{1/n} - \left(p_{O_2}^o \right)^{1/n} \right],$$ (4.17)

can be written as in Equation (4.18):

$$k_p'' = const. p_{O_2}^{1/n}.$$ (4.18)

The rate constant will vary with temperature according to the Arrhenius equation, thus the variation of k_p'' with oxygen partial pressure and temperature can be written as in Equation (4.19).

$$k_p'' = const'. p_{O_2}^{1/n} \exp(-Q/RT)$$ (4.19)

Figure 4.7 shows Mrowec and Przybylski's results plotted accordingly. From this figure it can be deduced that n varies from 3.4 at 950 °C to 3.96 at 1300 °C whereas Q, the apparent activation energy, increases from 159.6 kJ mol^{-1} at $p_{O_2} = 0.658$ atm to 174.7 kJ mol^{-1} at 6.58×10^4 atm. They further showed that these variations could be explained satisfactorily only by assuming that CoO contains defects arising intrinsically, i.e., Frenkel defects, as well as defects arising extrinsically as a result of deviations from stoichiometry.

The above work was carried out totally within the CoO range of existence. Earlier, Bridges *et al.*[36] studied the oxidation of cobalt over a wider range of oxygen pressures well into the Co_3O_4 range of existence and over the temperature range 950–1150 °C (see Figure 3.11). The results show much more scatter than those of Mrowec and Przybylski in the CoO field but also demonstrate clearly how the oxidation-rate constant ceases to vary with oxygen pressure once the two-layer $CoO–Co_3O_4$ scale, such as that shown in Figure 4.8, is formed. Hsu and Yurek[37] obtained similar results but suggested that transport was partially by grain-boundary diffusion.

A similar example of this is given for the case of copper,[5] demonstrating clearly the dependence of the parabolic rate constant on oxygen partial pressure so long as the scale is a single layer of Cu_2O (see Figure 3.12). As soon as an outer layer of CuO forms, the parabolic rate constant becomes independent of oxygen partial pressure.

Theory of multi-layered scale growth

As mentioned above, the formation of multiple-layer scales is common for metals such as Fe, Co, and Cu, as well as others.

Oxidation of pure metals

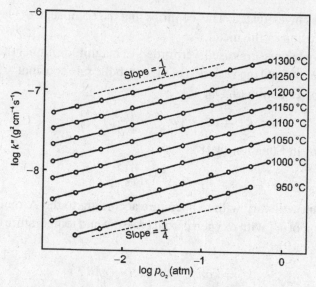

Figure 4.7 Parabolic rate constant for the oxidation of cobalt to CoO at various oxygen partial pressures and temperatures, determined by Mrowec and Przybylski.[35]

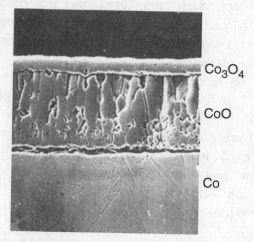

Figure 4.8 Two-layered scale formed on cobalt after 10 h in oxygen at 750 °C.

The theory of multi-layered scale growth on pure metals has been treated by Yurek *et al.*[38] The hypothetical system treated is shown in Figure 4.9. It is assumed that the growth of both scales is diffusion controlled with the outward migration of cations large relative to the inward migration of anions. The flux of cations in each oxide is assumed to be independent of distance. Each oxide exhibits predominantly

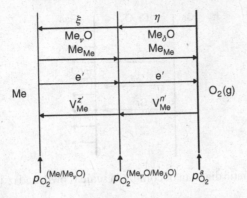

Figure 4.9 Schematic diagram of a hypothetical two-layered scale.

electronic conductivity and local equilibrium exists at the phase boundaries. The total oxidation reaction is shown in Equation (4.20),

$$(vw + \delta y)\,\mathrm{Me} + \left(\frac{w + y}{2}\right)\mathrm{O_2} \rightarrow w\mathrm{Me}_v\mathrm{O} + y\mathrm{Me}_\delta\mathrm{O},\tag{4.20}$$

where w and y are the partition fractions of the two oxides in the scale.

The rate of thickening of a single product layer of $\mathrm{Me}_v\mathrm{O}$ would be given by Equation (4.21),

$$\frac{d\xi}{dt} = J_{\mathrm{Me}} \left(\frac{V_{\mathrm{Me}_v\mathrm{O}}}{v}\right),\tag{4.21}$$

where J_{Me} is the outward flux of cations in mol $\mathrm{cm^{-2}\,s^{-1}}$ and $V_{\mathrm{Me}_v\mathrm{O}}$ is the molar volume of $\mathrm{Me}_v\mathrm{O}$. However, in the case of two-layered scale growth only a fraction $wv/(\delta y + vw)$ of the cations transported through $\mathrm{Me}_v\mathrm{O}$ leads to its growth with the remainder being transported further through the outer phase, $\mathrm{Me}_\delta\mathrm{O}$. Treatment of this case yields Equation (4.22),[37]

$$\frac{d\xi}{dt} = \frac{k_p^{\mathrm{Me}_v\mathrm{O}}}{\xi \left(1 + \dfrac{\delta y}{vw}\right)},\tag{4.22}$$

where k_p is the parabolic rate constant for exclusive growth of $\mathrm{Me}_v\mathrm{O}$ and the bracketed term accounts for the partitioning of cations. A similar treatment of the growth of the outer layer yields Equation (4.23):

$$\frac{d\eta}{dt} = \frac{k_p^{\mathrm{Me}_\delta\mathrm{O}}}{\eta \left(1 + \dfrac{w}{y}\right)}.\tag{4.23}$$

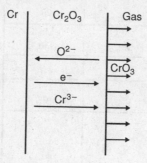

Figure 4.10 Schematic diagram of combined scale growth and oxide volatilization from Cr.

Combinations of Equations (4.22) and (4.23) allow calculation of the overall scale-growth rate and integration of the two equations allows the thickness ratio of the two layers to be calculated.

This theory has been shown by Garnaud[39] to describe the growth of CuO and Cu_2O on Cu, by Garnaud and Rapp[40] to describe the growth of Fe_3O_4 and FeO on Fe, and by Hsu and Yurek[37] to describe the growth of Co_3O_4 and CoO on Co.

Systems for which volatile species are important

Oxidation of chromium

The oxidation of pure Cr is, in principle, a simple process since a single oxide, Cr_2O_3, is observed to form. However, under certain exposure conditions, several complications arise, which are important both for the oxidation of pure Cr and for many important engineering alloys which rely on a protective Cr_2O_3 layer for oxidation protection. The two most important features are scale thinning by CrO_3 evaporation, and scale buckling as a result of compressive stress development.

The formation of CrO_3 by the reaction shown in Equation (4.24),

$$Cr_2O_3(s) + \frac{3}{2}O_2(g) = 2CrO_3(g), \tag{4.24}$$

becomes significant at high temperatures and high oxygen partial pressures. This is illustrated in Figures 2.6 and 2.7. The evaporation of CrO_3, shown schematically in Figure 4.10, results in the continuous thinning of the protective Cr_2O_3 scale so the diffusive transport through it is rapid. The effect of the volatilization on the oxidation kinetics has been analyzed by Tedmon.[41] The instantaneous change in scale thickness is the sum of two contributions: thickening due to diffusion and thinning due to volatilization, Equation (4.25),

$$\frac{dx}{dt} = \frac{k_d'}{x} - k_s', \tag{4.25}$$

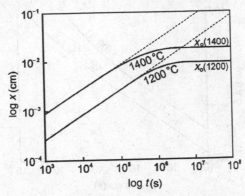

Figure 4.11 Scale thickness versus time for the oxidation of Cr. (The dashed lines correspond to oxide growth following parabolic kinetics for the case of negligible evaporation.)

where k'_d is a constant describing the diffusive process and k'_s describes the rate of volatilization. This equation can be rearranged to yield Equation (4.26),

$$\frac{dx}{\left(\dfrac{k'_d}{x} - k'_s\right)} = dt, \tag{4.26}$$

which, upon integration, yields Equation (4.27),

$$\frac{-x}{k'_s} - \frac{k'_d}{k'^2_s} \ln(k'_d - k'_s x) + C = t, \tag{4.27}$$

where C is an integration constant to be evaluated from the initial conditions. Taking $x = 0$ at $t = 0$ we obtain Equation (4.28):

$$t = \frac{k'_d}{k'^2_s} \left[-\frac{k'_s}{k'_d}x - \ln\left(1 - \frac{k'_s}{k'_d}x\right) \right]. \tag{4.28}$$

Initially, when the diffusion through a thin scale is rapid, the effect of CrO_3 volatilization is not significant but, as the scale thickens, the rate of volatilization becomes comparable and then equal to the rate of diffusive growth. This situation, *paralinear oxidation*, results in a limiting scale thickness, x_o, for which $dx/dt = 0$, which is shown schematically in Figure 4.11. Setting this condition in Equation (4.25) yields Equation (4.29):

$$x_o = \frac{k'_d}{k'_s}. \tag{4.29}$$

Since k'_d and k'_s have different activation energies, the value of x_o will depend on temperature. Figure 4.11 indicates a greater limiting thickness at higher oxidation temperature. This will be the case when the oxide growth process has a stronger

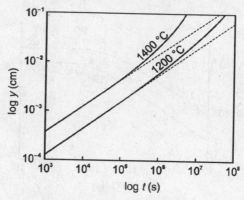

Figure 4.12 Metal recession versus time for the oxidation of Cr. (The dashed lines correspond to oxide growth following parabolic kinetics for the case of negligible evaporation.)

dependence on temperature than does the vaporization process, i.e., Q for diffusive growth is larger than that for the controlling surface process such as chemical reaction or gas-phase transport. This will be the case for most systems. However, one could envision systems where scale growth has a smaller activation energy. In this case, the limiting scale thickness would decrease with increasing oxidation temperature.

The occurrence of a limiting scale thickness implies protective behaviour but, in fact, the amount of metal consumed increases until a constant rate is achieved. This can be seen more clearly by considering the metal recession, y, rather than the scale thickness:

$$\frac{dy}{dt} = \frac{k_d}{y} + k_s.\tag{4.30}$$

Upon integration Equation (4.30) yields Equation (4.31):

$$t = \frac{k_d}{k_s^2}\left[\frac{k_s}{k_d}y - \ln\left(1 + \frac{k_s}{k_d}y\right)\right].\tag{4.31}$$

When $\log y$ is plotted against $\log t$, the curve shown in Figure 4.12 is obtained. Here it is clear that the metal consumption is greater than it would be for simple parabolic kinetics. This problem, which is more serious in rapidly flowing gases, is one of the major limitations on the high-temperature use of Cr_2O_3 forming alloys and coatings.

The second important factor associated with the oxidation of Cr is the apparent variation of the oxidation rate with different surface preparations and the buckling of the Cr_2O_3 scales due to compressive stress development.[42] The latter observation is illustrated schematically in Figure 4.13. These results were somewhat puzzling until the careful work of Caplan and Sproule.[43] In this study both electropolished

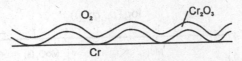

Figure 4.13 Schematic diagram of scale buckling on Cr.

and etched specimens of large-grained polycrystalline Cr were oxidized in 1 atm O_2 at 980, 1090, and 1200 °C. The electropolished specimens oxidized relatively rapidly and showed evidence of compressive stresses in the polycrystalline Cr_2O_3 scale that formed. The etched specimens showed similar behaviour for the scale formed on some grains, but on certain grains very thin, single crystalline Cr_2O_3 was observed, showing no evidence of compressive stresses. Caplan and Sproule concluded that the polycrystalline oxide grew by both outward transport of cations and inward transport of anions. The latter transport apparently occurs along oxide grain boundaries and results in compressive stresses by oxide formation at the scale–metal interface or within the scale. Subsequent work using ^{18}O tracers has indicated this to be the case with pure chromia growing primarily outward with approximately 1% of the growth being by inward anion transport.[44,45] The results of Caplan and Sproule would suggest that substantially more than 1% of the growth is inward. It should be noted that this transport is markedly affected by dopants such as Y and Ce. This will be discussed in Chapter 5. Polman *et al.*[46] oxidized Cr at 900 °C and found that k_P was independent of oxygen partial pressure, which would be consistent with Cr interstitials being the predominant defect. These authors propose that the defects responsible for grain-boundary diffusion are essentially the same as defects in the bulk lattice.

Oxidation of molybdenum and tungsten

The volatilization of oxides is particularly important in the oxidation of Mo and W at high temperatures and high oxygen pressures. Unlike Cr, which develops a limiting scale thickness, complete oxide volatilization can occur in these systems. The condensed and vapour species for the Mo–O and W–O systems have been reviewed by Gulbransen and Meier[47] and the vapour-species diagrams for a temperature of 1250 K are presented in Figures 4.14 and 4.15. The effects of oxide volatility on the oxidation of Mo have been observed by Gulbransen and Wysong[48] at temperatures as low as 475 °C and the rate of oxide evaporation above 725 °C was such that gas-phase diffusion became the rate-controlling process.[49] Naturally, under these conditions, the rate of oxidation is catastrophic. Similar behaviour is observed for the oxidation of tungsten, but at higher temperatures because of the lower vapour pressures of the tungsten oxides. The oxidation behaviour of tungsten has been reviewed in detail by Kofstad.[34]

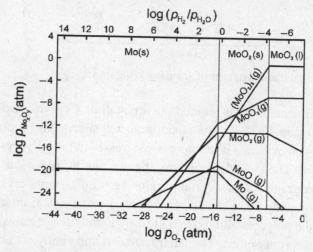

Figure 4.14 The Mo–O system volatile-species diagram for 1250 K.

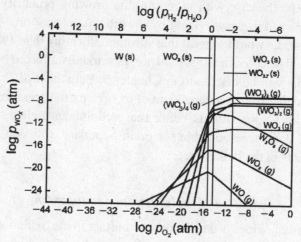

Figure 4.15 The W–O system volatile-species diagram for 1250 K.

Oxidation of platinum and rhodium

The oxidation of Pt, and Pt-group metals, at high temperatures is influenced by oxide volatility in that the only stable oxides are volatile. This results in a continuous mass loss. Alcock and Hooper[50] studied the mass loss of Pt and Rh at 1400 °C as a function of oxygen pressure. These results are presented in Figure 4.16 where it is seen that the mass loss is directly proportional to the oxygen partial pressure. The gaseous species were identified as PtO_2 and RhO_2. These results have an extra significance because Pt and Pt–Rh wires are often used to support specimens during high-temperature oxidation experiments. If these experiments involve mass-change

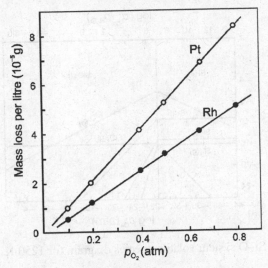

Figure 4.16 Mass loss of gaseous Pt and Rh oxides as a function of oxygen partial pressure at 1400 °C. After Alcock and Hooper.[50]

measurements, it must be recognized that there will be a mass loss associated with volatilization of oxides from the support wires.

Carol and Mann[51] have studied the oxidation of Rh in air at temperatures between 600 and 1000 °C. At 600 and 650 °C a crystalline oxide, identified as hexagonal Rh_2O_3, was observed on the surface. This film grew with logarithmic kinetics. As the oxidation temperature was increased, orthorhombic Rh_2O_3 was also observed and, above 800 °C, only the orthorhombic oxide was observed. This film grew with 'power law' kinetics which appear to be nearly parabolic. At 1000 °C, oxide volatility became significant and mass change versus time plots became paralinear, as described above for Cr oxidation.

Oxidation of silicon

The formation of SiO_2 on silicon, silicon-containing alloys and Si-based ceramics results in very low oxidation rates. However, this system is also one that can be influenced markedly by oxide vapour species. Whereas the oxidation of Cr is influenced by such species at high oxygen pressures, the effects for Si are important at low oxygen partial pressures. The reason for this may be seen from the volatile-species diagram for the Si–O system (see Figure 4.17). A significant pressure of SiO is seen to be in equilibrium with SiO_2 (s) and Si (s) at oxygen pressures near the dissociation pressure of SiO_2. This can result in a rapid flux of SiO away from the specimen surface and the subsequent formation of a non-protective SiO_2 smoke.

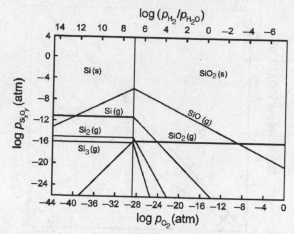

Figure 4.17 The Si–O system volatile-species diagram for 1250 K.

This formation of the SiO_2 as a smoke, rather than as a continuous layer, allows continued rapid reaction.

Wagner[52,53] has analyzed the conditions under which this "active" oxidation occurs. A simplified version[53] of this analysis follows. The fluxes of O_2 and SiO across a hydrodynamic boundary layer, taken to the be same thickness, δ, for both species are given by Equations (4.32) and (4.33),

$$J_O = \frac{p_{O_2} D_{O_2}}{\delta RT},$$
(4.32)

$$J_{SiO} = \frac{p_{SiO} D_{SiO}}{\delta RT},$$
(4.33)

where p_{SiO} is the pressure of SiO at the Si surface and p_{O_2} is the oxygen pressure in the bulk gas. Under steady-state conditions, the net transport rate of oxygen atoms must vanish. Therefore, we have Equation (4.34),

$$2J_{O_2} = J_{SiO},$$
(4.34)

which, upon substitution of Equations (4.32) and (4.33), and assuming $D_{SiO} \approx D_{O_2}$, yields Equation (4.35):

$$p_{SiO} = 2p_{O_2}.$$
(4.35)

If the equilibrium SiO pressure from the reaction given in Equation (4.36),

$$\frac{1}{2}Si\,(s,l) + \frac{1}{2}SiO_2\,(s) = SiO\,(g),$$
(4.36)

is greater than the SiO pressure in Equation (4.35), the Si surface will remain bare and Si will be continually consumed. Therefore, there is a critical oxygen partial pressure, Equation (4.37),

$$p_{O_2}(\text{crit}) \approx \frac{1}{2}p_{SiO}(\text{eq}),$$
(4.37)

below which 'active' oxidation will occur and the rate of metal consumption will be controlled by the SiO flux away from the surface:

$$J_{SiO} = \frac{2p_{O_2}D_{O_2}}{\delta RT}.$$ (4.38)

A more rigorous treatment of the problem[52] yields a slightly different expression for the critical oxygen pressure, Equation (4.39):

$$p_{O_2}(crit) = \frac{1}{2}\left(\frac{D_{SiO}}{D_{O_2}}\right)^{1/2} p_{SiO}(eq).$$ (4.39)

Gulbransen *et al.*[54] have tested Wagner's prediction by oxidizing Si over a range of temperatures and oxygen pressures and found good agreement. The rate of Si consumption was approximately 300 times faster for oxygen pressures below $p_{O_2}(crit)$ than those above it. Interestingly, this is a phenomenon which is rare in high-temperature oxidation but rather more common in aqueous corrosion, i.e., the rate of reaction being lower at higher driving forces (passive) than lower driving forces (active).

Metals with significant oxygen solubilities

Oxidation of titanium

The oxidation of Ti is quite complex because the Ti–O system exhibits a number of stable oxides and high oxygen solubility as seen from the phase diagram in Figure 4.18. The rate laws observed for the oxidation of Ti vary with temperature

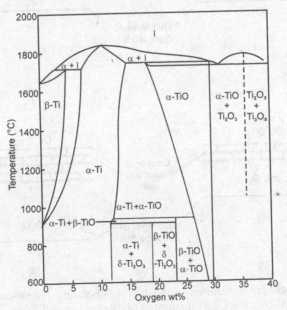

Figure 4.18 The Ti-rich portion of the Ti–O phase diagram.

as discussed by Kofstad.[34] However, in the temperature range of 600 to 1000 °C, the oxidation is parabolic but the rate is a combination of two processes, oxide scale growth and oxide dissolution into the metal. The mass change per unit area is given by Equation (4.40):

$$\frac{\Delta m}{A} = k_p(\text{oxide growth})t^{1/2} + k_p(\text{dissolution})t^{1/2}. \tag{4.40}$$

Similar behaviour is observed for the oxidation of Zr and Hf.[34]

Systems with significant scale cracking

Oxidation of niobium

The high-temperature oxidation of Nb is characterised by inward diffusion of oxygen through the scale. Initially, a protective layer is formed but, as the scale grows, the formation of oxide at the scale–metal interface stresses the oxide, resulting in scale cracking and a 'breakaway' linear oxidation. Roberson and Rapp[55] have shown this by an elegant experiment in which Nb was oxidized in sealed quartz tubes containing a mixture of Cu and Cu_2O powders at 1000 °C. The oxygen is transported from the powder by Cu_2O molecules and Cu is deposited at the reduction site. Figure 4.19 shows schematically the results of these experiments. Initially,

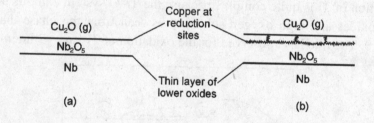

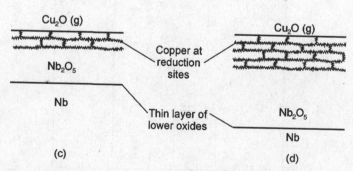

Figure 4.19 Schematic diagram of the high-temperature oxidation of Nb with oxygen supplied by a Cu/Cu_2O mixture. After Roberson and Rapp.[55]

the Cu was found at the scale–gas interface indicating the growth of a continuous layer of Nb_2O_5 over lower oxides of Nb by the inward migration of oxygen. With increasing time, the Cu was found to occupy cracks through the outer portion of the scale indicating that progressive scale cracking is the cause of the accelerated, linear, oxidation. The cracking of the Nb_2O_5 has also been observed *in situ* with acoustic emission measurements.[56] For example, oxidation at 600 °C resulted in an initial period during which the scale followed parabolic growth kinetics and no cracking was detected. After an incubation period of 15 min, significant cracking was detected by acoustic emission and the oxidation kinetics rapidly became linear. Similar oxidation behaviour is also observed for Ta. As a result, neither Nb nor Ta can be used for any extended period at elevated temperature without a protective coating.

References

1. F. S. Pettit, R. Yinger, and J. B. Wagner, *Acta Met.*, **8** (1960), 617.
2. J. C. Yang, M. Yeadon, B. Kolasa, and J. M. Gibson, *Scripta Mater.*, **38** (1998), 1237.
3. J. Romanski, *Corros. Sci.*, **8** (1968), 67; 89.
4. A. Bruckmann, in *Reaction Kinetics in Heterogeneous Chemical Systems*, ed. P. Barret, Amsterdam Elsevier, 1975, p. 409.
5. S. Mrowec and A. Stoklosa, *Oxid. Metals*, **3** (1971), 291.
6. O. Kubaschewski and O. van Goldbeck, *Z. Metallkunde*, **39** (1948), 158.
7. Y. Matsunaga, *Japan Nickel Rev.*, **1** (1933), 347.
8. W. J. Moore, *J. Chem. Phys.*, **19** (1951), 255.
9. E. A. Gulbransen and K. F. Andrew, *J. Electrochem. Soc.*, **101** (1954), 128.
10. W. Philips, *J. Electrochem. Soc.*, **110** (1963), 1014.
11. G. C. Wood and I. G. Wright, *Corr. Sci.*, **5** (1965), 841.
12. N. Birks and H. Rickert, *J. Inst. Metals*, **91** (1962), 308.
13. B. Ilschner and H. Pfeiffer, *Naturwissenschaft*, **40** (1953), 603.
14. L. Czerski and F. Franik, *Arch. Gorn. Hutn.*, **3** (1955), 43.
15. D. Caplan, M. J. Graham, and M. Cohen, *J. Electrochem. Soc.*, **119** (1978), 1205.
16. M. J. Graham, D. Caplan, and R. J. Hussey, *Can. Met. Quart.*, **18** (1979), 283.
17. S. Mrowec and T. Werber, *Gas Corrosion of Metals*, Springfield, VA, US Department of Commerce, National Technical Information Service, 1978, p. 383.
18. A. Atkinson and D. W. Smart, *J. Electrochem. Soc.*, **135** (1988), 2886.
19. K. Fueki and J. B. Wagner, *J. Electrochem. Soc.*, **112** (1965), 384.
20. C. Wagner and K. Grünewald, *Z. Phys. Chem.*, **40B** (1938), 455.
21. K. Wefers and C. Misra, AlCOA Technical Paper 19, Pittsburgh, PA, Aluminium Company of America, 1987.
22. P. Liu and J. Skogsmo, *Acta Crystallogr.*, **B47** (1991), 425.
23. K. Shimizu, K. Kobayashi, G. E. Thompson, and G. C. Wood, *Oxid. Met.*, **36** (1991), 1.
24. K. Shimizu, R. C. Furneaux, G. E. Thompson, *et al.*, *Oxid. Met*, **35** (1991), 427.
25. J. I. Eldridge, R. J. Hussey, D. F. Mitchell, and M. J. Graham, *Oxid. Met.*, **30** (1988), 301.
26. H. J. Engell, *Archiv. Eisenhüttenwesen*, **28** (1957), 109.

27. H. J. Engell, *Acta met.*, **6** (1958), 439.
28. M. H. Davies, M. T. Simnad, and C. E. Birchenall, *J. Met.*, **3** (1951), 889.
29. D. J. M. Deven, J. P. Shelton, and J. S. Anderson, *J. Chem. Soc.*, **2** (1948), 1729.
30. A. Bruckmann and G. Simkovich, *Corr. Sci.*, **12** (1972), 595.
31. W. Schwenk and A. Rahmel, *Oxid. Met.*, **25** (1986), 293.
32. R. H. Jutte, B. J. Kooi, M. A. J. Somers, and E. J. Mittemeijer, *Oxid. Met.*, **48** (1997), 87.
33. S. Mrowec and K. Przybylski, *Oxid. Met.*, **11** (1977), 383.
34. P. Kofstad, *High Temperature Oxidation of Metals*, New York, Wiley, 1966.
35. S. Mrowec and K. Przybylski, *Oxid. Met.*, **11** (1977), 365.
36. D. W. Bridges, J. P. Baur, and W. M. Fassel, *J. Electrochem. Soc.*, **103** (1956), 619.
37. H. S. Hsu and G. J. Yurek, *Oxid. Met.*, **17** (1982), 55.
38. G. J. Yurek, J. P. Hirth, and R. A. Rapp, *Oxid. Met.*, **8** (1974), 265.
39. G. Garnaud, *Oxid. Met.*, **11** (1977), 127.
40. G. Garnaud and R. A. Rapp, *Oxid. Met.*, **11** (1977), 193.
41. C. S. Tedmon, *J. Electrochem. Soc.*, **113** (1966), 766.
42. D. Caplan, A. Harvey, and M. Cohen, *Corr. Sci.*, **3** (1963), 161.
43. D. Caplan and G. I. Sproule, *Oxid. Met.*, **9** (1975), 459.
44. R. J. Hussey and M. J. Graham, *Oxid. Met.*, **45** (1996), 349.
45. C. M. Cotell, G. J. Yurek, R. J. Hussey, D. J. Mitchell, and M. J. Graham, *Oxid. Met.*, **34** (1990), 173.
46. E. A. Polman, T. Fransen, and P. J. Gellings, *Oxid. Met.*, **32** (1989), 433.
47. E. A. Gulbransen and G. H. Meier, Mechanisms of oxidation and hot corrosion of metals and alloys at temperatures of 1150 to 1450 K under flow. In Proceedings of 10th Materials Research Symposium, National Bureau of Standards Special Publications 561, 1979, p. 1639.
48. E. A. Gulbransen and W. S. Wysong, *TAIME*, **175** (1948), 628.
49. E. A. Gulbransen, K. F. Andrew, and F. A. Brassart, *J. Electrochem. Soc.*, **110** (1963), 952.
50. C. B. Alcock and G. W. Hooper, *Proc. Roy. Soc.*, **254A** (1960), 551.
51. L. A. Carol and G. S. Mann, *Oxid. Met.*, **34** (1990), 1.
52. C. Wagner, *J. Appl. Phys.*, **29** (1958), 1295.
53. C. Wagner, *Corr. Sci.*, **5** (1965), 751.
54. E. A. Gulbransen, K. F. Andrew, and F. A. Brassart, *J. Electrochem. Soc.*, **113** (1966), 834.
55. J. A. Roberson and R. A. Rapp, *TAIME*, **239** (1967), 1327.
56. R. A. Perkins and G. H. Meier, Acoustic emission studies of high temperature oxidation. In *High Temperature Materials Chemistry*, eds. Z. A. Munir and D. Cubicciotti, New York, NY, Electrochemical Society, 1983, p. 176.

5

Oxidation of alloys

Introduction

Many of the factors described for the oxidation of pure metals also apply to the oxidation of alloys. However, alloy oxidation is generally much more complex as a result of some, or all, of the following.

- The metals in the alloy will have different affinities for oxygen reflected by the different free energies of formation of the oxides.
- Ternary and higher oxides may be formed.
- A degree of solid solubility may exist between the oxides.
- The various metal ions will have different mobilities in the oxide phases.
- The various metals will have different diffusivities in the alloy.
- Dissolution of oxygen into the alloy may result in sub-surface precipitation of oxides of one or more alloying elements (internal oxidation).

This chapter describes the major effects occurring in alloy oxidation and their relation to the above factors. No attempt has been made to provide a complete survey of the extensive literature on this subject, rather, examples which illustrate the important fundamentals are presented. This is done by first classifying the types of reactions which occur, and then describing additional factors which have significant influences on the oxidation process. Subsequent chapters will describe the oxidation of alloys in complex environments, such as those involving mixed gases, liquid deposits, and erosive conditions and the use of coatings for oxidation protection. Previous reviews of alloy oxidation include those by Kubaschewski and Hopkins,[1] Hauffe,[2] Benard,[3] Pfeiffer and Thomas,[4] Birchenall,[5] Mrowec and Werber,[6] Kofstad,[7] and Beranger, Colson and Dabosi.[8]

Classification of reaction types

Wagner[9] has presented a lucid categorization of simple types of alloy oxidation according to the reaction morphologies. The classification presented here is a

101

modification of that given by Wagner in which alloys will be grouped as (a) noble parent with base alloying elements and (b) base parent with base alloying elements.

Noble parent with base alloying elements

This group includes alloy bases, such as Au, Ag, Pt, etc., which do not form stable oxides under normal conditions with alloying elements such as Cu, Ni, Fe, Co, Cr, Al, Ti, In, Be, etc., which form stable oxides. However, at reduced oxygen partial pressures, those metals, such as Cu, Ni, and Co, which have only moderately stable oxides (see Chapter 2) can behave as noble-parent metals.

A useful example of this alloy class is the Pt–Ni system. The important factors are a low solubility for oxygen in Pt and a small negative standard free energy of formation for NiO. When the alloy is first exposed to the oxidizing atmosphere, oxygen begins to dissolve in the alloy. However, as a result of the low oxygen solubility and only moderate stability of NiO, the oxide cannot nucleate internally. This results in the outward migration of Ni to form a continuous NiO layer on the alloy surface. Since Pt does not enter into oxide formation it is inevitable that it will concentrate at the scale–metal interface and, correspondingly, Ni will be denuded there. Schematic concentration gradients are shown in Figure 5.1. Here $N_{Ni}^{(o)}$ and $N_{Pt}^{(o)}$ are the bulk atom fractions in the alloy of Ni and Pt, respectively, and N_{Ni}' and N_{Pt}' are the interface concentrations. The reaction rate may be limited either by the supply of nickel from the alloy to the oxide or by the rate of diffusion of Ni ions across the oxide to the scale–gas interface where further oxide formation occurs. When the alloy nickel content is low, diffusion in the alloy is rate determining. When the nickel content is high, diffusion through the oxide is rate determining. This situation has been analyzed by Wagner[10] and the predictions agree well with experiment. Both processes are diffusion controlled and both, therefore, lead to

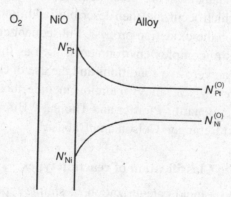

Figure 5.1 Schematic concentration profiles for the oxidation of Pt–Ni alloys.

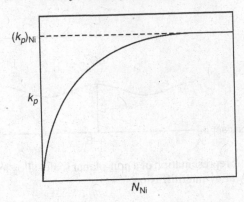

Figure 5.2 Effect of nickel on the parabolic rate constant for the oxidation of Pt–Ni alloys.

parabolic kinetics. As the nickel content is increased, the parabolic rate constant approaches values typical of pure nickel as shown schematically in Figure 5.2. The bulk nickel content below which diffusion in the alloy controls may be calculated by equating the fluxes arriving at the alloy–oxide interface to that required to maintain the growth of the oxide. Here we consider the general case of a system A–B where A represents the noble component and BO_v is the oxide being formed:

$$J_B = \frac{D_B}{V_m} \left(\frac{\partial N_B}{\partial x} \right)_{x=0} = \frac{1}{v} \cdot \frac{1}{2} \cdot \frac{k_p^{1/2}}{M_O} t^{-1/2}. \tag{5.1}$$

The concentration gradient in Equation (5.1) may be evaluated by differentiating Equation (A19) from Appendix A, with $N_B^{(S)}$ being taken as zero, leading to the result shown in Equation (5.2):

$$N_B^{(o)} = \frac{V_m}{32v} \left(\frac{\pi k_p}{D_B} \right)^{1/2}. \tag{5.2}$$

When the scale growth is controlled by diffusion in the alloy, the scale–alloy interface can become unstable as illustrated in Figure 5.3. Since any inward protuberance of the scale–alloy interface will shorten the diffusion distance across the zone depleted in B, such a protuberance will grow and result in a wavy scale–alloy interface. This problem has been described by Wagner.[11] Similar considerations apply to the scale–gas interface, which can develop oxide whiskers if gas-phase diffusion of oxidant to the surface controls the rate. Rickert[12] has shown this to be the case for the reaction of silver and sulphur using the experiment shown schematically in Figure 5.4. A silver specimen was placed in one end of a sealed, evacuated tube and a quantity of sulphur was placed in the other. The tube was heated to 400 °C in an inclined furnace. Sulphur transport through the vapour caused an Ag_2S layer to grow on the portion of silver nearest the sulphur. However, as the S was depleted

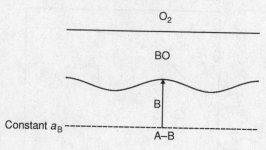

Figure 5.3 Schematic representation of a non-planar scale–alloy interface resulting from rate control by alloy interdiffusion.

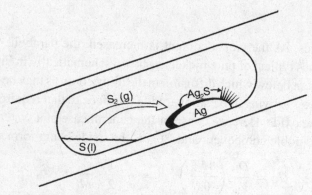

Figure 5.4 Whisker growth in the reaction between Ag and S resulting from gas-phase diffusion control.

from the vapour by this reaction, the rate of transport to the far end of the specimen became rate controlling so that Ag_2S whiskers grew on this end as indicated in Figure 5.4.

Silver-base alloys containing less noble solutes, such as In or Al, represent another useful case in this classification. In this case the solubility of oxygen in Ag is significant so that, for dilute alloys, internal oxidation can occur.

Internal oxidation

Internal oxidation is the process in which oxygen diffuses into an alloy and causes sub-surface precipitation of the oxides of one or more alloying elements. It has been the subject of reviews by Rapp,[13] Swisher,[14] Meijering,[15] and Douglass.[16]

The necessary conditions for the occurrence of the phenomenon are the following.

(1) The value of ΔG° of formation (per mole O_2) for the solute metal oxide, BO_{ν}, must be more negative than ΔG° of formation (per mole O_2) for the base metal oxide.

(2) The value of ΔG for the reaction $B + \nu\underline{O} = BO_{\nu}$ must be negative. Therefore, the base metal must have a solubility and diffusivity for oxygen which is sufficient to establish the required activity of dissolved oxygen $\underline{O}$ at the reaction front.

(3) The solute concentration of the alloy must be lower than that required for the transition from internal to external oxidation.

(4) No surface layer must prevent the dissolution of oxygen into the alloy at the start of oxidation.

The process of internal oxidation occurs in the following manner. Oxygen dissolves in the base metal (either at the external surface of the specimen or at the alloy–scale interface if an external scale is present) and diffuses inward through the metal matrix containing previously precipitated oxide particles. The critical activity product, $a_B a_O^{\nu}$, for the nucleation of precipitates is established at a reaction front (parallel to the specimen surface) by the inward-diffusing oxygen and the outward diffusion of solute. Nucleation of the oxide precipitate occurs and a given precipitate grows until the reaction front moves forward and depletes the supply of solute B arriving at the precipitate. Subsequent precipitate growth occurs only by capillarity-driven coarsening (Ostwald ripening).

Kinetics

The rate of internal oxidation will be derived here for a planar specimen geometry using the quasi-steady-state approximation. The results of similar derivations for cylindrical and spherical specimens will be presented. A more rigorous derivation of the kinetics of internal oxidation is presented in Appendix B.

Consider a planar specimen of a binary alloy A–B in which B is a dilute solute, which forms a very stable oxide. Assume the ambient oxygen pressure is too low to oxidize A but high enough to oxidize B (see Figure 5.5).

The quasi-steady approximation requires the assumption that the dissolved oxygen concentration varies linearly across the zone of internal oxidation. Therefore, the oxygen flux through the internal oxidation zone (IOZ) is given by Fick's first law as Equation (5.3),

$$J = \frac{dm}{dt} = D_O \frac{N_O^{(S)}}{X V_m} \ (\text{mol cm}^{-2}\text{s}^{-1}),\tag{5.3}$$

where $N_O^{(S)}$ is the oxygen solubility in A (atom fraction), V_m is the molar volume of the solvent metal or alloy (cm^3 mol^{-1}), and D_O is the diffusivity of oxygen in A (cm^2 s^{-1}). If counter-diffusion of solute B is assumed to be negligible, the amount of oxygen accumulated in the IOZ per unit area of reaction front is given by

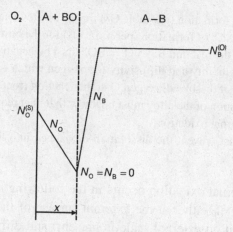

Figure 5.5 Simplified concentration profiles for the internal oxidation of A–B.

Equation (5.4),

$$m = \frac{N_B^{(o)} \nu X}{V_m} \ (\text{mol cm}^{-2})$$ (5.4)

where $N_B^{(o)}$ is the initial solute concentration. Differentiation of Equation (5.4) with respect to time gives an alternate expression for the flux:

$$\frac{dm}{dt} = \frac{N_B^{(o)} \nu}{V_m} \frac{dX}{dt}.$$ (5.5)

Equating Equations (5.3) and (5.5),

$$D_O \frac{N_O^{(S)}}{X V_m} = \frac{N_B^{(o)} \nu}{V_m} \frac{dX}{dt},$$ (5.6)

and rearranging Equation (5.6) gives Equation (5.7):

$$X dX = \frac{N_O^{(S)} D_O}{\nu N_B^{(o)}} dt.$$ (5.7)

Integration of Equation (5.7), assuming $X = 0$ at $t = 0$, yields Equation (5.8),

$$\frac{1}{2} X^2 = \frac{N_O^{(S)} D_O}{\nu N_B^{(o)}} t,$$ (5.8)

or Equation (5.9):

$$X = \left[\frac{2 N_O^{(S)} D_O}{\nu N_B^{(o)}} t \right]^{1/2}.$$ (5.9)

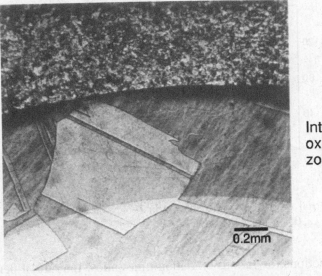

Internal oxidation zone

0.2mm

Figure 5.6 Optical micrograph showing the microstructure of a Cu–0.47 wt% Ti alloy internally oxidized at 800 °C for 97 h.

The latter gives the penetration depth of the internal oxidation zone as a function of oxidation time. The following points concerning Equation (5.9) are of interest.

(1) The penetration depth has a parabolic time dependence, $X \propto t^{1/2}$.
(2) The penetration depth for a fixed time is inversely proportional to the square root of the atom fraction of solute in the bulk alloy.
(3) Careful measurements of the front penetration as a function of time for an alloy of known solute concentration can yield a value for the solubility–diffusivity product, $N_O^{(S)} D_O$ (permeability) for oxygen in the matrix metal.

The results of a similar derivation for a cylindrical specimen[14] yield Equation (5.10),

$$\frac{(r_1)^2}{2} - (r_2)^2 \ln\left(\frac{r_1}{r_2}\right) + \frac{1}{2} = \frac{2N_O^{(S)} D_O}{\nu N_B^{(o)}} t, \qquad (5.10)$$

and for a spherical specimen,[14] Equation (5.11),

$$\frac{(r_1)^2}{3} - (r_2)^2 + \frac{2}{3}\frac{(r_2)^3}{(r_1)} = \frac{2N_O^{(S)} D_O}{\nu N_B^{(o)}} t, \qquad (5.11)$$

where r_1 is the specimen radius and r_2 is the radius of the unoxidized alloy core.

Figure 5.6 shows a polished and etched cross-section of an internally oxidized Cu–Ti cylinder showing the penetration of the internal oxidation zone.[17] The TiO$_2$ particles are too small to resolve in the optical microscope. Measurements of r_2

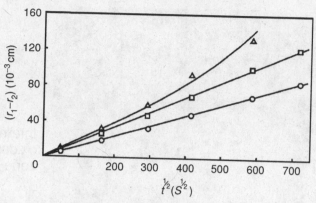

Figure 5.7 Internal oxidation zone penetration for cylindrical Cu–Ti alloys at 850 °C: ○, 0.91; □, 0.47; and △, 0.25 wt%.

from such cross-sections for various oxidation times allow the rate of penetration to be measured. Note that, for this geometry, the penetration depth is expressed as $(r_1 - r_2)$. Figure 5.7 shows the results of such measurements for Cu–Ti alloys at 850 °C. The curvature of the plot is predicted by Equation (5.10) and can be understood, qualitatively, on the basis of the flux of oxygen feeding into a continually decreasing volume as the front penetrates. Plotting the left hand side of Equation (5.10) versus time yields a straight line with a slope of $2N_O^{(S)} D_O / \nu N_B^{(o)}$ from which $N_O^{(S)} D_O$ may be calculated.[17]

Precipitate morphology
Many of the effects of internal oxidation, both on the overall corrosion process for an alloy and on influencing the mechanical, magnetic, etc., properties of the alloy, are intimately related to the morphology of the oxide precipitates. The following is a rather qualitative discussion of the factors, which influence the size, shape, and distribution of internal oxides.

In general, the oxide particle size is determined by a competition between the rate of nucleation as the internal oxidation front passes, and the subsequent growth and coarsening rates of the particles. The longer the time a nucleated particle has to grow by arrival of oxygen and solute at its surface before the next nucleation event 'which depletes the supply of solute' the larger will be the particle. Subsequent coarsening may occur, but this phenomenon will be temporarily neglected. Therefore, those factors which favour high nucleation rates will decrease the particle size and those which favour high growth rates will increase the size.

If the nucleation of precipitates is assumed to be controlled by the velocity of the oxidation front (i.e., the rate at which a critical supersaturation may be built up) and the growth is controlled by the length of time available for growth of

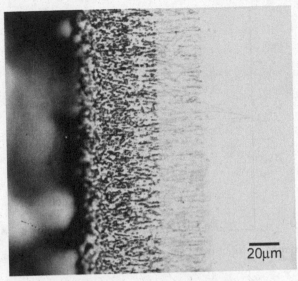

20μm

Figure 5.8 Optical micrograph of a Co–5 wt%Ti alloy internally oxidized for 528 h at 900 °C.

a precipitate, the size should be inversely proportional to the front velocity, i.e., $r \propto 1/\vec{v}$. Rewriting Equation (5.7) gives the velocity for a planar specimen as in Equation (5.12):

$$\vec{v} = \frac{dX}{dt} = \frac{N_O^{(S)} D_O}{\nu N_B^{(o)}} \frac{1}{X}. \tag{5.12}$$

Therefore, other things being equal, one would predict the particle size to increase proportionally to (a) X, (b) $N_B^{(o)}$, and (c) $1/N_O^{(S)} \propto (1/p_{O_2})^{1/2}$.

The effect of temperature is difficult to evaluate but, apparently, the nucleation rate is affected much less by increasing temperature than is the growth rate so that particle size increases with increasing temperature of internal oxidation. As the front penetration increases, the velocity eventually decreases to a point where diffusion of solute to a growing particle occurs rapidly enough to prevent nucleation of a new particle. This results in the growth of elongated or needle-like particles,[18] e.g., Figure 5.8. The formation of elongated particles can also be promoted by inward interfacial diffusion of oxygen along the alloy oxide interfaces.

The effect of particle-matrix interfacial free energy is often overlooked but is particularly important in the nucleation and coarsening of internal oxides. Consider the classical nucleation problem of forming a spherical nucleus. If strain is neglected, the free energy of formation of a nucleus of radius r is given by Equation (5.31),

$$\Delta G = 4\pi r^2 \sigma + \frac{4}{3}\pi r^3 \Delta G_V, \tag{5.13}$$

Oxidation of alloys

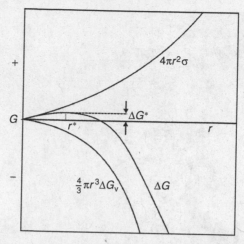

Figure 5.9 Schematic plot of free energy versus radius of a nucleating particle.

where σ is the specific interfacial free energy and ΔG_V is the free-energy change per unit volume of precipitate formed during the reaction. The surface and volume terms, along with ΔG, are plotted as a function of r in Figure 5.9. Nuclei with radii greater than r^* will tend to grow spontaneously. At r^* we have Equation (5.14),

$$\frac{d\Delta G}{dr} = 8\pi r^*\sigma + 4\pi r^{*2}\Delta G_V = 0, \tag{5.14}$$

which may be solved for r^* as in Equation (5.15):

$$r^* = \frac{-2\sigma}{\Delta G_V}. \tag{5.15}$$

Substitution into Equation (5.13) yields ΔG^*, the height of the activation barrier to nucleation:

$$\Delta G^* = 4\pi \frac{4\sigma^3}{\Delta G_V^2} - \frac{4}{3}\pi \frac{8\sigma^3}{\Delta G_V^3}\Delta G_V$$

$$= \left(16 - \frac{32}{3}\right)\pi \frac{\sigma^3}{\Delta G_V^2} = \frac{16}{3}\pi \frac{\sigma^3}{\Delta G_V^2}. \tag{5.16}$$

It is this value (ΔG^*) which enters into the nucleation rate as calculated from absolute reaction-rate theory, Equation (5.17),

$$J = \omega C^* = \omega C_o \exp(-\Delta G^*/RT), \tag{5.17}$$

where C_o is the concentration of reactant molecules, C^* is the concentration of critical nuclei, and ω is the frequency with which atoms are added to the critical nucleus. Clearly, since the nucleation rate is proportional to the exponential of a quantity, which in turn depends on the cube of the surface free energy, this parameter will have a drastic effect on the nucleation, and particle size will increase with increasing σ. This influence will be felt further in the coarsening of the particles since the interfacial free energy is the driving force for this process.

In a similar manner the relative stability of the internal oxide, $\Delta G°$, will influence ΔG_V with more stable oxides resulting in higher nucleation rates and smaller particles.

In summary, the size of internal oxides will be determined by a number of factors with larger particles forming for:

(1) deeper front penetrations, X
(2) higher solute concentrations, $N_B^{(o)}$
(3) lower oxygen solubilities, i.e., lower p_{O_2}
(4) higher temperature
(5) higher particle-matrix interfacial free energies
(6) less stable oxides.

These predictions have been born out in a number of studies of particle morphology available in the literature.[17-20] However, the internal oxidation process is quite complex and many systems show deviations from the simple analysis provided above. Douglass[16] has described these phenomena which include the following.

(1) Accelerated transport along the interfaces between internal oxides and the alloy matrix producing oxide needles and deeper penetrations than those predicted from Equation (5.9).
(2) Preferential formation of internal oxides on alloy grain boundaries often with zones adjacent to the boundary which are denuded of precipitates.
(3) Alternating bands of internal oxide particles running parallel to the specimen surface.
(4) Extrusion of parent metal from the internal oxidation zone to form pure metal nodules on the surface.
(5) Formation of solute-oxygen clusters containing more than the stoichiometric oxygen concentration.

Transition from internal to external oxidation
Consideration of Equation (5.12) shows that the penetration velocity of the internal oxidation zone decreases with increasing $N_B^{(o)}$ and decreasing $N_O^{(S)}$. Thus there will be a limiting concentration of solute in the alloy above which the outward diffusion of B will be rapid enough to form a continuous blocking layer of BO_ν and stop internal oxidation (shown schematically in Figure 5.10). This transition to external oxidation is the basis for the design of Fe-, Ni-, and Co-base engineering alloys. These contain a sufficiently high concentration of a solute (e.g., Cr, Al, or Si) to produce an external layer of a stable, slowly growing oxide (e.g., Cr_2O_3, Al_2O_3, or SiO_2), which prevents oxidation of the parent metal. This process, known as *selective oxidation*, will be discussed further but at this point the fundamentals of the transition from internal to external oxidation must be considered. The analysis of

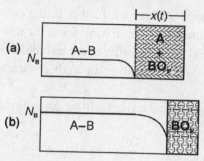

Figure 5.10 Schematic diagram of the transition from internal oxidation of a solute B to the formation of an external layer of BO_y. An alloy undergoing internal oxidation is presented in (a), and a more concentrated alloy, where the oxide forms an external layer, is presented in (b).

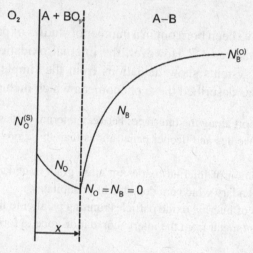

Figure 5.11 Concentration profiles for internal oxidation of A–B.

this problem has been carried out by Wagner.[21] Consider the concentration profiles in Figure 5.11. The penetration of the zone of internal oxidation may be expressed as in Equation (5.18) (see Appendix B):

$$X = 2\gamma(D_O t)^{1/2}. \tag{5.18}$$

Fick's second law for the counter-diffusion of B is shown in Equation (5.19),

$$\frac{\partial N_B}{\partial t} = D_B \frac{\partial^2 N_B}{\partial x^2}, \tag{5.19}$$

with initial conditions;

$$t = 0, \quad N_B = 0 \quad \text{for} \quad x < 0;$$
$$N_B = N_B^{(o)} \quad \text{for} \quad x > 0;$$

and boundary conditions:

$$t = t, \quad N_B = 0 \quad \text{for} \quad x = X;$$
$$N_B = N_B^{(o)} \quad \text{for} \quad x = \infty.$$

The solution to Equation (5.19), therefore, becomes that given in Equation (5.20),

$$N_B(x, t) = N_B^{(o)} \left[1 - \frac{\text{erfc}\left(\dfrac{x}{2(D_B t)^{1/2}} \right)}{\text{erfc}(\theta^{1/2}\gamma)} \right], \tag{5.20}$$

where $\theta = D_O/D_B$.

If f represents the mole fraction of BO_ν in the internal oxidation zone and V_m the molar volume of the alloy, then f/V_m will be the concentration in moles per volume and the number of moles in a volume element, AdX, will be $(f/V_m)AdX$, where A is the cross-sectional area for diffusion. This quantity must be equal to the number of moles of B arriving at $x = X$ in the time dt by diffusion from within the sample, i.e., from $x > X$. Therefore, we obtain Equation (5.21):

$$\frac{f\,AdX}{V_m} = \left[\frac{A D_B}{V_m} \frac{\partial N_B}{\partial x} \right] dt, \tag{5.21}$$

Substitution of Equations (5.18) and (5.20) into Equation (5.21) yields an enrichment factor of α, Equation (5.22):

$$\alpha = \frac{f}{N_B^{(o)}} = \frac{1}{\gamma \pi^{1/2}} \left(\frac{D_B}{D_O} \right)^{1/2} \frac{\exp(-\gamma^2 \theta)}{\text{erfc}(\gamma \theta^{1/2})}. \tag{5.22}$$

For the case of interest here, $N_O^{(S)} D_O \ll N_B^{(o)} D_B$, i.e., the oxygen permeability is significantly less than the corresponding product for B, the product $\gamma \theta^{1/2} \ll 1$ and γ is given by Equation (B20) from Appendix B. Therefore, α may be written as in Equation (5.23):

$$\alpha \approx \frac{2\nu}{\pi} \left[\frac{N_B^{(o)} D_B}{N_O^{(S)} D_O} \right]. \tag{5.23}$$

When α is large, we expect the accumulation and lateral growth of the internal oxides to form a continuous layer, i.e., the transition to external oxidation. Figure 5.12 shows this occurring in a Co–7.5 wt% Ti alloy.[18] Wagner states that, when the volume fraction of oxide, $g = f(V_{ox}/V_m)$, reaches a critical value, g^*, the transition from internal to external scale formation should occur. Insertion of f, in terms of g^*, in α in Equation (5.23) then gives the criterion for external oxidation as shown

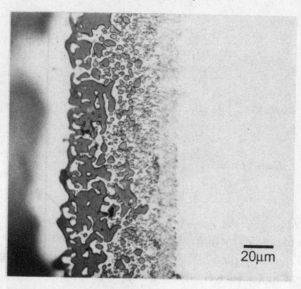

Figure 5.12 Optical micrograph showing transition from internal to external oxidation in a Co–7.5 wt% Ti alloy oxidized for 528 h at 900 °C.

in Equation (5.24):

$$N_{B}^{(o)} > \left[\frac{\pi \, g^*}{2 \nu} N_{O}^{(S)} \frac{D_O V_m}{D_B V_{ox}} \right]^{1/2}.$$ (5.24)

The latter enables prediction of how changing exposure conditions will affect the concentration of solute required to produce an external scale. Those conditions that decrease the inward flux of oxygen, e.g., lower $N_{O}^{(S)}$ (i.e. lower p_{O_2}) and those that increase the outward flux of B, e.g., cold working the alloy (increasing D_B by increasing the contribution of short-circuit diffusion) will allow the transition to external oxidation to occur at lower solute concentrations. Rapp[22] has tested the Wagner theory for Ag–In alloys. It was found that at $p_{O_2} = 1$ atm and $T = 550$ °C the transition occurred for $g^* = 0.3$ ($N_{In} = 0.16$). Rapp then used this value to predict the effect of reduced p_{O_2} (reduced $N_{O}^{(S)}$) on the In concentration required for the transition. The agreement between experiment and Equation (5.24) was excellent. Rapp's experiments also gave an indirect verification of the effect of increased D_B, caused by cold work, on the transition. Specimens containing only 6.8 at% In were observed to undergo the transition to external In_2O_3 formation around scribe marks under exposure conditions that required 15 at% In to produce the transition on undeformed surfaces.

An additional factor which can have a pronounced influence on the transition is the presence of a second solute whose oxide has a stability intermediate between that of A and B. If the p_{O_2} is great enough to form the oxide of the second solute this will, in effect, decrease the inward flux of oxygen and allow external formation of

BO_ν at lower concentrations of B.[23] This concept is often referred to as *secondary gettering*. Wagner observed the effect for Cu–Zn–Al alloys[23] and Pickering obtained similar results for Ag–Zn–Al alloys.[24] In the latter experiments Ag–3 at% Al alloys were found to oxidize internally while, under the same conditions, a Ag–21 at% Zn–3 at% Al alloy produced an external Al_2O_3 scale. The secondary getter concept is based essentially on the second solute decreasing $N_O^{(S)}$. However, explanation of the effect should actually consider the effect on all of the terms in Equation (5.24).[25,26] Nevertheless, the concept has proved useful in the design of high-temperature alloys since the element which provides protection by selective oxidation (e.g., Al) often has detrimental effects on other alloy properties, particularly mechanical properties, and should, in these cases, be held at as low a concentration as is feasible. This concept has led to the development of the M–Cr–Al (M = Fe, Ni, or Co) alloys, which will be discussed in the next section.

The above systems have been considered since they allow some of the features of alloy oxidation to be examined without the complication of the simultaneous formation of two oxides. In principle, any alloy system can be examined in this way, provided the ambient oxygen pressure is below that required to form the lowest oxide of the more noble component.

Selective oxidation

The process described above in which a solute oxidizes preferentially to the parent element and forms a continuous layer on the surface is referred to as *selective oxidation*. The selective oxidation of elements which form a slowly growing, protective layer is the basis for the oxidation protection of all alloys and coatings used at high temperature. The only elements which consistently result in protective scales are Cr (chromia scale), Al (alumina scale), and Si (silica scale). Therefore, much research has been directed at finding alloy and coating compositions, which meet other property (e.g., mechanical) requirements and also form one of these scales.

Base parent with base-alloying elements

This classification represents the systems most commonly used and the formation of two or more oxides must be considered. The principles of oxidation of this class of alloys will be developed through specific examples.

Oxidation of nickel-, iron-, and cobalt-base alloys

These alloys form the basis for most commercial high-temperature alloys and consist of a base of Ni, Fe, or Co, which forms a moderately stable oxide, and an alloying element Cr, Al, or Si, which forms a highly stable oxide.

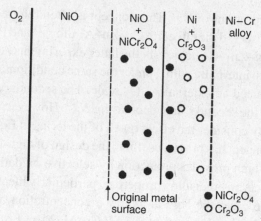

Figure 5.13 Schematic diagram of the oxidation morphology of dilute Ni–Cr alloys.

Chromia-forming alloys

(i) Nickel–chromium alloys

Extensive studies of the oxidation behaviour of Ni–Cr alloys have been published by a number of investigators.[27,28] Alloys in this system with low Cr contents show internal oxidation of Cr forming Cr_2O_3 islands within a matrix of almost pure Ni. An outer scale of NiO is formed with an inner layer, sometimes porous, of NiO containing $NiCr_2O_4$ islands as shown in Figure 5.13.

The NiO of the inner duplex scale layer contains Cr ions in solution in equilibrium with the $NiCr_2O_4$ second phase. This provides cation vacancies, thus increasing the mobility of Ni ions in this region. This doping effect, explained in Appendix C, results in an increase in the observed oxidation rate constant, compared with pure Ni, which is also increased, on a mass-gain basis, due to the extra oxygen tied up in Cr_2O_3 in the internal oxidation zone, Figure 5.14.

As the outer scale advances into the metal, the Cr_2O_3 internal oxide islands are engulfed by NiO and a solid-state reaction ensues to form $NiCr_2O_4$:

$$NiO + Cr_2O_3 = NiCr_2O_4. \tag{5.25}$$

(Some $NiCr_2O_4$ actually starts to form by oxygen moving in advance of the NiO front.) These $NiCr_2O_4$ islands remain as a second phase in the scale in the cases of all but the most dilute Cr alloys and their extent marks the position of the original metal surface.

Since cation diffusion is much slower through the $NiCr_2O_4$ spinel than it is through NiO, the spinel islands in the scale act as diffusion blocks for outward-migrating Ni ions. Consequently, as the Cr content of the alloy is increased, the increasing volume fraction of spinel reduces the total Ni flux through the scale and the rate constant begins to fall. Eventually as the Cr content is increased further,

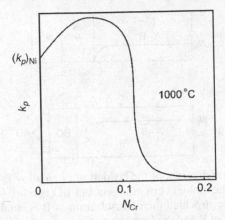

Figure 5.14 Dependence of the parabolic rate constant for the oxidation of Ni–Cr alloys on Cr content.

to about 10 wt% at 1000 °C, the mode of oxidation changes, as with Ag–In alloys, to give a complete external scale of Cr_2O_3. At this and higher Cr concentrations the oxidation rate falls abruptly to values more typical of Cr than of Ni. This is illustrated in Figure 5.14.

The above discussion of the effect of Cr content applies to steady-state oxidation. In fact, when a clean alloy surface, even one with a Cr concentration in excess of the critical value, is exposed to an oxidizing atmosphere, oxides containing Ni as well as Cr will form initially. Since the Ni-containing oxides grow much more rapidly than Cr_2O_3 a significant amount of NiO and $NiCr_2O_4$ can form before a continuous Cr_2O_3 layer can form. This phenomenon, known as *transient oxidation*,[29–31] occurs for almost all alloy systems for which the oxides of more than one component have negative free-energy changes for their formation in the given atmosphere. This is shown schematically in Figure 5.15b for the transition from internal to external oxidation of an element B (e.g., Cr, Al), which forms a more stable oxide than A (e.g., Ni, Fe) but where some AO forms before BO becomes continuous. The nature of the transient oxidation often affects the ultimate selective oxidation process in which only the most stable oxide is formed. For example, the critical solute content for the transition from internal to external oxidation increases as the parabolic rate constant for the growth of the transient oxide increases.[32] The extent of the transient period is decreased by those factors which promote selective oxidation. For example, for Ni–Cr alloys the amount of Ni-rich oxide is observed to be less for higher alloy Cr contents, reduced oxygen pressures, and exposure of cold-worked alloys.

With Cr contents of around the critical value, the protection conferred by the scale of Cr_2O_3 may not be permanent, since the selective oxidation of Cr to form the scale may denude the alloy below the scale of Cr to the point where its concentration is

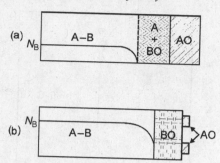

Figure 5.15 Schematic diagram of the oxidation of an alloy A–B for which both oxides are stable in the ambient environment but BO is more stable than AO. (a) Illustration of an alloy in which the concentration of B is small, which results in the internal oxidation of B and the formation of a continuous external layer of AO. (b) Illustration of an alloy, which has sufficient B to form a continuous external layer, but since both oxides can form, initially *transient* AO forms along with BO and grows until the BO becomes continuous.

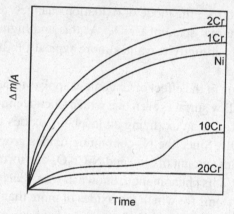

Figure 5.16 Mass change versus time for the oxidation of Ni with various Cr contents.

below the critical one. Therefore, any rupturing, or mechanical damage, of the scale will expose a lower Cr alloy which will undergo internal oxidation and NiO scale formation with a corresponding increase in reaction rate as illustrated in Figure 5.16. In essence the system is returned to the transient oxidation period.

The phenomenon of depletion of an element being selectively oxidized is a common one already mentioned for Pt–Ni alloys and is illustrated for Ni–50 wt% Cr alloys in Figure 5.17. Since these alloys contain two phases, a Ni-rich fcc matrix and Cr-rich bcc precipitate, the Cr depletion is delineated by the depth to which the Cr-rich second phase has dissolved. The extent of depletion of an element being selectively oxidized depends on a number of factors; its concentration, the rate of scale growth, and the alloy interdiffusion coefficient being the most important.

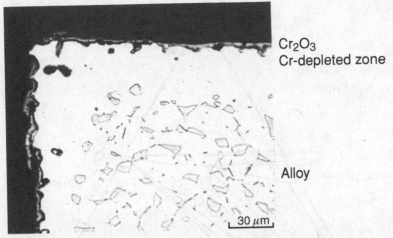

Figure 5.17 Scanning electron micrograph showing the chromium depleted zone for a Ni–50 wt% Cr alloy oxidized for 21 h at 1100 °C in oxygen.

If the alloy is depleted below the concentration that will allow diffusion in the alloy to provide a sufficient flux of Cr to the alloy–oxide interface to maintain the stability of the chromia layer, relative to Ni-containing oxides, the chromia layer can break down without fracturing. The condition for this is described by Equation (5.2). Because of this depletion effect, oxidation-resistant alloys based on the Ni–Cr system usually contain at least 18–20 wt% Cr.

(ii) Iron–chromium alloys

As before, it is convenient to consider this system in order of increasing Cr content. At low Cr contents, no internal oxidation zone is seen. This is because the rate at which the external scale is formed is so rapid that the thickness of the internal oxidation zone is negligibly small. The system is best described in terms of the Fe–Cr–O phase diagram shown schematically in Figure 5.18. The rhombohedral oxides Fe_2O_3 and Cr_2O_3 show a continuous series of solid solutions. The iron- and chromium-oxides react to form spinels which form solid solutions with Fe_3O_4.

At low Cr contents, both chromium-rich oxides and iron oxides form on the surface. Some Cr will enter solution in the FeO phase but, due to the stability of spinel, the solubility is limited as in the case of Ni–Cr alloys (see Figure 5.18). Furthermore, the effect of a few more vacancies is not observable in the highly defective p-type FeO. Thus, an increase in rate constant is rarely observed. On increasing the Cr content, Fe^{2+} ions are progressively blocked by the $FeCr_2O_4$ islands and the FeO layer correspondingly becomes thinner relative to the Fe_3O_4-layer thickness (Figure 5.19). In these stages the reaction rate is still quite high and typical of pure iron. When the Cr content is increased further, a scale of mixed spinel $Fe(Fe,Cr)_2O_4$ is produced and the parabolic rate constant is lower. Apparently, iron

Oxidation of alloys

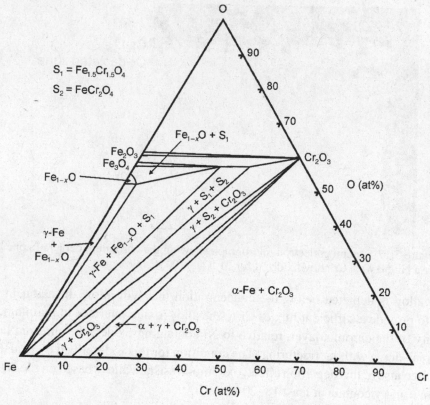

Figure 5.18 Isothermal section of the Fe–Cr–O phase diagram at 1200 °C.

ions are much more mobile in this oxide than Cr^{3+} ions since at longer times quite pure iron oxides can be found at the outer surface of the scale when the oxidation rate is controlled by the diffusion of iron ions through the inner mixed-spinel layer.

On exceeding the Cr concentration N_{Cr}^* an outer scale is formed initially of Cr_2O_3 with a corresponding reduction in the parabolic rate constant. A permanent protective scale cannot be achieved unless N_{Cr}^* is exceeded and most Fe–Cr systems designed for heat resistance have in excess of 20 wt% Cr. It should be emphasized that the popular 'stainless steels' 8 wt% Ni–18 wt% Cr were developed for corrosion resistance to aqueous environments only and should not be regarded as oxidation resistant at high temperatures – a common error. In fact, the Fe–Cr system is not a good system to use as the basis of high-temperature oxidation resistant alloys for long-term exposures because of the range of solid solutions formed by the Fe_2O_3–Cr_2O_3 rhombohedral oxides. Even with high Cr levels, iron ions will dissolve in and diffuse rapidly through the Cr_2O_3 scale and eventually an outer layer of fairly pure iron oxides will result. This is illustrated in Figure 5.20 for Fe–25 wt% Cr which has been oxidized at 1150 °C for 24 h. Protrusions of Fe_3O_4 (A) are seen projecting

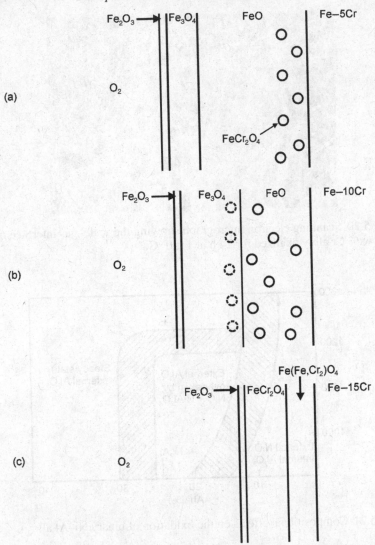

Figure 5.19 Schematic diagrams of the scale morphologies on (a) Fe–5 wt% Cr, (b) Fe–10 wt% Cr, and (c) Fe–15 wt% Cr.

through the Cr_2O_3 scale (B). The use of Cr contents higher than 20–25 wt% cannot be achieved without other alloying additions to avoid sigma-phase formation with its attendant embrittlement of the alloy.

(iii) Cobalt–chromium alloys
The oxidation of Co–Cr alloys is qualitatively similar to that of Ni–Cr alloys. However, the critical Cr content required to develop an external chromia layer on Co–Cr alloys is substantially higher because of a low alloy interdiffusion coefficient

Oxidation of alloys

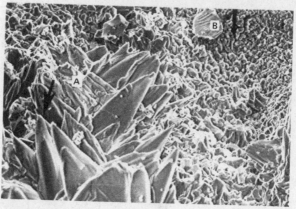

Figure 5.20 Scanning electron micrograph showing the scale–gas interface of an Fe–25 wt% Cr alloy oxidized for 24 h at 1150 °C.

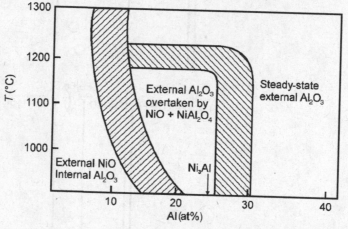

Figure 5.21 Compositional effects on the oxidation of binary Ni–Al alloys.

and somewhat more rapid transient oxidation.[33] As a result, Co-base alloys designed for oxidation resistance contain on the order of 30 wt% Cr.

Alumina-forming alloys

Figure 5.21 presents oxidation data for binary Ni–Al alloys as a function of temperature and composition.[34] The behaviour is characterized by three regions: (I) 0–6 wt% (0–13 at%) Al, internal Al_2O_3 (+ $NiAl_2O_4$) + NiO external scales; (II) 6–17 wt% (13–31 at%) Al, external Al_2O_3 forms initially but cannot be sustained because of an inadequate supply of Al, Equation (5.2), and is overtaken by a rapidly growing NiO + $NiAl_2O_4$ + Al_2O_3 mixture; (III) ≥17 wt% (31 at%), the external Al_2O_3 scale is maintained by a sufficient supply of Al. Higher temperatures extend this region to lower Al contents. The crosshatched region indicates where variable results can be obtained depending upon surface conditions. In fact, in such regions, a given specimen may develop different oxide scales at different locations upon its surface. This diagram indicates that NiAl should form and maintain protective alumina under all conditions while Ni_3Al (γ') is a 'marginal' alumina former at temperatures below 1200 °C.

The stable form of alumina is always the α (corundum) structure. However, under some conditions it is preceded by metastable forms.[35] The oxide scales formed on Ni_3Al at oxygen partial pressures on the order of 1 atm in the temperature range 950–1200 °C consist mainly of Ni-containing transient oxides, NiO and $NiAl_2O_4$, over a layer of columnar α-alumina. Schumann and Rühle[36] studied the very early stages of transient oxidation of the (001) faces of Ni_3Al single crystals in air at 950 °C using cross-section transmission electron microscopy (TEM). After 1 min oxidation simultaneous formation of an external NiO scale and internal oxidation of γ' were observed. The internal oxide particles were identified as γ-Al_2O_3, which possesses a cube-on-cube orientation with respect to the Ni matrix. After 6 min oxidation a continuous γ-Al_2O_3 layer had formed between the IOZ and the γ' single crystal. Oxidation for 30 min resulted in a microstructure similar to the 6 min oxidation but the Ni in the two-phase zone was oxidized to NiO. Oxidation for 50 h resulted in a scale consisting of an outer layer of NiO, an intermediate layer of $NiAl_2O_4$, and an inner layer of γ-Al_2O_3 in which α-Al_2O_3 grains had nucleated at the alloy–oxide interface. The spinel was presumed to have formed by a solid-state reaction between NiO and γ-Al_2O_3. A crystallographic orientation relationship was found between the α- and γ-alumina, whereby $(0001)[1\bar{1}00]_\alpha \| [(111)[1\bar{1}0]_\gamma$, i.e., close-packed planes and close-packed directions of α are parallel to close-packed planes and directions in γ.

The oxidation of NiAl is somewhat unique in that, at temperatures of 1000 °C and above, there are negligible amounts of Ni-containing transient oxides. The transient oxides are all metastable phases of Al_2O_3 (γ, δ, and/or θ).[35] The transition of these metastable phases to the stable α-Al_2O_3 results in significant decreases in the scale growth rate and a 'ridged' oxide morphology which is distinct from the columnar morphology observed on Ni_3Al.[35] The transition aluminas have been shown to grow primarily by outward migration of cations while α-alumina grows primarily by inward transport of oxygen along oxide grain boundaries. The effect of oxidation time and temperature on the phases present in the scales has been studied by several authors. Rybicki and Smialek[37] identified θ-alumina as the transient oxide on Zr-doped NiAl and found that the transition to α occurred at longer times at lower oxidation temperatures, e.g., scales consisted entirely of θ after 100 h at 800 °C while it transformed to α in about 8 h at 1000 °C. Only α was observed at 1100 and 1200 °C. Pint and Hobbs[38] observed only α on undoped NiAl after 160 s at 1500 °C. Brumm and Grabke[39] observed two transformations at 900 °C for undoped NiAl. The scale consisted initially of γ-alumina which transformed to θ-alumina

Oxidation of alloys

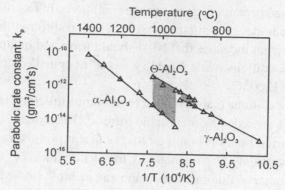

Figure 5.22 Arrhenius plot of k_p for the oxidation of pure NiAl. After Brumm and Grabke.[39]

after approximately 10 h. The transformation of θ to α occurred at much longer times but accelerated as the oxidation temperature was increased. The oxidation kinetics of NiAl, as a function of temperature, are illustrated in Figure 5.22.[39] The scale transformations have also been found to be sensitive to the presence of third elements in the alloy. Additions of Cr accelerate the transformation to α[39] by a proposed mechanism involving transient Cr_2O_3, which is isostructural with α-alumina, providing nucleation sites. This results in a finer-grained α which grows somewhat faster than the α formed on binary NiAl. A study of oxide-dispersed NiAl found that the θ to α transformation was slowed by Y, Zr, La, and Hf and accelerated by Ti.[40] The proposed explanation of these results is that the transformation is slowed by large ions, which can enter the more open lattices of the transition aluminas.

The addition of Cr to Ni–Al alloys results in a remarkable synergistic effect which is of great technological importance. For example, chromium additions of about 10 wt% can enable external Al_2O_3 formation on alloys having aluminium levels as low as 5 wt%. This phenomenon has allowed the design of more ductile alloys and coatings.

The compositional control of oxidation is most easily summarized in *oxide maps*[41,42] where oxidation data are superposed on the ternary composition triangle, as illustrated in Figure 5.23. These maps are not thermodynamic diagrams but are based upon kinetic processes which take place during scale development. Three primary regions of oxidation can be seen corresponding to (I) NiO external scales + Al_2O_3/Cr_2O_3 internal oxides, (II) Cr_2O_3 external scale + Al_2O_3 internal oxides, and (III) external scales of only Al_2O_3.

The role of chromium in producing Al_2O_3 scales at much lower aluminium contents than in binary Ni–Al alloys can be described by considering the transient oxidation phenomena for a Ni–15 wt% Cr–6 wt% Al shown schematically in

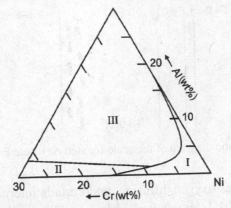

Figure 5.23 Compositional effects on the oxidation of Ni–Cr–Al ternary alloys. (I) External NiO, internal $Cr_2O_3/Al_2O_3/Ni(Al,Cr)_2O_4$; (II) external Cr_2O_3, internal Al_2O_3; (III) external Al_2O_3.

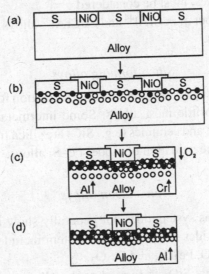

Figure 5.24 Synergistic effects of Cr on the formation of external Al_2O_3 scales during the transient oxidation of Ni–15 wt% Cr–6 wt% Al at 1000 °C. After (a) 1 min; (b) 5 min; (c) 40 min; and (d) >40 min; S = $Ni(Cr,Al)_2O_4$, • = Cr_2O_3, ∘ = Al_2O_3.

Figure 5.24.[43] The initial oxide (a) contains all the cations of the alloy surface resulting in 15 wt% NiO–85 wt% $Ni(Cr,Al)_2O_4$ coverage, (b) subscale formation of Cr_2O_3 occurs because it is stable at the low oxygen activity defined by the NiO–alloy equilibrium and internal oxidation of Al occurs ahead of this front since Al_2O_3 is stable at the even lower oxygen activities here. The high chromium content results in a Cr_2O_3 subscale, which may be continuous (c) and defines a lower scale–alloy

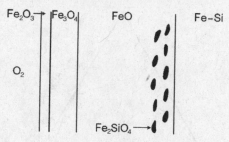

Figure 5.25 Schematic diagram of the scale formed on dilute Fe–Si alloys.

oxygen activity, reduces oxygen diffusion, and curtails internal Al_2O_3 formation. Further, $NiO–Ni(Cr,Al)_2O_4$ growth is also blocked. Eventually, the Al_2O_3 subscale becomes continuous and rate controlling. The effect of Cr was originally described in purely thermodynamic terms as 'gettering';[23] however, it is now clear that the effect of Cr on all the parameters in Equations (5.2) and (5.24), as well as the growth rate of the transient oxides, must be considered.[25,26]

The oxidation behaviour of FeCrAl[44] and CoCrAl[45] alloys is qualitatively similar to that for NiCrAl alloys.

Silica-forming alloys

There are no structural alloys which rely on silica formation for protection primarily because Si tends to embrittle most alloys. Some intermetallic compounds (e.g., $MoSi_2$), silicide coatings, and ceramics (e.g., SiC) are silica formers. The following is a brief discussion of the oxidation of model Fe–Si alloys.

Iron–silicon alloys

As with Fe–Cr alloys, this system does not generally show internal oxidation for the same reasons. The oxides involved in this system include SiO_2 which forms a silicate, Fe_2SiO_4, and FeO, Fe_3O_4, and Fe_2O_3.

At low silicon contents, SiO_2 is formed at the alloy surface. It is apparently distributed very finely since, instead of being surrounded by the advancing iron-oxide scale, it is swept along and accumulated at the alloy surface. Simultaneously it reacts with FeO to form fayalite, Fe_2SiO_4, and, when the particles become larger, they are engulfed by the scale. These islands lie in the FeO layer as markers but do not give an accurate indication of the position of the original metal–scale interface. The fayalite is indeed frequently seen to lie as stringers parallel to the metal surface (see Figure 5.25).

The silicon content can be increased to provide a continuous, protective SiO_2 scale.[46] However, at the concentrations required, the formation of intermetallic compounds make the alloys unstable mechanically.

Both Fe–Cr and Fe–Si alloys can be made to show internal oxidation by reducing or preventing the inward movement of the outer scale. This can be demonstrated by oxidizing in air to form a scale, then encapsulating the sample under vacuum. Alternatively, internal oxidation can be observed at areas where scale–metal separation has occurred, i.e., usually at corners and edges.

In the systems discussed so far, the alloying element has been much more reactive towards oxygen than the parent metal and has consequently tended to segregate preferentially to the oxide formed. Where the alloy features internal oxidation as a reaction characteristic, the segregation is almost complete resulting in a matrix heavily deficient in the alloying element. Where an external scale only has formed although the segregation is severe, it does not generally lead to heavy denudation of the alloying component in the metal phase unless the alloy interdiffusion coefficient is very small. Furthermore, under conditions where internal oxidation would be expected a very high rate of external scale formation may reduce the thickness of the internal oxidation zone to negligible proportions and thus the denuded layer in the alloy does not develop.

Oxidation of other selected alloys

Iron–copper alloys

A further classification in terms of behaviour and resulting morphology comes from the case when the alloying element is more noble than the parent element. A good example of this with commercial significance is given by the iron–copper system. When Fe–Cu alloys are oxidized, the copper does not enter the oxide phase but stays behind and enriches in the metal leading to the formation of a copper-rich rim at the scale–metal interface. This enrichment proceeds with continued scaling until the copper solubility limit in iron at that temperature is exceeded, when a second, metallic, copper-rich phase precipitates at the scale–metal interface. If the oxidation temperature is below the melting point of copper, this second phase is solid and, due to its low solubility for iron, may act as a barrier to iron diffusion, resulting in slower rates of oxidation.

If, however, the scaling temperature is above the melting point of copper, the new phase is precipitated as a liquid. This liquid phase then penetrates inward along alloy grain boundaries. This situation can develop in the reheating of Cu-bearing steels before rolling. Copper may be present either as an impurity from the reduction of traces of copper minerals in iron ore in the blast furnace or remelting of Cu-bearing scrap or, intentionally, as an element to improve corrosion resistance. When such a steel is subsequently rolled, the liquid-infiltrated grain boundaries, unable to support a shear or tensile stress, open up and the steel surface appears to be crazed. The resulting slab, billet, or ingot cannot usefully be processed and must be scrapped or returned for surface conditioning. This condition is known as 'hot shortness'. Hot shortness can also be encouraged by the presence in the steel of other relatively noble elements, such as Sn, As, Bi, or Sb, which also enrich with copper and lower the melting point of the second phase, thus aggravating the situation.

The problem of hot shortness can be approached in several ways. Initially, the copper level of the charge to a steelmaking furnace can be monitored and, if necessary, diluted with hot metal or pig iron. Secondly, the time–temperature history of the reheating programme can be scheduled to expose the material to sensitive temperatures for the shortest possible time. Thirdly, it is possible that heating at very high temperatures may result in back-diffusion of Cu being more rapid than scaling so the Cu concentration at the surface does not build up significantly. This is not a popular technique due to the excessive losses in yield due to scaling. Finally, the presence of nickel, which also enriches, serves to increase the solubility of copper in the enriched surface layers thus delaying precipitation. Alternatively, the phase precipitated may be nickel-rich and not liquid. Of these alternatives the first two, i.e., prevention at source and strictly controlled reheating procedures, are the most advantageous and, usually, the only economically feasible options.

Niobium–zirconuim-type alloys

Alloys of Nb with small additions of Zr exhibit internal oxidation of Zr under an external scale of Nb-rich oxides. This class of alloy is somewhat different from those such as dilute Ni–Cr alloys in that the external Nb-rich scale grows at a linear, rather than parabolic rate. The kinetics of this process have been analyzed by Rapp and Colson.[47] The analysis indicates the process should involve a diffusion-controlled internal oxidation coupled with the linear scale growth, i.e., a paralinear process. At steady state, a limiting value for the penetration of the internal zone below the scale–metal interface is predicted. Rapp and Goldberg[48] have verified these predictions for Nb–Zr alloys.

Nickel–cobalt alloys

Alloys of this type oxidize in a manner similar to pure nickel since CoO is only slightly more stable than NiO and the two oxides form a single-phase solid solution scale.[49] However, the rate of oxidation is somewhat faster than that of pure nickel and a segregation of cations is observed across the scale. This segregation is illustrated in Figure 5.26 which shows the concentration profiles for Co in the alloy and scale.[49] The increase in Co concentration near the scale–gas interface is the result of a higher mobility for Co ions than Ni ions through the oxide lattice. The higher mobility of

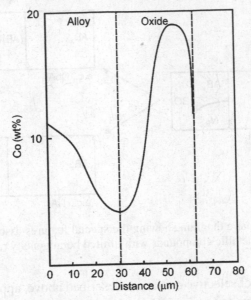

Figure 5.26 Concentration profiles of Co across a (Ni,Co)O scale formed on a Ni–10.9% Co alloy after 24 h in 1 atm O_2 at 1000 °C.

Co is the result of two effects: a higher cation-vacancy concentration in CoO than NiO, and a lower activation energy for the motion of the Co ions. The latter factor is consistent with calculations based on the difference in energy between a cation on an octahedral or tetrahedral site in a close-packed oxygen-anion lattice.[50] The most feasible diffusion path from one octahedral site to a vacant one is thought to be via an intervening tetrahedral site, which indicates why this energy difference affects the activation energy for cation-motion. Since the Co-rich oxide has a higher cation-vacancy concentration, as well as a lower activation energy for motion, the increase in CoO concentration near the scale–gas interface is responsible for the higher growth rate of the scale.

The partitioning of cations through a single-phase scale due to different mobilities and resultant effect of alloy composition on scale growth rate have been analyzed by Wagner.[51] Bastow *et al.*[52] have measured concentration profiles and growth rates for a number of Ni–Co alloys and found close agreement with Wagner's model.

Oxidation of intermetallic compounds

The oxidation behaviour of intermetallic compounds has been the subject of several reviews.[35,53–56] This section briefly points out the features of selective oxidation which are peculiar to intermetallic compounds and compares them with the behaviour of conventional Ni- and Fe-base alloys, described above.

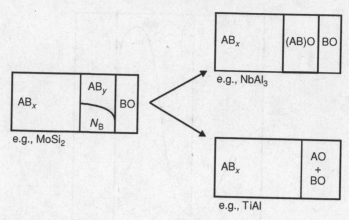

Figure 5.27 Schematic diagram showing the special features associated with the oxidation of intermetallic compounds with limited homogeneity ranges.

The fundamentals of selective oxidation, described above, apply to intermetallic compounds just as they do to alloys. However, some quantitative differences often arise.

Oxygen solubilities

Few quantitative data on oxygen solubilities in intermetallic compounds are available but indirect evidence indicates that most intermetallics have negligible oxygen solubilities. As a result, internal oxidation is rarely observed in intermetallic compounds. Exceptions to this rule are Ti_3Al and Ni_3Al, which dissolve substantial amounts of oxygen.

Compound stoichiometry

Most intermetallic compounds exist over only narrow ranges of composition. Therefore, the selective oxidation of one component from the compound can result in the formation of a lower compound immediately below the oxide as illustrated in Figure 5.27. This situation means that the properties of the lower compound determine the ability of the alloy to maintain the growth of the protective oxide, Equation (5.2). Apparently some compounds, such as Mo_5Si_3, provide a sufficient flux of Si to the oxide–alloy interface to maintain the growth of a silica layer on $MoSi_2$. This is not the case, however, for many systems as is illustrated for the oxidation of $NbAl_3$. The initial exposure of this compound results in the formation of a continuous alumina layer on the surface but the depletion of Al causes the next lower aluminide, Nb_2Al to form beneath the oxide. This compound has been shown to be incapable of forming continuous alumina[57] and, indeed, the alumina breaks down and then begins to reform under the Nb_2Al layer and envelops it. This

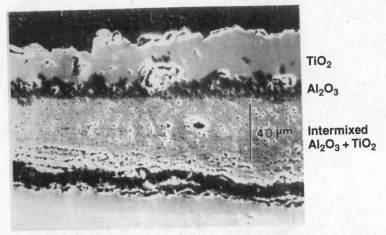

TiO$_2$

Al$_2$O$_3$

40 µm

Intermixed
Al$_2$O$_3$ + TiO$_2$

Figure 5.28 Scanning electron micrograph showing the cross-section of Ti$_3$Al after oxidation at 900 °C for 165 h in 1 atm O$_2$.

enveloped metal is eventually converted to a Nb–Al oxide in the scale. Repetition of these processes results in a layered scale, which grows substantially faster than alumina.

Relative stability of oxides

The selective oxidation of an alloy component, e.g., Al or Si, requires the alumina or silica to be more stable than the oxides of the other components in the alloy. Figure 2.5 indicates this condition would be met for compounds such as nickel aluminides and molybdenum silicides. However, in the case of Nb- or Ti-base compounds the oxides of the base metal are nearly as stable as those of Al or Si. This can result in conditions for which selective oxidation is impossible. This situation exists for titanium aluminides containing less than 50 at% Al as illustrated in Figure 5.27. In this case a two-phase scale of intermixed Al$_2$O$_3$ and TiO$_2$ is generally observed. It should be emphasized that the determination of which oxide is more stable must take into account the prevailing metal activities.

The actual oxidation morphologies are often more complex than those diagrammed in Figure 5.27. This is illustrated in Figure 5.28 for the compound Ti$_3$Al which was oxidized for 165 h in O$_2$ at 900 °C. The major portion of the scale is a two-phase mixture of α-Al$_2$O$_3$ and TiO$_2$ (rutile). However, on top of this layer is a layer consisting of discrete alumina crystals and on top of this is a layer of virtually pure TiO$_2$. The development of this complex microstructure clearly involves inward transport of oxygen and outward transport of Al and Ti through the intermixed layer. The alloy beneath the scale has a significant amount of oxygen in solution which results in severe embrittlement.

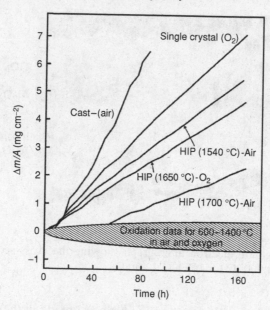

Figure 5.29 Oxidation rates at 500 °C in air for $MoSi_2$ in various fabrication forms (cast polycrystal, single crystal and hot isostatically pressed (HIP) at 1540, 1650, and 1700 °C. The rates for the temperature range 600–1400 °C fall in the crosshatched region.

Growth rate of transient oxides

The formation of transient oxides can influence the ability of an alloy to form a protective oxide layer, even when the protective oxide is the thermodynamically stable oxide, as described above. This is particularly true for some intermetallic compounds such as those involving the refractory metals. Figure 5.29 shows the oxidation rates of $MoSi_2$ as a function of temperature.[58] At temperatures above 600 °C the oxidation kinetics are very slow as the result of the formation of a continuous silica scale. However, at 500 °C the kinetics are greatly accelerated and a mixed oxide of SiO_2 and MoO_3 forms. At high temperatures the transient molybdenum oxides evaporate (see Figure 4.14), as shown in Figure 5.30, and the silica regions grow laterally into a continuous layer. At low temperatures the reduced volatility of the molybdenum oxides and the slower growth of silica prevent the development of a continuous silica layer. Rapid inward growth of MoO_3 results in continuous incorporation of the silica nuclei and produces the intermixed layer of MoO_3 and SiO_2. In polycrystalline material formation of this mixed oxidation product can form along alloy grain boundaries and result in fragmentation or 'pesting' of the alloy.[58] Similar observations have been made for $TaSi_2$ and $NbSi_2$ which undergo accelerated oxidation to temperatures above 1000 °C because the transient tantalum and niobium oxides are not volatile.[59]

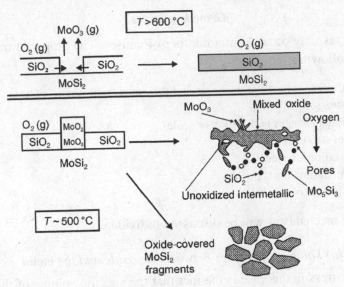

Figure 5.30 Schematic diagram showing the oxidation mechanism for temperatures of 600 °C and above (top) and for temperatures near 500 °C (bottom).

Additional factors in alloy oxidation

Stress development and relief in oxide films

The discussion of alloy oxidation in this chapter and the discussion of pure metal oxidation in the previous one have indicated that resistance to high-temperature oxidation requires the development of an oxide barrier which separates the environment from the substrate. Continued resistance requires the maintenance of this protective barrier. Therefore, stress generation and relief in oxide films and the ability of an alloy to reform a protective scale, if stress-induced spalling or cracking occurs, are important considerations in the high-temperature oxidation of alloys. This subject has been discussed in reviews by Douglass,[60] Stringer,[61] Hancock and Hurst,[62] Stott and Atkinson,[63] and Evans.[64]

Stress generation

Stresses can, of course, result from an applied load on the component undergoing oxidation. However, additional stresses are generated by the oxidation process. These are *growth stresses*, which develop during the isothermal formation of the scale, and *thermal stresses*, which arise from differential thermal expansion or contraction between the alloy substrate and the scale.

Growth stresses

Growth stresses may occur from a number of causes. The proposed mechanisms include the following

(1) Volume differences between the oxide and the metal from which it forms.
(2) Epitaxial stresses.
(3) Compositional changes in the alloy or scale.
(4) Point defect stresses.
(5) Recrystallization stresses.
(6) Oxide formation within the scale.
(7) Specimen geometry.

Each of these mechanisms will be discussed individually.

(a) Volume differences between the oxide and the metal

The cause of stress in this case is the fact that the specific volume of the oxide is rarely the same as that of the metal which is consumed in its formation. The sign of the stress in the oxide may be related to the Pilling–Bedworth ratio (PBR):[65]

$$\text{PBR} = \frac{V_{\text{ox}}}{V_{\text{m}}}. \tag{5.26}$$

Table 5.1 lists the PBRs for a number of systems.[62] The oxide is expected to be in compression if the PBR is greater than unity (which is the case for most metals)

Table 5.1 *Oxide–metal-volume ratios of some common metals*

Oxide	Oxide–metal-volume ratio
K_2O	0.45
MgO	0.81
Na_2O	0.97
Al_2O_3	1.28
ThO_2	1.30
ZrO_2	1.56
Cu_2O	1.64
NiO	1.65
FeO (on α-Fe)	1.68
TiO_2	1.70–1.78
CoO	1.86
Cr_2O_3	2.07
Fe_3O_4 (on α-Fe)	2.10
Fe_2O_3 (on α-Fe)	2.14
Ta_2O_5	2.50
Nb_2O_5	2.68
V_2O_5	3.19
WO_3	3.30

and in tension if the PBR is less than unity. Generally those systems which develop tensile stresses in the oxide, e.g. K, Mg, Na, cannot maintain protective films. The oxides on most metals and alloys form in compression as expected from the PBR. However, the development of compressive stresses would only seem to be feasible if the oxide were growing at the scale–metal interface by inward migration of oxide ions. In fact the development of compressive stresses during the growth of an oxide depends upon the experimental conditions. If the oxide is growing at the scale–metal interface by the inward migration of oxide ions, compressive stress should develop providing the oxide scale cannot move outward away from the metal, as in the case of small coupon specimens. Scales forming at the oxide–gas interface on a planar specimen should not develop stresses because the volume difference between metal and oxide is simply reflected in the scale thickness. Also, the magnitudes of the stresses are not always in the order expected from the PBR.[62] Clearly, while the volume ratio is a cause of growth stresses in some cases, other mechanisms must also be operative in many systems.

(b) Epitaxial stresses

Nucleation considerations may dictate that the first oxide to form will have an epitaxial relationship with the substrate. This constraint will result in stress development because of the difference in lattice parameter between the metal and the oxide. This mechanism would only seem to generate significant stresses when the oxide is very thin, i.e., for short oxidation times and low oxidation temperatures. However, there are proposals that the action of intrinsic dislocations, in what amounts to a semi-coherent interface, in annihilating vacancies during cationic scale growth can lead to sizeable stresses.[66] However, there are also interface dislocation structures proposed, which could annihilate vacancies without generating significant stress.[67]

(c) Compositional changes in the alloy or scale

Compositional changes can result in growth stresses by several means. Changes in the lattice parameter of the alloy as one or more elements are depleted by selective oxidation can produce stresses as can changes in scale composition. Dissolution of oxygen in metals such as Nb, Ta, and Zr, which have high oxygen solubilities, can result in stress development. In a similar manner the volume changes associated with internal oxide or carbide formation can result in stresses in some alloys. Internal oxidation of Ag–In[68] and Ag–Mg[69] alloys resulted in the formation of Ag nodules on the specimen surface. Large compressive stresses in the internal oxidation zone lead to diffusional transport of Ag to the surface. This was explained in terms of dislocation pipe diffusion of Ag. Nodules of pure Ni were also observed during the internal nitriding of Ni–Cr–Al alloys.[16]

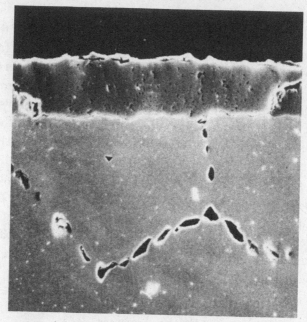

Figure 5.31 Scanning electron micrograph showing the grain-boundary voids formed in Ni by vacancy condensation during oxidation at 950 °C.

(d) Point-defect stresses

Stresses can be generated in scales which exhibit large deviations from stoichiometry, e.g., FeO, due to the gradient in point defects, which will result in a lattice-parameter variation across the scale. Also metals which oxidize by the outward migration of cations may develop a vacancy gradient through the substrate because of vacancy injection from the scale. The lattice-parameter variation produced by this gradient by relaxation around the vacancies could result in stresses in the substrate. However, as pointed out by Hancock and Hurst,[62] these vacancies may also be a source of stress relief by enhancing creep within the metal. However, the effects of this vacancy supersaturation are likely to be small, since it is difficult to maintain, and the vacancies tend to precipitate as voids at the scale–metal interface and grain boundaries in the substrate. This can be seen in the case of Ni in Figure 5.31. The most important effect of vacancies appears to be that of decreasing the contact area between the scale and metal by void formation.

(e) Recrysallization stresses

Recrystallization in the oxide scale has been suggested as a cause of stresses.[70,71] However, this phenomenon would seem to relieve growth stresses rather than generate them. One similar case in which stress generation has been reported is for the oxidation of fine-grained Fe–Cr alloys.[72] Grain growth in the alloy was observed to disrupt the Cr_2O_3 scale locally and produce thick nodules of Fe-rich oxide. Oxidation of a coarse-grained alloy produced a continuous Cr_2O_3 scale with no nodules. However, this should be contrasted with the oxidation behaviour of Ni–Cr alloys where the recrystallization zone at the alloy surface, resulting from prior deformation, promoted the development of a continuous Cr_2O_3 scale.[73] In this latter case, however, stresses were not a factor, since the recrystallized structure was proposed to affect transport of chromium to the surface of the alloy.

(f) Oxide formation within the scale

The inward migration of oxidant along oxide grain boundaries and through microcracks can lead to compressive stress generation within the oxide scale if it results in oxide formation at sites within the scale. Grain-boundary transport was proposed by Rhines and Wolf[74] to explain the lengthening of Ni rods and the increase in area of Ni sheet during oxidation, both of which suggest compressive growth stresses in the scale.

The possibility of new oxide forming within a scale is a matter of some controversy. It has been questioned if this process can satisfy the equilibrium between point-defect concentrations within the scale.[75] Whether new oxide is formed or consumed within the scale has been shown to be dependent on the types of point defects involved.[76] Apparently, depending on the type of point defects, oxide can be formed internally, consumed internally, or neither of the conditions may arise. When grain-boundary diffusion is dominant, these conditions are still valid, but knowledge of the point defects in the grain boundaries is required.[76] Experiments using secondary-ion mass spectrometry (SIMS),[77] in which the oxidizing gas has been switched from ^{16}O to ^{18}O during the growth of alumina, have indicated the presence of ^{18}O within the scale, which would be consistent with new oxide forming within the scale. Caplan and Sproule[78] found that areas of single-crystal oxide formed on Cr were smooth, giving no indication of large stresses, whereas polycrystalline areas of oxide on the same specimen were buckled, indicating large compressive stresses within the scale. Lillerud and Kofstad[79] invoked new oxide formation along grain boundaries in chromia scales to explain oxide buckling during Cr oxidation and Golightly et al. [80] explained buckling of alumina scales on Fe–Cr–Al alloys in a similar manner. There is, thus, substantial evidence that this mechanism is operative in some systems.

(g) Specimen geometry

The previously discussed mechanisms have applied to large planar specimens. However, an important source of growth stresses arises due to the finite size of specimens and the resultant curvature. The nature of the stress developed will

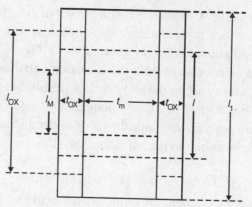

Figure 5.32 Schematic section through an oxidized sheet of metal showing the dimensions needed to calculate the thermal stress in the oxide.

depend on the curvature and the mechanism of scale growth. These effects, for the common situation of the PBR > 1, have been divided into four categories by Hancock and Hurst:[62] cationic oxidation on convex surfaces, anionic oxidation on convex surfaces, cationic oxidation on concave surfaces, and anionic oxidation on concave surfaces.

Thermal stresses

Even when no stress exists at the oxidation temperature, stresses will be generated during cooling because of the difference in thermal-expansion coefficient of the metal and oxide. A derivation for the stresses generated in a bimaterial strip subjected to a change in temperature, ΔT, was originally developed by Timoshenko.[81] The following derivation is based on Timoshenko's, for the specific case of an oxide on both the top and the bottom of the metal, and thus no bending of the oxide–metal system. The oxidized specimen is shown in one dimension in Figure 5.32. At the oxidation temperature (T_H) both metal and oxide have length l_1 and, after cooling to a lower temperature (T_L), they have length l. If they were not bonded, on cooling to T_L, the metal and oxide would experience free thermal strains as shown in Equations (5.27) and (5.28):

$$\varepsilon_{\text{thermal}}^{\text{metal}} = \alpha_M \Delta T, \tag{5.27}$$

$$\varepsilon_{\text{thermal}}^{\text{oxide}} = \alpha_{Ox} \Delta T, \tag{5.28}$$

where α_M and α_{Ox} are the (assumed constant) linear thermal-expansion coefficients for the metal and oxide, respectively, and $\Delta T = T_L - T_H$ and is thus negative in sign. Because they are bonded, the oxide and metal also experience mechanical strains due to residual stress, see Equations (5.29) and (5.30):

$$\varepsilon_{\text{mechanical}}^{\text{metal}} = \frac{\sigma_M(1 - \nu_M)}{E_M}, \tag{5.29}$$

$$\varepsilon_{\text{mechanical}}^{\text{oxide}} = \frac{\sigma_{Ox}(1 - \nu_{Ox})}{E_{Ox}}, \tag{5.30}$$

where Hooke's law for an equal biaxial state of stress has been used. Because the oxide and metal are bonded, their total axial strains must be the same. In other words, we obtain Equation (5.31):

$$\varepsilon_{\text{thermal}}^{\text{metal}} + \varepsilon_{\text{mechanical}}^{\text{metal}} = \varepsilon_{\text{thermal}}^{\text{oxide}} + \varepsilon_{\text{mechanical}}^{\text{oxide}}. \tag{5.31}$$

Performing a force balance on the specimen, where t_M and t_{Ox} are the thickness of metal and a single-oxide scale, respectively yields Equation (5.32):

$$\sigma_M t_M + 2\sigma_{Ox} t_{Ox} = 0. \tag{5.32}$$

Combining these equations yields Equation (5.33), or (5.34):

$$\alpha_M \Delta T - \frac{2\sigma_{Ox} t_{Ox}(1 - \nu_M)}{t_M E_M} = \alpha_{Ox} \Delta T + \frac{\sigma_{Ox}(1 - \nu_{Ox})}{E_{Ox}}; \tag{5.33}$$

$$\sigma_{Ox} = \frac{-(\alpha_{Ox} - \alpha_M)\Delta T}{\dfrac{2t_{Ox}(1 - \nu_M)}{t_M E_M} + \dfrac{(1 - \nu_{Ox})}{E_{Ox}}}. \tag{5.34}$$

If there is no Poisson ratio mismatch between the metal and oxide, the formula simplifies to Equation (5.35):

$$\sigma_{Ox} = \frac{-E_{Ox}(\alpha_{Ox} - \alpha_M)\Delta T}{(1 - \nu)\left(1 + 2\dfrac{t_{Ox} E_{Ox}}{t_M E_M}\right)}. \tag{5.35}$$

Note that because ΔT is assumed negative and α_{Ox} is typically less that α_M, the stress in the oxide is typically negative. For the case of a thin oxide on only one side of a thick metal substrate, it may still be reasonable to assume no bending deformation of the oxide–metal laminate. In such cases, the Equation (5.35) can be used if the factor of 2 in the last term of the denominator is removed. In cases where the oxide is very thin relative to the thickness of the metal, the last term in the denominator of Equation (5.35) can be neglected, yielding Equation (5.36),

$$\sigma_{Ox} = \frac{-E_{Ox}(\alpha_{Ox} - \alpha_M)\Delta T}{(1 - \nu)}, \tag{5.36}$$

which corresponds to the physical case of residual stresses in the oxide inducing essentially no deformation in the metal. Equations (5.35) and (5.36) have been widely used for calculating thermal stresses in oxides.[82]

Table 5.2 *Linear coefficients of thermal expansion of metals and oxides*[62]

System	Oxide coefficient $\times 10^{-6}$	Metal coefficient $\times 10^{-6}$	Ratio
Fe–FeO	12.2	15.3	1.25
Fe–Fe$_2$O$_3$	14.9	15.3	1.03
Ni–NiO	17.1	17.6	1.03
Co–CoO	15.0	14.0	0.93
Cr–Cr$_2$O$_3$	7.3	9.5	1.30
Cu–Cu$_2$O	4.3	18.6	4.32
Cu–CuO	9.3	18.6	2.00

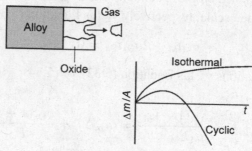

Figure 5.33 Schematic plot of mass change versus oxidation time comparing typical behaviour during isothermal and cyclic oxidation.

The presence of thermal stresses often results in spalling of protective oxides. This generally makes exposures in which the temperature is cycled more severe than isothermal exposures, as shown schematically in Figure 5.33. As mentioned, the thermal expansion coefficient for the oxide will generally be less than that for the metal, as seen in Table 5.2, so that compressive stresses will develop in the oxide during cooling. The magnitude of the stresses will be proportional to $\Delta\alpha = \alpha_M - \alpha_{Ox}$. This is consistent with the observation that scales formed on Ni and Co tend to remain adherent after cooling whereas those on Cu and Cr do not.

Measurement of stresses

Stresses in oxide layers have been measured by a variety of techniques, which include mechanical tests where bending of a specimen is observed, when the oxide is formed on only one side, or a change in specimen length can be detected.[64] Usually the most accurate techniques are those that directly measure strain in the oxide lattice.

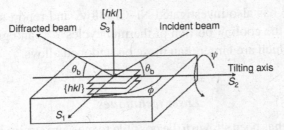

Figure 5.34 Schematic diagram of the tilting technique for measuring the strains in an oxide layer.

X-Ray diffraction

The strain in an oxide film can be measured by X-ray diffraction (XRD) as a change in lattice spacing with inclination, with respect to the surface of a sample. This strain is usually expressed as in Equation (5.37),

$$\varepsilon_{\phi\psi} = \frac{d_{\phi\psi} - d_o}{d_o} \tag{5.37}$$

in which d_o is the stress-free lattice spacing of the selected (hkl) planes and $d_{\phi\psi}$ is the lattice spacing of these (hkl) planes in the stressed-film for a given tilt ψ. The geometry of the commonly used 'tilting technique' is presented in Figure 5.34. It can be shown that this strain is expected to be proportional to $\sin^2 \psi$. For an equal biaxial stress in the irradiated layer, Equation (5.37) can be expressed as in Equation (5.38):[83]

$$\frac{d_{\phi\psi} - d_o}{d_o} = \frac{1+\nu}{E}\sigma_{\phi} \sin^2 \psi - \frac{\nu}{E}(\sigma_1 + \sigma_2). \tag{5.38}$$

If a plot of ε versus $\sin^2 \psi$ curve is linear, as is expected from an isotropic surface layer which is polycrystalline and not textured, the stress can be accurately calculated from the slope of the d vs. $\sin^2 \psi$ line and values for E, ν, and d_o. This technique and variations thereof have been used to measure the stresses in alumina films both at room temperature and during oxidation at high temperature (see, for example, ref. 84). Growth stresses in alumina can vary from negligible values to greater than 1 GPa compression. The residual stress, which includes the effects of growth and thermal stresses, can exceed 5 GPa compression for oxidation temperatures on the order of 1100 °C. The stresses generated in chromia films have been investigated by several groups using XRD techniques. Hou and Stringer[85] measured the residual stress in chromia on Ni–25 wt% Cr–1 wt% Al and Ni–25 wt% Cr–0.2 wt% Y after oxidation at 1000 °C and found the stresses to be compressive and of similar magnitude for both alloys. Comparison with calculated thermal stresses led to the conclusion that growth stresses were negligible. Shores and coworkers,[86] on the other hand, have reported substantial compressive growth stresses for chromia on Cr. This group has also investigated Ni–Cr alloys and report significant stress relaxation during the cooling portion of thermal cycles[86–88] and growth strains on Y-doped alloys which are larger than those on undoped alloys.[89]

Laser techniques

In some systems it has been shown to be possible to measure strains in oxide films by laser Raman spectroscopy[90] and, in alumina films, by photostimulated chromium luminescence spectroscopy.[91]

Response to stress

The growth and thermal stresses generated during oxidation may be accommodated by a number of mechanisms. The most important are the following:

(1) Cracking of the oxide.
(2) Spalling of the oxide from the alloy substrate.
(3) Plastic deformation of the substrate.
(4) Plastic deformation of the oxide.

All of these mechanisms have been observed to operate in various systems. The particular mechanisms will depend in a complicated manner on virtually all of the variables controlling the oxidation process. Mechanisms (1) and (2) tend to produce the most severe consequences because they can expose fresh metal to the oxidizing environment.

Oxide cracking occurs when the oxide is put into tension. This is the situation described for the oxidation of Nb in Chapter 4 which results from the growth mechanism of Nb_2O_5. Oxides forming on the alkali metals also form under tension because the PBR is less than unity for these systems (Table 5.1).

The oxides on most engineering alloys are in compression because the growth stresses tend to be compressive and, particularly, the thermal stresses, when they develop on cooling, are compressive because of the sign of the thermal expansion mismatch between the alloy and oxide (Table 5.2). The spalling of a compressively stressed protective oxide scale will occur when the elastic strain energy stored in the intact scale exceeds the fracture resistance, G_c, of the interface. Denoting E and v as the elastic modulus and Poisson's ratio of the scale, h as the scale thickness, and σ as the equal biaxial residual stress in the film, the elastic strain energy stored in the scale per unit area is $(1 - v)\sigma^2 h/E$. The criterion for failure then becomes that given in Equation (5.39):[92,93]

$$(1 - v)\sigma^2 h/E > G_c. \tag{5.39}$$

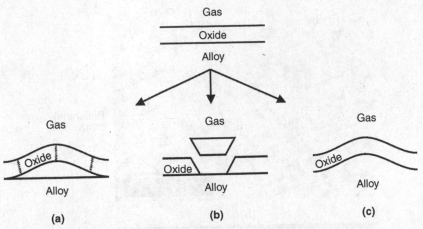

Figure 5.35 Schematic diagram of responses of an oxide which is loaded in compression. (a) Buckling of the oxide, (b) shear cracking of the oxide, and (c) plastic deformation of the oxide and alloy.

According to this criterion, which is a necessary but not sufficient condition, a scale will spall if the stress is high, the scale is thick, or the interfacial free energy is high (the work of adhesion is low). Decohesion of films under compression, such as most oxide scales, however, require either a buckling instability or development of a wedge crack in order to spall.[64] According to elastic mechanics, buckling of a thin film under biaxial compression to form an axisymmetric buckle of radius a will occur at a critical stress, σ_c, given by Equation (5.40):

$$\sigma_c = 1.22 \frac{E}{1 - v^2} \left(\frac{h}{a}\right)^2 .$$

(5.40)

However, such a buckle is stable and will not propagate to cause decohesion failure by delamination unless the strain-energy release rate also satisfies Equation (5.39). Figure 5.35(a) shows a schematic diagram of the buckling of a scale and Figures 5.36 and 5.37 show the results of this type of behaviour for alumina scales formed on Fe–Cr–Al and Fe–Cr–Al–Ti alloys, respectively.

The buckling stress, Equation (5.40), increases as the square of the scale thickness such that, for thick scales, buckling may not be feasible. In this case, shear cracks can form in the oxide and, if Equation (5.39) is satisfied, lead to scale spallation by a 'wedging mechanism,' which is shown schematically in Figure 5.35(b).

In some situations, if G_c is high and the alloy is relatively weak, the compressive stresses can be accommodated by simultaneous deformation of the scale and alloy without spallation. This phenomenon is shown schematically in Figure 5.35(c) and illustrated for an alumina scale in Figure 5.38.

(a)

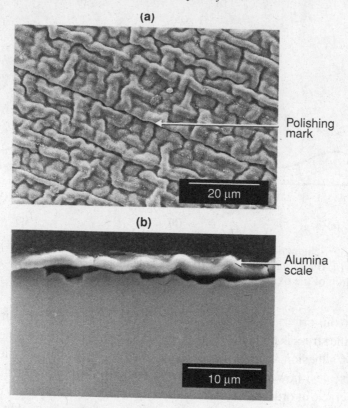

Polishing mark

20 μm

(b)

Alumina scale

10 μm

Figure 5.36 Scanning electron micrographs of (a) the surface and (b) the cross-section of an Fe–Cr–Al alloy which was oxidized at 1000 °C and cooled to room temperature. Buckling of the alumina scale is evident.

Reactive element effects

It is now well established that small additions of reactive elements, such as yttrium, hafnium, and cerium, substantially improve the adherence of chromia and alumina films to alloy substrates.[94] This is illustrated in Figure 5.39 where comparison of the cyclic oxidation kinetics of the Fe–Cr–Al (normal S) and Fe–Cr–Al–Y alloys shows the substantial influence of the Y addition. Figure 5.40 shows that the alumina scale on a Fe–Cr–Al–Y alloy is still completely adherent after the exposure. While the effects produced by the reactive elements are widely known, the mechanisms whereby they improve adherence are not completely understood. Over the last fifty years a number of mechanisms have been proposed. These include: reactive elements acting as vacancy sinks to suppress void formation at the alloy–oxide interface,[95,96] formation of oxide pegs at the alloy–oxide interface,[97] alteration of the growth mechanism of the oxide resulting in reduced growth stresses,[80] reactive-element segregation to the alloy–oxide interface to form a graded seal[98] or otherwise

(a)

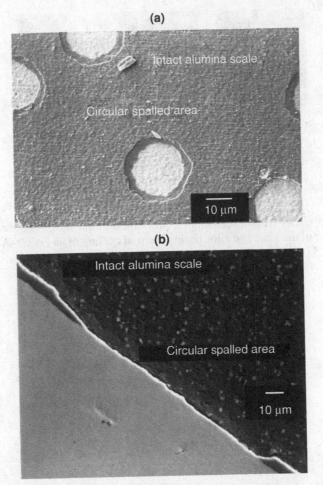

(b)

Figure 5.37 Scanning electron micrographs of (a) the surface and (b) the cross-section of an Fe–Cr–Al–Ti alloy which was oxidized at 1100 °C and cooled to room temperature. Circular buckles have formed in the alumina scale, which has spalled from these regions.

strengthen the alloy–oxide bond,[99] increased oxide plasticity,[100] and reactive elements tying up sulphur in the alloy to prevent it from segregating to the alloy–oxide interface and weakening an otherwise strong bond.[101,102]

The importance of the latter mechanism has been illustrated in experiments where hydrogen annealing of nickel-base single crystals[103,104] and Fe–Cr–Al alloys[105] lowered the sulphur contents to very low levels and resulted in dramatic improvements in the adherence of alumina films to the alloys. This is illustrated in Figure 5.39 where the cyclic oxidation behaviour of the Fe–Cr–Al (low S) alloy, which is the same alloy as Fe–Cr–Al (normal S) that has been desulphurized by hydrogen annealing, is comparable to that of the Y-doped alloys.

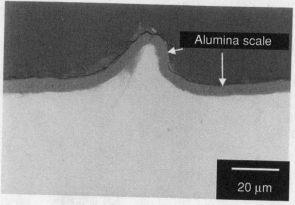

Figure 5.38 Cross-section scanning electron micrograph showing the concurrent deformation of the scale and alloy during thermal cycling of an Fe–Cr–Al–Ti alloy from 1100 °C to room temperature.

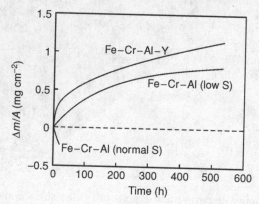

Figure 5.39 Cyclic oxidation kinetics for several Fe–Cr–Al alloys exposed at 1100 °C.

The low-sulphur theory is generally accepted, but the mechanism by which sulphur disrupts the alumina scale is not well defined. Grabke *et al.*,[106] using Auger electron spectroscopy (AES) depth-profiling techniques, have concluded that there is no equilibrium segregation of sulphur to the metal–oxide interface. They argue that, from thermodynamic considerations, the size and charge effect of a sulphur atom would make the segregation process impossible, and that a void must first form; then sulphur segregation to the surface of voids lowers surface energy. However, Hou and Stringer[107] found sulphur at the scale–metal interface in the absence of voids. Grabke *et al.*[108] comment that the amount of sulphur observed at apparently

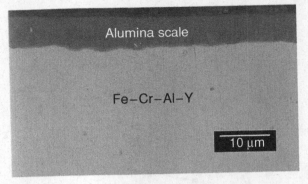

Figure 5.40 Cross-section scanning electron micrograph showing the alumina scale formed on a Y-doped Fe–Cr–Al alloy after cyclic oxidation at 1100 °C for 525 h.

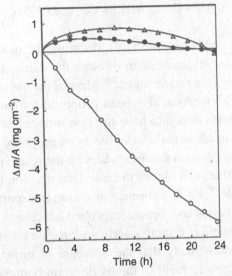

Figure 5.41 Effect of cerium content on the cyclic oxidation mass loss of Ni–50 wt% Cr in air at 1100 °C: ∘, 0% Ce; △, 0.01% Ce; •, 0.08% Ce.

intact alloy–oxide interfaces is small and could result from segregation at defects too small to be resolved by Auger techniques, e.g., microvoids, dislocations, etc.

High-temperature stainless steels, most polycrystalline superalloys, and chromized coatings rely on the formation of a surface layer of chromia for oxidation protection. The effects of reactive element additions are often more dramatic in the case of chromia-forming alloys than alumina formers in that, in addition to improving adherence (Figure 5.41), they decrease the amount of transient oxidation, reduce

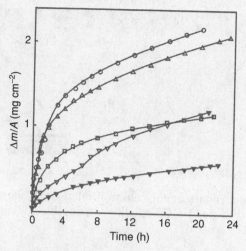

Figure 5.42 Effect of cerium content on the oxidation rate of Ni–50 wt% Cr in oxygen: ○, 0% Ce; △, 0.004 Ce; □, 0.010% Ce; ▽, 0.030% Ce; ▼, 0.080 Ce.

the growth rate of the oxide (Figure 5.42), decrease the oxide grain size (Figure 5.43) and alter the transport mechanism through the oxide. This subject has been extensively reviewed by Hou and Stringer.[109] Most of the mechanisms for improved adherence mentioned above have also been proposed for chromia formers; however, sulphur has not been shown to have the first-order effect it does for alumina formers.[109] Recently, an additional model has been proposed to explain the effect of reactive elements on chromia formers whereby the reactive element segregates to the alloy oxide interface and blocks the interface reaction for the incorporation of cations into the scale.[66,110] The absence of cation transport is proposed to lead to decreased growth stresses and, hence, improved adherence.

The reactive-element effect is more poorly understood for chromia formers than alumina formers. It seems likely that the decrease in growth rate is caused by segregation of the reactive element to the oxide grain boundaries as proposed by Ecer and Meier.[111] This, coupled with the fact that the growth direction is changed by the reactive-element additions from primarily outward to primarily inward, raises the complication of the latter two factors possibly being responsible for the better adhesion. However, Al additions have been shown to improve chromia adhesion to Ni–Cr alloys without affecting either growth rate or direction.[85]

Another complication with chromia formers is that surface application of yttria (and ceria) has been shown to produce the same effects as a reactive element incorporated into the alloy,[112,113] but this precludes the operation of many of the mechanisms proposed for the reactive-element effect. The application of MgO has been shown to be ineffective, at least with regard to altering the growth rate of the chromia.[112]

(a)

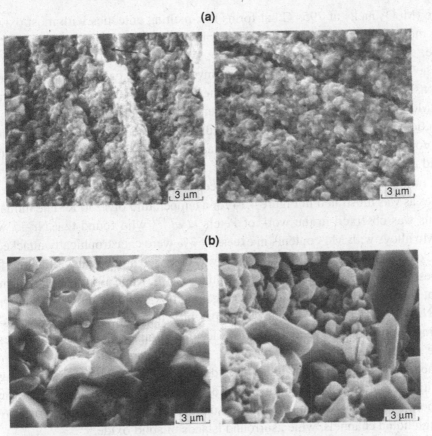

(b)

Figure 5.43 Scanning electron micrographs showing the effect of cerium on the Cr_2O_3 grain size during oxidation of Ni–50 wt% Cr in oxygen at 1100 °C. (a) After 1 min the oxide grain size on Ni–50 wt% Cr (left) is similar to that on Ni–50 wt% Cr–0.09 wt% Ce (right) but after 21 h (b) the average grain size of the oxide on Ni–50 wt% Cr (left) is much larger than that on Ni–50% Cr–0.09 wt% Ce (right).

Catastrophic oxidation caused by refractory-metal additions

The oxidation literature contains several examples of catastrophic attack induced by high-refractory-metal contents in Fe- and Ni-base alloys.[114] Leslie and Fontana[115] found that Fe–25 wt% Ni–16 wt% Cr–6 wt% Mo alloys underwent catastrophic oxidation at 900 °C in static air even though the oxidation was good in flowing air. Brenner[116] found that binary Ni–Mo and Fe–Mo alloys containing as much as 30 wt% Mo did not oxidize catastrophically at 1000 °C. However, a number of Fe–Ni–Mo and Fe–Cr–Mo alloys did undergo catastrophic attack.[117] The mechanism postulated for this attack consists of the formation of a layer of MoO_2 at the scale–alloy interface, which is oxidized to liquid MoO_3 following cracking of the scale (MoO_3 melts at 795 °C but forms low-melting eutectics with most oxides). The molten oxide may result in the dissolution and disruption of the protective scale.

The postulate of a liquid phase being involved in the catastrophic degradation is consistent with the earlier work of Rathenau and Meijering,[118] which indicated that Mo-induced accelerated attack commenced at temperatures approximating the eutectic temperatures for the oxides formed on the alloys and molybdenum oxides. The corrosion of Cr_2O_3-forming alloys was associated with Cr_2O_3 dissolution by liquid molybdenum oxides.

A further important feature of Mo-induced attack is the high volatility of the Mo oxides as was illustrated in Figure 4.14 for a temperature of 1250 K. The influence of this was observed in the work of Peters *et al.*[119] who found that Ni–15 wt% Cr–Mo alloys with Mo contents in excess of 3% were catastrophically attacked in static oxygen at 900 °C. A Mo-rich oxide was observed at the scale–alloy interface suggesting the importance of MoO_3 accumulation. The same alloy, however, formed a protective scale in rapidly flowing oxygen and oxidized at about the same rate as a Ni–15 wt% Cr alloy without Mo. Apparently, the evaporation of MoO_3 in the flowing atmosphere prevented sufficient accumulation of molybdenum oxides to cause severe corrosion.

The oxidation kinetics when liquid oxides are formed often exhibit two stages: an initial period of rapid parabolic oxidation followed by linear kinetics. The rapid parabolic period has been modeled by assuming diffusion of metal and/or oxygen through liquid channels, which surround islands of solid oxide.[120]

Tungsten additions have not been observed to cause catastrophic oxidation of Ni- and Co-base alloys, perhaps because of the higher melting temperatures of the tungsten oxides as compared with molybdenum oxides, e.g., $T_{mp} = 1745$ K for WO_3. However, W additions have been observed to induce some scale breakdown on Ni–Cr alloys.[121] Additions of W to Co–Cr alloys appear to be beneficial in decreasing the transient oxidation period and promoting the formation of a continuous Cr_2O_3 layer.[122]

Relatively little work has been done on the effects of refractory-metal additions on the oxidation of Al_2O_3-forming alloys. In superalloys, where the refractory-metal concentration is rarely above 10 weight percent, three effects have been observed.[114] One effect is beneficial and arises since these elements can be considered to be oxygen getters in comparison to the base metals (Ni and Co) and they can, therefore, promote the selective oxidation of aluminium and chromium. The other two effects are deleterious. One occurs because the refractory elements decrease the diffusion coefficients of the elements needed to be selectively oxidized. The other arises because of the relatively poor protectiveness of the oxides of the refractory metals, which makes their presence in the external scale undesirable. Refractory metals have been observed to initiate severe corrosion when a molten sulphate deposit is on the alloy.[123] This phenomenon will be discussed in Chapter 8.

Oxidation and decarburization of steels

The oxidation of carbon and low-alloy steels follows oxidation mechanisms similar to those described for the oxidation of pure iron in Chapter 4. At low temperatures ($\approx 500\,°C$), typical of service conditions, the alloys form layered scales consisting of Fe_3O_4 and Fe_2O_3. The oxidation of Fe–C alloys containing 0.1, 0.5 and 1.0 wt% C at $500\,°C$ in 1 atm oxygen has been studied by Caplan *et al.*[124] It was found that the oxide scales grew thicker over pearlite colonies than over single-phase ferrite for annealed alloys as the result of a finer grained Fe_3O_4 layer forming over the pearlite. Cold-worked alloys oxidized substantially faster than annealed alloys. No carburization or decarburization was observed for any of the alloys. This was postulated to result from C diffusing out through the inner portion of the Fe_3O_4 layer and being oxidized to CO which escaped through channels and pores in the outer part of the Fe_3O_4 and the Fe_2O_3 layer. Decarburization has been reported at $500\,°C$ for oxygen partial pressures on the order of 10 torr.[125]

Oxidation at high temperatures, typical of reheating conditions during steel processing, can lead to substantial decarburization. This can have a large effect on the properties of components. The tensile strength of a heat-treated steel depends primarily on the carbon content. The maximum stress on a component in bending occurs at the surface. Clearly, if excessive failure of steel components in service, particularly under reversing bending stresses, is to be avoided, the specified carbon content must be maintained within the surface layers of the component. This is particularly necessary with rotating shafts and stressed threads in bolts, etc.

Unfortunately, there is a strong tendency for carbon to be lost from the surface of steel when it is reheated for hot working or for heat treatment. The loss of carbon, decarburization, is one of the oldest, most persistent problems in ferrous production metallurgy.

During normal reheating which is usually carried out in a large furnace where the steel stock is in contact with the products of combustion of a fuel (usually oil or gas), the steel oxidizes to form a scale and, therefore, undergoes simultaneous decarburization and scaling. The mechanisms by which the decarburization of steels occurs are well understood,[126,127] particularly in the case of plain carbon and low-alloy steels to which the following discussion is confined.

Basically, in the oxidizing atmosphere of the furnace, a scale of iron oxide forms and grows. At the scale–metal interface, carbon interacts with the scale to form

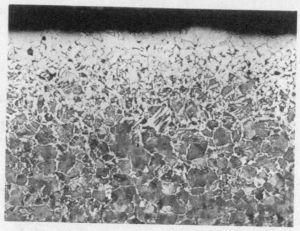

Figure 5.44 Microstructure of 0.6 wt% C steel after heating in air for 30 min at 1100 °C.

carbon monoxide by the reaction shown in Equation (5.41):

$$\underline{C} + FeO = Fe + CO \quad \Delta \dot{G}^\circ = 147\,763 - 150.07T \text{ J} \qquad (5.41)$$

(NB, an underlined element indicates it is in solution in the steel.)

This reaction can only proceed if the carbon monoxide reaction product can escape through the scale. In general, porous scales are produced, particularly under industrial conditions, and removal of carbon monoxide is not a problem. However, it has been shown[128] that very careful heating can produce a non-porous scale or a scale of greatly reduced permeability to carbon monoxide. As a result, instead of showing a decarburized surface, the steel shows carbon enrichment at the surface. This clearly demonstrates that removal of the carbon monoxide gas is vital for decarburization to occur.

Figure 5.44 shows the decarburized surface of a 0.6% C steel heated in air at 1100 °C for 30 min. Several points should be noted. The decarburization occurs in the surface layers, but carbon is clearly withdrawn rapidly from the grain-boundaries, and grain-boundary decarburization persists well into the specimen. Although decarburization is a surface phenomenon, there is no inner limit of the decarburized zone. This latter point is important since measurements of the 'depth of decarburization' are used commercially to describe the condition of the steel. However, this measurement is not precise, requiring judgement of the position of the inner limit, thus introducing a human element into the measurement. Further difficulties are introduced by the effect of cooling rate and alloy content, such as manganese, on the precipitation of proeutectoid ferrite during cooling and the eventual composition of the pearlite that forms as the eutectoid structure. This is not the place to go into these factors in detail, but the above warning note should be borne in mind.

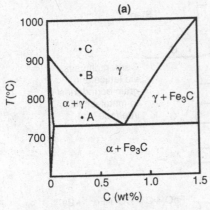
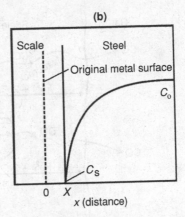

Figure 5.45 (a) Iron–carbon phase diagram with different heating conditions A, B, and C indicated; α = ferrite, γ = austenite. (b) Carbon profile and model for the decarburization of plain carbon steel above 910 °C, corresponding to condition C in (a).

Simultaneous isothermal scaling and decarburization

The various conditions likely to be met are shown in Figure 5.45. Figure 5.45(a) shows the various temperatures for reheating a 0.4 wt% C steel and it should be noticed that conditions A and B result in the formation of a surface layer of ferrite, as carbon is removed, which, due to its low solubility for carbon, restricts the outward diffusion of carbon. Under these conditions, decarburization is restricted to a very thin layer of total decaburization at the surface. Although this may be significant in the final heat treatment of finished components, it is negligible when reheating for hot working, in which case the reheating is carried out well into the austenite range and corresponds to condition C in Fig. 5.45(a). Under these conditions the austenite structure is maintained, even at zero carbon content, and carbon can diffuse rapidly out of the steel forming a deep rim of decarburization. The relevant carbon profile is shown in Fig. 5.45(b). The mechanism of decarburization is shown in Figure 5.46.

The temperature below which a ferrite surface layer forms on decarburization depends on the nature and concentration of other alloying elements in the steel, but, for plain carbon steels, can be taken to be 910 °C as for pure iron. Prediction of the effect of alloying elements on this temperature is complicated by the fact that the alloying elements will undergo either denudation or concentration at the surface during scaling.

In the model shown for the carbon profile in Figure 5.45(b), C_o is the original carbon content, C_s is the carbon content at the scale–metal interface, x is distance measured from the original metal surface, and X is the position of the scale–metal interface.

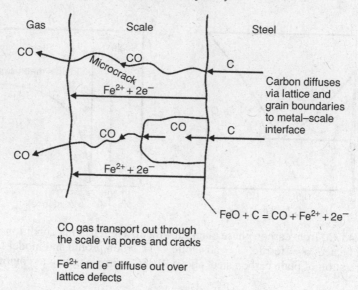

Figure 5.46 Mechanism of decarburization of plain carbon steel with simultaneous scaling.

The shape of the carbon profile can be obtained by solving Fick's second law for the case of a semi-infinite slab, assuming that the diffusion coefficient of carbon, D_C, is constant; the general solution for a semi-infinite slab is (Appendix A) given in Equation (5.42):

$$C = A + B_{\mathrm{erf}}\left(\frac{x}{2\sqrt{D_C t}}\right). \tag{5.42}$$

Here C represents the carbon concentration in wt%. The constants A and B are eliminated by considering the initial and steady-state boundary conditions shown in Equation (5.43),

$$C = C_o \quad \text{for } x > 0, \quad t = 0, \tag{5.43}$$

i.e., the carbon concentration was initially uniform, and in Equation (5.44),

$$C = C_S \quad \text{for } x = X, \quad t > 0, \tag{5.44}$$

i.e., the carbon concentration at the metal–scale interface, C_S, is held in equilibrium with the scale. The carbon profile at constant temperature is given by Equation (5.45):

$$\frac{C_o - C}{C_o - C_S} = \frac{\mathrm{erfc}\left(\dfrac{x}{2\sqrt{D_C t}}\right)}{\mathrm{erfc}\left(\dfrac{k_c}{2D_C}\right)^{1/2}}. \tag{5.45}$$

where k_c is the parabolic rate constant for the oxidation of the steel given by Equation (5.46),

$$k_c = \frac{X^2}{2t},\qquad(5.46)$$

where X is the depth of metal consumed by scale formation at time t. Equation (5.45) can be simplified by setting $C_S = 0$, since this value is extremely low (≈ 0.01 wt%), yielding Equation (5.47):

$$C = C_0\left[1 - \frac{\mathrm{erfc}\left(\dfrac{x}{2\sqrt{D_C t}}\right)}{\mathrm{erfc}\left(\dfrac{k_c}{2D_C}\right)^{1/2}}\right].\qquad(5.47)$$

Using a steel of 0.85 wt% C, 0.85 wt% Mn, and 0.18 wt% Si, data for k_c were obtained by direct measurement[129] and found to obey the relationship shown in Equation (5.48):

$$k_c = 57.1\exp\left(\frac{-21\,720}{T}\right)\ \mathrm{mm^2\,s^{-1}}.\qquad(5.48)$$

In practice, the value of D_C varies with carbon content, but values for low carbon content were found to give good agreement between calculated and measured profiles.[128] Extrapolating the data of Wells[130] to zero carbon content yielded[128] Equation (5.49):

$$D_C = 24.6\exp\left(\frac{-17\,540}{T}\right)\ \mathrm{mm^2\,s^{-1}}.\qquad(5.49)$$

Combining Equations (5.47), (5.48), and (5.49), carbon profiles can be calculated and, as shown in Figure 5.47, these give excellent agreement with measured profiles.

The actual value of the depth of decarburization is more difficult to establish since the carbon profile is smooth and it is difficult to define an inner limit. Although this could, and perhaps should, be set by defining a carbon content below which the mechanical properties of the steel are below specification, current practice relies on the choice of this position 'by eye' by an experienced metallographer. By comparing reported depths of decarburization with the carbon profiles, the inner limit was established as the position where the carbon content is 92% of the original uniform carbon content of the steel. This is also indicated in Figure 5.47.

Denoting the position of the inner limit as $x = x^*$, where $C = C^* = 0.92C_0$, the depth of decarburization, d, can now be established as shown in Equation (5.50),

$$d = x^* - X = x^* - (2k_c t)^{1/2};\qquad(5.50)$$

it can be seen that the value of d depends on the value of X or k_c, and in atmospheres where scaling is rapid k_c and X will be high and d will be correspondingly low.

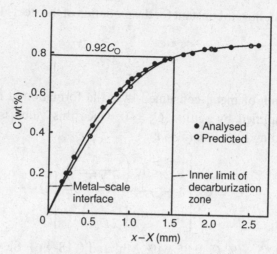

Figure 5.47 Comparison of analyzed and predicted profiles for 0.85 wt% C steel heated at 1100 °C for 90 min showing the position of the inner decarburization limit.

Equation (5.50) has been evaluated for a eutectoid steel and the depth of decarburization over the temperature range 900–1300 °C has been shown to be given by Equation (5.51)[129,131]

$$d = 10.5 \exp\left(\frac{-8710}{T}\right) t^{1/2} \, \text{mm} \tag{5.51}$$

where t is in s. Equation (5.51) allows the observed depth of decarburization at a constant temperature, T, between 900 and 1300 °C, to be calculated for a 0.85 wt% C steel. For other steels a similar equation would be valid, the difference being mainly in the pre-exponential term.

It should also be noted that, in steels of lower carbon content than the eutectoid composition, the presence of ferrite in the microstructure makes the judgement of the inner limit of decarburization more difficult and open to greater error. Thus, the agreement between prediction and observation is expected to deteriorate at lower carbon contents.

Effects of scaling rate on decarburization

One of the most common fallacies is to attempt to reduce decarburization by reducing the oxygen potential in the furnace. Since the carbon content at the metal–scale interface is constant, so long as FeO remains in contact with the steel, the driving force for carbon diffusion is also constant in the presence of a scale. However, by reducing the oxygen potential of the atmosphere, the scaling rate can be reduced and this will affect the observed depth of decarburization.

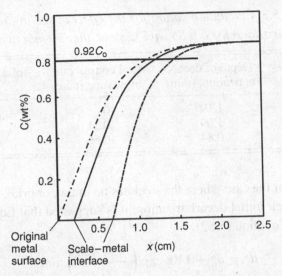

Figure 5.48 Effect of scaling rate on depth of decarburization, calculated for 0.85 wt% C steel heated for 90 min at 1050 °C; $k_c = 0$ (—·—), 4.1×10^{-6} (—), or 4.1×10^{-5} mm s^{-1} (---).

To illustrate this, calculated carbon profiles are plotted in Figure 5.48 for an 0.85 wt% C steel heated for 90 min at 1050 °C for various assumed values of k_c, the corrosion constant.[131] The profiles are plotted relative to the original metal surface and refer to the following: values of k_c 4.1×10^{-6} mm^2 s^{-1}, which is a realistic value for a furnace atmosphere; a hypothetical value of 4.1×10^{-5} mm^2 s^{-1} to represent a very high scaling rate; and 0 mm^2 s^{-1}, representing the case where the atmosphere just fails to form a scale but still reduces the surface carbon content to very low values.

It can be seen from Figure 5.48 that reducing the scaling rate reduces the total metal wastage but increases the observed depth of decarburization. This can be seen from the values of metal lost due to scaling, X, and the observed depth of decarburization, d, given in Table 5.3.

If the product is to be machined to remove decarburization, less metal is wasted if less aggressive atmospheres are used. Alternatively, reheating an awkwardly shaped product, such as wire, in a very aggressive atmosphere (by injecting steam) may cause the observed depth of decarburization to be reduced. Obviously, the question of cost and economy must carefully be considered.

Effect of prior decarburization

Most steels already have a decarburized surface layer, particularly at the billet stage, before they are subjected to reheating. Equation (5.51), in fact, predicts the depth of

Table 5.3 *Calculated values for the effect of scaling rate on decarburization for a 0.85 wt% C steel after 90 min at 1050°C*

Rate constant, k_c (mm^2 s^{-1})	Depth of decarburization, (mm)	Metal consumption by scaling (mm)	Total depth affected (mm)
0	1.19	0	1.19
4.1×10^{-6}	1.09	0.21	1.30
4.1×10^{-5}	0.84	0.66	1.50

decarburization for the case where the steel has no initial, or prior, decarburization. To account for such initial decarburization, it is suggested that Equation (5.51) be rearranged as in Equation (5.52),

$$d^2 = d_o^2 + 110t \exp\left(\frac{-17\,420}{T}\right) \text{mm}^2 \qquad (5.52)$$

where d_o is the initial depth of decarburization at $t = 0$. Thus, from Equation (5.51), if a steel with no initial decarburization is reheated at 1200°C for 20 mins, the value of $d^2 = 0.965$ mm^2 is given, corresponding to a depth of decarburization of 0.98 mm. If the steel already had an initial depth of decarburization of 0.5 mm before this treatment then subsequent to reheating at 1200°C the depth $d = (0.5^2 + 0.98^2)^{1/2}$ mm $= 1.10$ mm is expected.

Simultaneous non-isothermal scaling and decarburization

All of the previous discussion refers to isothermal conditions. Very few commercial reheating cycles can be adequately represented by an isothermal process. Consequently for the non-isothermal case that is normally encountered in industrial practice it would be necessary to take account of the existence of temperature gradients in the steel, grain-boundary diffusion of the carbon and variability of carbon diffusivity with both carbon content and temperature. This may be possible using a suitable computer program; meanwhile an approximate attempt to solve this problem by dividing the heating profile into discrete isothermal steps has been attempted.[127,129]

Conclusions

The quantitative treatment presented of the decarburization of steels contains many assumptions, despite which it produces acceptable results in the isothermal case. To extend the treatment to the non-isothermal case, certain factors that have been overlooked should be considered. For example, the existence of a temperature gradient in the steel has been ignored and both grain-boundary diffusion and the variation of carbon diffusion with carbon content have not been accounted for. Nevertheless, the model can produce useful results for commercial application and a more comprehensive, computerised version should be capable of more accurate predictions.

References

1. O. Kubaschewski and B. E. Hopkins, *Oxidation of Metals and Alloys*, London, Butterworth, 1962.
2. K. Hauffe, *Oxydation von Metallen und Metallegierungen*, Berlin, Springer, 1957.
3. J. Benard, *Oxydation des Métaux*, Paris, Gauthier-Villars, 1962.
4. H. Pfeiffer and H. Thomas, *Zunderfeste Legierungen*, Berlin, Springer, 1963.
5. C. E., Birchenall, Oxidation of alloys. In *Oxidation of Metals and Alloys*, ed. D. L. Douglass, Metals Park, Ohio, ASM, 1971, ch. 10.
6. S. Mrowec and T. Werber, *Gas Corrosion of Metals*, Washington, DC, National Bureau of Standards and National Science Foundation (translated from Polish), 1978.
7. P. Kofstad, *High Temperature Corrosion*, London, Elsevier Applied Science, 1988.
8. G. Beranger, J. C. Colson, and F. Dabosi, *Corrosion des Materiaux à Haute Température*, Les Ulis, les Éditions de Physique, 1987.
9. C. Wagner, *Ber. Bunsenges. Phys. Chem.*, **63** (1959), 772.
10. C. Wagner, *J. Electrochem. Soc.*, **99** (1956), 369.
11. C. Wagner, *J. Electrochem. Soc.*, **103** (1956), 571.
12. H. Rickert, *Z. Phys. Chem. NF* **21** (1960), 432.
13. R. A. Rapp, *Corrosion*, **21** (1965), 382.
14. J. H. Swisher, Internal oxidation. In *Oxidation of Metals and Alloys*, ed. D. L. Douglass, Metals Park, Ohio, ASM, 1971, ch. 12.
15. J. L. Meijering, Internal oxidation in alloys. In *Advances in Materials Research*, ed. H. Herman, New York, Wiley, 1971, Vol. 5, pp. 1–81.
16. D. L. Douglass, *Oxid. Met.*, **44** (1995), 81.
17. S. Wood, D. Adamonis, A. Guha, W. A. Soffa, and G. H. Meier, *Met. Trans.*, **6A** (1975), 1793.
18. J. Megusar and G. H. Meier, *Met. Trans.*, **7A** (1976), 1133.
19. G. Bohm and M. Kahlweit, *Acta met.*, **12** (1964), 641.
20. P. Bolsaitis and M. Kahlweit, *Acta met.*, **15**, (1967), 765.
21. C. Wagner, *Z. Elektrochem.*, **63** (1959), 772.
22. R. A. Rapp, *Acta met.*, **9** (1961), 730.
23. C. Wagner, *Corr. Sci.*, **5** (1965), 751.
24. H. R. Pickering, *J. Electrochem. Soc.*, **119** (1972), 64.
25. G. H. Meier, *Mater. Sci. Eng.*, **A120** (1989), 1.
26. F. H. Stott, G. C. Wood, and J. Stringer, *Oxid. Met.*, **44** (1995), 113.
27. C. S. Giggins and F. S. Pettit, *TAIME*, **245** (1969), 2495.
28. N. Birks and H. Rickert, *J. Inst. Met.*, **91** (1962–63), 308.
29. G. C. Wood, *Oxid. Met.*, **2** (1970), 11.
30. G. C. Wood, I. G. Wright, T. Hodgkiess, and D. P. Whittle, *Werkst. Korr.*, **21** (1970), 900.
31. G. C. Wood and B. Chattopadhyay, *Corr. Sci.*, **10** (1970), 471.
32. F. Gesmundo and F. Viani, *Oxid. Met.*, **25** (1986), 269.

33. G. C. Wood and F. H. Stott, *Mater. Sci. Tech.*, **3** (1987), 519.
34. F. S. Pettit, *Trans. Met. Soc. AIME*, **239** (1967), 1296.
35. J. Doychak, in *Intermetallic Compounds*, eds. J. H. Westbrook, R. L. Fleischer, New York, Wiley, 1994, p. 977.
36. E. Schumann and M. Rühle, *Acta metall. mater.*, **42** (1994), 1481.
37. G. C. Rybicki and J. L. Smialek, *Oxid. Met.*, **31** (1989), 275.
38. B. A. Pint and L. W. Hobbs, *Oxid. Met.*, **41** (1994), 203.
39. M. W. Brumm and H. J. Grabke, *Corr. Sci.*, **33** (1992), 1677.
40. B. A. Pint, M. Treska, and L. W. Hobbs, *Oxid. Met.*, **47** (1997), 1.
41. C. S. Giggins and F. S. Pettit, *J. Electrochem. Soc.*, **118** (1971), 1782.
42. G. R. Wallwork and A. Z. Hed, *Oxid. Met.*, **3** (1971), 171.
43. B. H. Kear, F. S. Pettit, D. E. Fornwalt, and L. P. Lemaire, *Oxid. Met.*, **3** (1971), 557.
44. G. R. Wallwork, *Rep. Prog. Phys.*, **39** (1976), 401.
45. C. S. Giggins and F. S., Pettit, Final Report to Aerospace Research Laboratories, Dayton, OH, Wright-Patterson AFB, contract NF33615–72-C-1702, 1976.
46. T. Adachi and G. H. Meier, *Oxid. Met.*, **27** (1987), 347.
47. R. A. Rapp and H. Colson, *Trans. Met. Soc. AIME*, **236** (1966), 1616.
48. R. A. Rapp and G. Goldberg, *Trans. Met. Soc. AIME*, **236** (1966), 1619.
49. G. C. Wood, in *Oxidation of Metals and Alloys*, ed. D. L. Douglass, Metals Park, OH, ASM, 1971, ch. 11.
50. M. G. Cox, B. McEnaney, and V. D. Scott, *Phil. Mag.*, **26** (1972), 839.
51. C. Wagner, *Corr. Sci.*, **10** (1969), 91.
52. B. D. Bastow, D. P. Whittle, and G. C. Wood, *Corr. Sci.*, **16** (1976), 57.
53. G. H. Meier, Fundamentals of the oxidation of high temperature intermetallics. In *Oxidation of High Temperature Intermetallics*, eds. T. Grobstein and J. Doychak, Warrendate, PA, TMS, 1989, p. 1.
54. G. H. Meier, N. Birks, F. S. Pettit, R. A. Perkins, and H. J. Grabke, Environmental behavior of intermetallic materials. In *Structural Intermetallics*, 1993, p. 861.
55. G. H. Meier, *Mater. Corr.*, **47** (1996), 595.
56. M. P. Brady, B. A. Pint, P. F. Tortorelli, I. G. Wright, and R. J. Hanrahan, Jr., High temperature oxidation and corrosion of intermetallics. In *Materials Science and Technology: A Comprehensive Review*, eds. R. W. Cahn, P. Haasen, and E. J. Kramer, Wiley–VCH Verlag, 2000, Vol. II, ch. 6.
57. R. Svedberg, Oxides associated with the improved air oxidation performance of some niobium intermetallics and alloys. In *Properties of High Temperature Alloys*, eds. Z. A. Foroulis and F. S. Pettit, NewYork, NY, The Electochemical Society, 1976, p. 331.
58. H. J. Grabke and G. H. Meier, *Oxid. Met.*, **44** (1995), 147.
59. D. A. Berztiss, R. R. Cerchiara, E. A. Gulbransen, F. S. Pettit, and G. H. Meier, *Mater. Sci. Eng.*, **A155** (1992), 165.
60. D. L. Douglass, Exfoliation and the mechanical behavior of scales. In *Oxidation of Metals and Alloys*, ed. D. L. Douglass, Metals Park, OH, ASM, 1971.
61. J. Stringer, *Corr. Sci.*, **10** (1970), 513.
62. P. Hancock and R. C., Hurst, The mechanical properties and breakdown of surface oxide films at elevated temperatures. In *Advances in Corrosion Science and Technology*, eds. R. W. Staehle and M. G. Fontana, New York, NY, Plenum Press, 1974, p. 1.
63. F. H. Stott and A. Atkinson, *Mater. High Temp.*, **12** (1994), 195.
64. H. E. Evans, *Int. Mater. Rev.*, **40** (1995), 1.

65. N. B. Pilling and R. E. Bedworth, *J. Inst. Met.*, **29** (1923), 529.
66. B. Pieraggi and R. A. Rapp, *Acta met.*, **36** (1988), 1281.
67. J. Robertson and M. J. Manning, *Mater. Sci. Tech.*, **4** (1988), 1064.
68. G. Guruswamy, S. M. Park, J. P. Hirth, and R. A. Rapp, *Oxid. Met.*, **26** (1986), 77.
69. D. L. Douglass, B. Zhu, and F. Gesmundo, *Oxid. Met.*, **38** (1992), 365.
70. W. Jaenicke and S. Leistikow, *Z. Phys. Chem.*, **15** (1958), 175.
71. W. Jaenicke, S. Leistikow, and A. Stadler, *J. Electrochem. Soc.*, **111** (1964), 1031.
72. S. Horibe and T. Nakayama, *Corr. Sci.*, **15** (1975), 589.
73. C. S. Giggins and F. S. Pettit, *Trans. Met. Soc. AIME*, **245** (1969), 2509.
74. F. N. Rhines and J. S. Wolf, *Met. Trans.*, **1** (1970), 1701.
75. M. V. Speight and J. E. Harris, *Acta met.*, **26** (1978), 1043.
76. A. Atkinson, *Corr. Sci.*, **22** (1982), 347.
77. R. Prescott and M. J. Graham, *Oxid. Met.*, **38** (1992), 233.
78. D. Caplan and G. I. Sproule, *Oxid. Met.*, **9** (1975), 459.
79. K. P. Lillerud and P. Kofstad, *J. Electrochem. Soc.*, **127** (1980), 2397.
80. F. A. Golightly, F. H. Stott, and G. C. Wood, *J. Electrochem. Soc.*, **126** (1979), 1035.
81. S. P. Timoshenko, *J. Opt. Soc. Amer.*, **11** (1925), 233.
82. J. K. Tien and J. M., Davidson, Oxide spallation mechanisms. In *Stress Effects and the Oxidation of Metals*, ed. J. V. Cathcart, New York, AIME, 1975, p. 200.
83. I. C. Noyan and J. B. Cohen, *Residual Stresses*, Berlin, Springer-Verlag, 1987.
84. C., Sarioglu, J. R. Blachere, F. S. Pettit, and G. H. Meier, Room temperature and *in-situ* high temperature strain (or stress) measurements by XRD techniques. *Microscopy of Oxidation 3*, eds., S. B. Newcomb and J. A. Little, London, The Institute of Materials. 1997, p. 41.
85. P. Y. Hou and J. Stringer, *Acta metall. mater.*, **39** (1991), 841.
86. J. H. Stout, D. A. Shores, J. G. Goedjen, and M. E. Armacanqui, *Mater. Sci. Eng.*, **A120** (1989), 193.
87. J. J. Barnes, J. G. Goedjen, and D. A. Shores, *Oxid. Met.*, **32** (1989), 449.
88. J. G. Goedjen, J. H. Stout, Q. Guo, and D. A. Shores, *Mater. Sci. Eng.*, **A177** (1994), 15.
89. Y. Zhang, D. Zhu, and D. A. Shores, *Acta metall. mater.*, **43** (1995), 4015.
90. D. J. Gardiner, Developments in Raman spectroscopy and applications to oxidation studies. In *Microscopy of Oxidation 2*, eds. S. B. Newcomb and M. J. Bennett, London, Institute of Materials, 1993, p. 36.
91. M. Lipkin and D. R. Clarke, *J. Appl. Phys.*, **77** (1995), 1855.
92. U. R. Evans, *An Introduction to Metallic Corrosion*, London, Edward Arnold, 1948, p. 194.
93. H. E. Evans and R. C. Lobb, *Corr. Sci.*, **24** (1984), 209.
94. D. P. Whittle and J. Stringer, *Phil. Trans. Roy. Soc. Lond.*, **A295** (1980), 309.
95. J. Stringer, *Met. Rev.*, **11** (1966), 113.
96. J. K. Tien and F. S. Pettit, *Met. Trans.*, **3** (1972), 1587.
97. E. J. Felten, *J. Electrochem. Soc.*, **108** (1961), 490.
98. H. Pfeiffer, *Werks. Korr.*, **8** (1957), 574.
99. J. E. McDonald and J. G. Eberhardt, *Trans. TMS-AIME*, **233** (1965), 512.
100. J. E. Antill and K. A. Peakall, *J. Iron Steel Inst.*, **205** (1967), 1136.
101. A. W. Funkenbusch, J. G. Smeggil, and N. S. Bornstein, *Met. Trans.*, **16A** (1985), 1164.

102. J. G. Smeggil, A. W. Funkenbusch, and N. S. Bornstein, *Met. Trans.*, **17A** (1986), 923.

103. B. K. Tubbs and J. L., Smialek, Effect of sulphur removal on scale adhesion to PWA 1480. In *Corrosion and Particle Erosion at High Temperatures*, eds. V. Srinivasan and K. Vedula, Warrendate, PA, TMS, 1989, p. 459.

104. R. V. McVay, P. Williams, G. H. Meier, F. S. Pettit, and J. L. Smialek, Oxidation of low sulphur single crystal nickel-base superalloys. In *Superalloys 1992*, eds. S. D. Antolovich, R. W. Stusrud, R. A. MacKay, D. L. Anton, T. Khan, R. D. Kissinger, and D. L. Klarstrom, Warrendale, PA, TMS, 1992, p. 807.

105. M. C. Stasik, F. S. Pettit, G. H. Meier, A. Ashary, and J. L. Smialek, *Scripta met. mater.*, **31** (1994), 1645.

106. H. J. Grabke, D. Wiener, and H. Viefhaus, *Appl. Surf. Sci.*, **47** (1991), 243.

107. P. Y. Hou and J. Stringer, *Oxid. Met.*, **38** (1992), 323.

108. H. J. Grabke, G. Kurbatov, and H. J. Schmutzler, *Oxid. Met.*, **43** (1995), 97.

109. P. Y. Hou and J. Stringer, *Mater. Sci. Eng.*, **A202** (1995), 1.

110. R. A. Rapp and B. Pieraggi, *J. Electrochem. Soc.*, **140** (1993), 2844.

111. G. M. Ecer and G. H. Meier, *Oxid. Met.*, **13** (1979), 159.

112. L. B. Pfeil, UK Pat. No. 459848, 1937.

113. G. M. Ecer, R. B. Singh, and G. H. Meier, *Oxid. Met.*, **18** (1982), 53.

114. F. S. Pettit and G. H., Meier, The effects of refractory elements on the high temperature oxidation and hot corrosion properties of superalloys. In *Refractory Alloying Elements in Superalloys*, eds. J. K Tien and S. Reichman, Metals Park, OH, ASM, 1984, p. 165.

115. W. C. Leslie and M. G. Fontana, *Trans. ASM*, **41** (1949), 1213.

116. S. S. Brenner, *J. Electrochem. Soc.*, **102** (1955), 7.

117. S. S. Brenner, *J. Electrochem. Soc.*, **102** (1955), 16.

118. G. W. Rathenau and J. L. Meijering, *Metallurgia*, **42** (1950), 167.

119. K. R. Peters, D. P. Whittle, and J. Stringer, *Corr. Sci.*, **16** (1976), 791.

120. V. V. Belousov and B. S. Bokshtein, *Oxid. Met.*, **50** (1998), 389.

121. M. E. El-Dashan, D. P. Whittle, and J. Stringer, *Corr. Sci.*, **16** (1976), 83.

122. M. E. El-Dashan, D. P. Whittle, and J. Stringer, *Corr. Sci.*, **16** (1976), 77.

123. J. A. Goebel, F. S. Pettit, and G. W. Goward, *Met. Trans.*, **4** (1973), 261.

124. D. Caplan, G. I. Sproule, R. J. Hussey, and M. J. Graham, *Oxid. Met.*, **12** (1968), 67.

125. W. E. Boggs and R. H. Kachik, *J. Electrochem. Soc.*, **116** (1969), 424.

126. K. Sachs and C. W. Tuck, ISI Publication 111, London, The Iron and Steel Institute, 1968, p. 1.

127. N. Birks and W. Jackson, *J. Iron Steel Inst.*, **208** (1970), 81.

128. J. Baud, A. Ferrier, J. Manenc, and J. Bénard, *Oxid. Met.*, **9** (1975), 1.

129. N. Birks and A. Nicholson, ISI Publication 123, London, The Iron and Steel Institute, 1970, p. 219.

130. C. Wells, *Trans. Met. Soc. AIME*, **188** (1950), 553.

131. N. Birks, ISI Publication 133, London, The Iron and Steel Institute, 1970, p. 1.

6

Oxidation in oxidants other than oxygen

Introduction

Most metals are thermodynamically unstable with respect to many of their compounds, particularly metal halides and sulphides, in addition to oxides. This is reflected in the extensive processing that is required to extract metals from their various ores. This chapter is concerned with the corrosion reactions by which metals return to their stable oxidized state, by reaction with sulphur, carbon, nitrogen, and the halogens. Often the corrosive gases encountered in practice contain oxygen, as well as one of these components. This problem of 'mixed-oxidant corrosion' will be addressed in the next chapter.

Reactions with sulphur

Metals react readily with sulphur, by mechanisms that parallel those involved in reactions with oxygen. The sulphides of iron and nickel show metallic conduction, i.e., they have high electrical conductivities, which decrease with increasing temperature and are nearly independent of sulphur partial pressure and stoichiometry (H. J. Grabke, personal communication).[1] The departure from stoichiometry in sulphides is generally greater than in the case of oxides, diffusion is faster in sulphides, which are also more plastic than the corresponding oxides and generally have lower melting points. In particular, low-melting metal-sulphide eutectics are common.

Compared with oxidation, the sulphidation of a metal proceeds more rapidly and can be studied at lower temperatures and over shorter times. Sulphide scales tend to be quite plastic but their adherence is not good because most grow by outward cation diffusion and form voids at the scale–metal interface. Rickert and coworkers[2-4] performed extensive studies on sulphidation and used a mechanical load to press the sulphide and metal together.

Reactions in metal–sulphur systems have been studied, as model systems, to verify mechanisms that are proposed for the high-temperature oxidation processes. The excellent correspondence between the initial theories of Wagner[5] and observation in the case of sulphidation established confidence in the theory, which was then applied widely to the more practically important metal-oxidation processes.

The standard free energy of formation of a sulphide is usually less negative than that of the corresponding oxide. As a result, the partial pressure of sulphur in equilibrium with a mixture of metal and its sulphide, the decomposition pressure of the sulphide is usually higher than the decomposition pressure of the corresponding oxide. Consequently, reaction mechanisms that involve the transfer of sulphur within the pores of a scale can be studied quite readily and the results applied, by comparison, to the oxidation process. In this way, it has been possible to emphasize the role of sulphide and oxide decomposition in the formation of inner porous layers of a scale, and that these processes proceed more quickly at grain boundaries than on grain surfaces,[6] as shown in Figure 4.2.

In addition to illustrating mechanistic principles, metal–sulphur systems are studied for their role in the attack of metal components in the reforming of oil and in the gasification of coal. In these applications, the oxygen potentials of the atmospheres are usually very low and, consequently, the sulphur potentials can be high, if a sufficient concentration of sulphur species is present. These atmospheres represent an interesting case in materials application, where a metal is selected for use at high temperature based on its oxidation resistance, only to fall prey to the more aggressive sulphidation attack that prevails in the low oxygen potentials of such atmospheres. A detailed consideration of these features is beyond the scope of this book, but good reviews of the processes are given[7–9] for sulphur and hydrogen–hydrogen sulphide mixtures. Usually the reaction kinetics in H_2S or H_2–H_2S mixtures are found to be controlled by a surface reaction, the transfer of S from H_2S. This has been illustrated for the growth of FeS in a wide range of conditions.[10,11]

The subject of the sulphidation of metals was reviewed earlier by Strafford,[12] who has also carried out substantial research into the subject, with particular emphasis on the sulphidation of refractory metals. In the case of the refractory metals, the sulphides are very stable and form scales of tight stoichiometry and, consequently, low rates of growth.

Sulphides generally deviate from stoichiometry to a much greater extent than oxides. This is true for all 'common' metals including chromium.[13,14] This means that the sulphides contain much higher defect concentrations than oxides. Therefore, for a given mobility of defects, diffusion through sulphides is expected to be substantially more rapid than through oxides, for a given potential gradient. Mrowec and Fueki and coworkers[13–15] compared the chemical diffusion coefficients of oxides and sulphides and found that they were very similar, within an

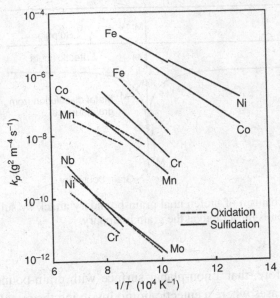

Figure 6.1 Comparison of typical sulphidation and oxidation rates of several transition and refractory metals. (Most of the data are from Mrowec.[13])

order of magnitude. They concluded that the higher self-diffusion of cations in metal sulphides, compared with oxides, is due to the higher defect concentrations in the sulphides.

In contrast to the sulphides of most of the transition metals, sulphides of the refractory metals have quite tight stoichiometry, similar to Cr_2O_3, although, in the cases of the refractory-metal sulfides and oxides, the defects appear on the anion sub-lattice.[16] Figure 6.1 compares the rates of oxidation and sulphidation for several of the transition and refractory metals. The low rates of sulphidation of the refractory metals are thought to be due to the low concentrations of defects in the sulphide structures.[13]

Many of the transition-metal–sulphur systems show low-temperature eutectics, for instance $Ni–Ni_3S_2$ (645 °C), $Fe–FeS$ (985 °C) and $Co–Co_4S_3$ (880 °C). Hancock[17] exposed nickel to hydrogen sulphide at 900 °C and found extremely rapid corrosion involving the formation of the liquid phase, as confirmed by the observation of dendritic features.

The formation of a liquid scale would be expected to give rise to the rapid attack that can be associated with rapid diffusion through the liquid. If the liquid is also a eutectic between the sulphide and the metal, then the reaction at the scale–metal interface involves a dissolution step. Such dissolution is capable of occurring more rapidly from some surface orientations than from others. In addition the dissolution is likely to be very rapid at the disordered areas of the grain boundaries. It is to

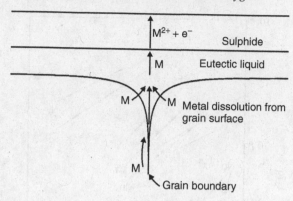

Figure 6.2 Mechanism of preferential grain-boundary attack by sulphide liquid. Metal dissolution is most rapid at the grain boundary.

be expected, therefore, that a non-planar surface with grain-boundary incursions would develop in cases where a eutectic liquid involving the metal is formed. The resulting grain-boundary attack produces 'notch' features that would accelerate processes such as fatigue and creep, in a component under load. The mechanism of the preferential grain-boundary attack is shown in Figure 6.2.

Examples of the use of sulphidation to verify oxidation theory has been provided also by Wagner.[18] In Chapter 3, the parabolic rate constant was derived from first principles using the simplifying assumption that cations and electrons were the only mobile species involved. It was assumed that the anions migrated so slowly that they could be considered to be immobile. Wagner[5,19] has considered the mobilities of all three species (cations, anions, and electrons) and expressed the 'rational' rate constant in terms of the self-diffusion coefficients of the cation and anion species, as in Equation (6.1),

$$k_r = C \int_{a'_X}^{a''_X} \left(\frac{Z_c}{Z_a} D_M + D_X \right) d \ln a_X, \qquad (6.1)$$

where C is the concentration of ions in equivalents per cubic centimeter, Z_c and Z_a and D_M and D_X are the valencies and self-diffusion coefficients of the cation and anion species, respectively. Equation (6.1) is the same as Equations (3.48) and (3.49) except that here k_r is used, rather than k', and both ions are presumed to be mobile. This equation can be differentiated, for the condition where the migration of anions can be neglected and for constant a'_X to give Equation (6.2):

$$\frac{dk_r}{(d \ln a''_X)_{a'_X}} = 2 \left(\frac{A}{Z_a \overline{V}} \right)^2 \frac{Z_c}{Z_a} D_M. \qquad (6.2)$$

Mrowec[16] measured the parabolic rate constant for the sulphidation of iron at different sulphur activities in the atmosphere at 800 °C. By plotting the parabolic rate constant against the logarithm of the sulphur activity the the tangent to the curve can be used to obtain the cation diffusion coefficient as a function of the sulphur activity. The cation diffusion coefficient was found to be independent of the sulphur activity with a value of 1.8×10^{-7} cm^2 s^{-1}. Radio active tracer experiments by Condit[20] gave a value of 8.6×10^{-8} cm^2 s^{-1} in good agreement, confirming Wagner's model for high-temperature tarnishing reactions.

The usefulness and the hazards of marker experiments have also been demonstrated using sulphide systems. When using markers, a chemically inert material is placed on the metal surface before exposure. After oxidation or sulphidation, the position of the marker with respect to the scale–metal interface should give evidence regarding whether the scale grew predominantly by cation migration, anion migration, or both. If the scale is compact and free of pores, then a marker positioned at the scale–metal interface infers that the scale has grown by outward diffusion of cations. Conversely, a marker found at the scale–gas interface, infers that scale growth occurred by inward anion diffusion. It follows that, if cations and anions both are mobile to the same extent, the markers would be expected to be found in the middle of the scale. Unfortunately, markers found in the middle of a scale pose a problem, since most compounds of which scales are formed usually show predominant migration of one of the ionic species only. The location of markers in the middle of scales is usually evidence that the scale has separated from the metal substrate during growth or that species can be transported by mechanisms other than pure lattice diffusion. This has been discussed by several workers,[19–23] whose publications should be consulted before undertaking marker experiments.

The defect concentrations found in sulphides are so high that some interaction of defects is almost inevitable. In the case of iron sulphide, FeS, if there were localized electronic defects, the progressive ionization of defects could be expressed by Equations (6.3)–(6.5):

$$\frac{1}{2}S_2(g) = S_S^X + V_{Fe},\tag{6.3}$$

$$\frac{1}{2}S_2(g) = S_S^X + V'_{Fe} + h^{\cdot},\tag{6.4}$$

$$\frac{1}{2}S_2(g) = S_S^X + V''_{Fe} + 2h^{\cdot}.\tag{6.5}$$

According to which expression prevails, the concentration of iron ion vacancies would be expressed in terms of the sulphur partial pressure as shown in

Equations (6.6)–(6.8):

$$[V_{Fe}] = const.\, p_{S_2}^{1/2},$$
(6.6)

$$[V'_{Fe}] = const.\, p_{S_2}^{1/4},$$
(6.7)

$$[V''_{Fe}] = const.\, p_{S_2}^{1/6}.$$
(6.8)

Mrowec,[16] using existing data[24-26], calculated that the concentration of iron-ion vacancies was proportional to the sulphur partial pressure to the power of 1/1.9 at 800 °C. This result is consistent with the metallic nature of FeS,[1] i.e., there are an insignificant number of localized electronic defects.

One final interesting point about sulphidation is that, because the free energies of formation of the sulphides of iron, nickel, cobalt, chromium, and aluminium are quite similar, selective sulphidation of chromium and aluminium is not favoured as in the case of oxidation. Lai[9] has reviewed the literature of the sulphidation of alloys and has made several generalizations as follows.

(1) Alloys of iron, nickel, and cobalt sulphidize at similar rates – within an order of magnitude.
(2) Alloys containing less than 40% Cr may show improved sulphidation resistance by forming an inner layer of ternary sulphide or chromium sulphide. It has been found that Cr concentrations in excess of 17 wt% Cr provide some sulphidation protection in steels.[27]
(3) Alloys containing more than 40% Cr could form a single, protective, outer layer of Cr_2S_3. At the higher Cr levels, all three alloy systems showed similar resistance.
(4) The presence of Al in Fe and Co alloys could be beneficial due to the formation of an aluminium-based sulphide. In the case of Ni-based alloys the results were erratic.
(5) However, formation of Al_2S_3 scales is not particularly useful because they react with H_2O, even at room temperature.

On the other hand, refractory-metal sulphides are both very stable and slow growing. This has been addressed by Douglass and coworkers,[28-35] who demonstrated that the addition of Mo and Mo plus Al to nickel substantially reduced the parabolic rate constant for sulphidation, by about five orders of magnitude for the composition Ni–30 wt% Mo–8 wt% Al; and, in the case of iron alloys, by six orders of magnitude for the composition Fe–30 wt% Mo– 5 wt% Al. In these cases the scales formed were $Al_{0.5}Mo_2S_4$, which gave excellent resistance to sulphidation, but would form molybdenum oxides in oxidizing atmospheres.[13]

Reactions with halogens

The reactions of metals to form stable halides are important for various reasons. The metal halides generally have low boiling points and high volatility. For this reason, they are used in several important processes for the production and refining of metals, such as the reactive metals titanium and zirconium. These metals are produced using the Kroll process, in which the metal oxide is converted to metal chloride or fluoride, which is then reduced to metal. This route avoids several formidable difficulties involved in the reduction of the oxides of these metals. Details of these processes can be found in extractive-metallurgy textbooks.

Although stability diagrams may be prepared relating halide and oxide phase fields, there exists a strong possibility that stable and volatile oxyhalide species may also form, such as $CrOCl_2$ and WO_2Cl_2. In these cases, the stability diagram must be altered to accommodate these phases, as discussed in Chapter 2.

Volatile metal halides, usually chlorides and fluorides, also form the heart of several processes used to produce surface layers, rich in aluminium, chromium, or silicon, or combinations of these. In these processes, the workpiece to be coated is buried in a powder bed and heated to reaction temperature. The bed consists of a mixture of inert alumina filler, a master alloy powder that contains the aluminium, etc., and an activator such as ammonium chloride. Basically, at about 630 °C, the activator volatilizes and the aluminium chloride vapour reacts with the master alloy to produce a volatile aluminium chloride, which then reacts with the workpiece surface to deposit aluminium. The deposited aluminium proceeds to diffuse into the surface layers of the workpiece to produce a diffusion coating. The process is driven basically by the difference in aluminium activity between the master alloy and the workpiece. These processes are well documented in principle, but their execution to provide reproducible and reliable results still involves considerable experience, or rule of thumb. These processes will be described in detail in Chapter 10. Finally, a chlorination treatment is used to remove tin from tin-plated steel.[36] This uses a normally deleterious reaction to advantage and profit in the recovery of both tin and steel for recycling. Fluorination is used in the manufacture of polymers and fluorocarbon; consequently, materials suitable for construction of these plants must be resistant to fluorine attack.

In principle, the formation of halide scales on metals exposed to halide atmospheres will follow the same mechanisms found to be valid for both oxide and sulphide formation. A comprehensive review of the reactions of metals with halogens has been given by Daniel and Rapp.[37] In fact, many metal halides are almost pure ionic conductors so that the growth of halide scales is expected to be controlled by electron migration rather than by ionic transport. In this case, the halide scale will grow laterally over the metal surface more readily than it will thicken.

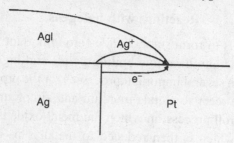

Figure 6.3 Growth of silver iodide on silver as it proceeds to cover a piece of platinum in electrical contact with the silver.

In fact a halide scale will grow beyond the reacting metal and over the surface of an inert metal attached to it, by the mechanism proposed by Ilschner-Gench and Wagner.[38] In this mechanism, the ions migrate through the halide and the electrons migrate through the inert metal to the reaction site. This mechanism is shown in Figure 6.3 for the case of the iodization of silver and indicates clearly how the low electronic conductivity of the halide emphasizes surface coverage rather than scale thickening. The ionic nature of PbI_2 scales growing on Pb has been demonstrated by Kuiry *et al.*,[39] who showed that scale growth rate was much more rapid when the electrons were shorted across the scale.

In common with oxide scales, halide scales can also suffer mechanical damage such as cracks, etc. In addition, the vapour pressure of the halide can be quite high, such that evaporation of the halide from the scale–gas interface can be substantial. This leads to simultaneous formation and evaporation of the halide scale and has also been treated by Daniel and Rapp.[37] The rate of thickening of a scale forming under these conditions then follows paralinear kinetics[40] as described for the combined scale growth and evaporation of chromium oxide in Chapter 4.

Lai[9] indicates that the addition of Cr and Ni to iron alloys improves their chlorination resistance. Consequently, nickel alloys are used for exposure to chlorine-bearing atmospheres. Most metals appear to resist attack up to a certain temperature, above which rapid attack ensues. This critical temperature for iron is about 250 °C, and 500 °C for nickel.[9]

Many environments contain oxygen and chlorine, in which case there is competition to form a solid oxide layer or a volatile chloride. The tendency to form the volatile chloride increases with chlorine partial pressure and temperature, as shown by Maloney and McNallen for cobalt.[41] In general, alloys that contain molybdenum and tungsten do not perform well in halogen-bearing atmospheres because of the formation of their volatile oxychlorides MoO_2Cl_2 and WO_2Cl_2.[42] In the presence of oxygen and chlorine any reduction in reaction rate depends on whether or not a protective oxide scale can form. According to Lai,[9] this is possible in many engineering alloys above 900 °C, when aluminium is present and capable of forming an alumina scale.

Fluorine attack follows the mechanisms mentioned for the case of chlorine. Namely, the reaction is slow at lower temperatures when a fluoride protective layer is formed; but at higher temperatures, the formation of liquid and, especially, volatile fluorides leads to rapid attack of the metal.

More detail of the response of alloys to atmospheres based on Cl_2, H_2–H_2S, O_2–Cl_2, and corresponding fluorine atmospheres is given by Lai.[9]

The halogenation of metals provides opportunities to study scaling reactions in which the reaction product is a pure ionic conductor, which can also be highly volatile. These systems can be studied at relatively low temperature; they are important in their own right in terms of the mechanisms of extraction and chemical-vapour-deposition processes that employ gaseous halides as intermediate compounds.

Reactions with carbon

In many processes, carbon is present as the primary or secondary oxidant. Where carbon is the primary oxidant, the principal reaction is that of carburization or the dissolution of carbon into the metal matrix. The solubility of carbon in metals varies widely, being very low in Ni, Cu, Co, and ferritic iron but quite substantial in austenitic iron. Carburization of alloys, principally steels, is a common treatment for developing a hard and strong surface on components that are exposed to contact wear during service. The theory and techniques for this are clarified in the literature.[43]

Carbon may be introduced into the metal surface using several gas systems involving reactions such as those shown in Equations (6.9)–(6.11):

$$CO\,(g) + H_2(g) = C + H_2O\,(g); \quad \Delta G_9^o, \tag{6.9}$$

$$2CO\,(g) = C + CO_2\,(g); \quad \Delta G_{10}^o, \tag{6.10}$$

$$CH_4\,(g) = C + 2H_2\,(g); \quad \Delta G_{11}^o. \tag{6.11}$$

In each case, the carbon activity of the gas is given using the equilibrium constant of the reactions given in Equations (6.12)–(6.14):

$$a_C = \frac{p_{CO}\,p_{H_2}}{p_{H_2O}} \exp\left(\frac{-\Delta G_9^o}{RT}\right), \tag{6.12}$$

$$a_C = \frac{p_{CO}^2}{p_{CO_2}} \exp\left(\frac{-\Delta G_{10}^o}{RT}\right), \tag{6.13}$$

$$a_C = \frac{p_{CH_4}}{p_{H_2}^2} \exp\left(\frac{-\Delta G_{11}^o}{RT}\right). \tag{6.14}$$

Standard references[9,43] contain diagrams that give the carbon activities of various atmospheres over a range of temperatures. The time of treatment is controlled by the diffusion of carbon into the component and depends upon the required surface concentration, depth of carburized rim, temperature, and the diffusion coefficient of carbon. Treatments of this simple process, in cases of complete diffusion control involve the solution of Fick's second law[44] for planar, cylindrical, or spherical coordinates, depending on the specimen dimensions and geometry.

Under most practical carburization conditions for steels the reaction rate is under mixed control of surface reaction and diffusion.[45] Whereas the carburization of simple steels, for instance, is straightforward, and hardness is achieved by the formation of high-carbon martensite on heat treatment, the presence of carbon in stainless steels and superalloys results in the formation of carbides based on chromium and other alloying elements. Excessive carburization can result in the removal from solution of protective elements such as chromium. This can seriously reduce the corrosion resistance of the component, particularly at grain boundaries.

Although carburization can enhance the performance of certain components, in cases such as reformer tubes in the treatment of oil and hydrocarbons, carburization of stainless steels is deleterious and life limiting. In this case internal carbides form with kinetics analogous to those for internal oxidation.[46] Consequently, alloys that are resistant to carburization are developed, primarily by alloying with nickel to reduce the diffusion coefficient of carbon, and with silicon and aluminium, which are thought to impart some protection by the formation of impervious silica and alumina surface films in the low-oxygen-potential atmospheres.[10]

Metal dusting is another carbon interaction that is harmful.[47,48] In this reaction, the carbon activity in the gas must be greater than unity, and it appears that the carbon from the atmosphere gas species dissolves into the metal faster than it can nucleate as soot on the metal surface. This produces high carbon activities in the metal and, for the case of iron-base alloys, leads to the growth of metastable carbides, which subsequently decompose to a powdery product. These reactions occur typically in the temperature range 450–800 °C.

Much of the work into the dusting attack of alloys has been carried out on nickel- and iron-based stainless systems similar to those used in reforming reactions where the phenomenon is a nuisance.[47–49] The reaction is also dangerous in that it can lead to the perforation of tubes very rapidly, within a few hours, and it is found generally where metal components are in contact with carbonaceous gas mixtures of low oxygen potential, typically $CO-H_2$ mixtures, or atmospheres containing organic gases. Because the carbon activity of the atmosphere exceeds unity and because of the difficulty in nucleating graphite on the metal surface, without a catalyst, it is possible for metastable carbides of the type M_3C and $M_{23}C_6$ to form. Following nucleation of graphite, these metastable carbides decompose to metal and carbon. Thus, the corrosion products of metal dusting attack are powdery mixtures of fine metal powder and carbon. The alloy is quite uniformly attacked in the case of low alloy steels, but involves pitting in the cases of alloy steels and nickel-base alloys. The degradation of Ni-base alloys has been reported to occur by the formation of graphite without the intermediate formation of metastable carbides.[50]

Metal dusting of iron proceeds in five stages.[51,52]

(1) Carbon dissolves in the metal to a degree of supersaturation.
(2) Cementite precipitates at the metal surface and in grain boundaries.
(3) Graphite deposits from the atmosphere on to the cementite.
(4) The cementite begins to decompose forming graphite and metal particles.
(5) Further graphite deposition proceeds, catalyzed by the metal particles.

Similar mechanisms are thought to be valid for low alloy steels.

In the case of high alloy steels that contain high concentrations of chromium, a protective scale of chromium oxides is formed. Where this scale fails locally, the dusting sequence can proceed since carbon can only penetrate the scale at defects. This results in the formation of a pitted surface, beginning with small pits, which widen to larger hemispherical sites with time. It is feasible that, if the chromia scale could be made more secure, the steel would show improved resistance to metal dusting. It has been observed that surface working is effective in improving the protectiveness of the chromia scales.[53,54] It should be recognized, however, that this increased resistance is not stable and that failure would be swift, once the sequence of dusting is begun.

In practice, it is found that the presence of sulphur in the atmosphere in the form of H_2S effectively suppresses metal dusting. This is thought[54–57] to result from the suppression of graphite nucleation by the sulphur, which thereby halts the chain of dusting reactions.

Reactions with nitrogen

There have been a number of studies directed towards the effects of nitrogen on high-temperature corrosion of metals.[9,58–63] Pettit *et al.*[58] oxidized pure chromium in air at 1200 °C and found an external scale of chromia to be formed together with a layer of chromium nitride, Cr_2N, that formed immediately on top of the metal. This structure is consistent if the nitrogen is transported through the chromia layer to form nitride at the metal surface. Perkins[59] observed very similar results for a series of chromium alloys oxidized in air at 1150 °C. Most of the alloys also showed internal nitridation. It was found that alloy surface preparation could influence the rate of transport of nitrogen through the chromia scales.

When iron-, nickel-, and cobalt-base alloys containing chromium were oxidized in air at 900 °C for 25 h[60] only oxide scales were formed and nitride formation was observed only on a Co–35 wt% Cr alloy. Presumably the inward diffusion of nitrogen through the chromia scale occurred in all cases but only the Co–35 wt% Cr alloy has a high enough Cr activity to form the nitride. The presence of nitrogen did not alter the isothermal behaviour of the alloys, but extensive internal precipitation occurred on cyclic oxidation of Ni–25 wt% Cr–6 wt% Al and Co–25 wt% Cr–6 wt% Al alloys. It appears that the inclusion of nitrogen into the metal substrate, with resultant nitride precipitation and, perhaps, embrittlement, may be the main causes of a detrimental effect of nitrogen.

References

1. H. Rau, *J. Phys. Chem. Solids*, **37** (1976), 425.
2. H. Rickert, *Z. Phy. Chem. NF*, **23** (1960), 355.
3. S. Mrowec and H. Rickert, *Z. Phy. Chem. NF*, **28** (1961), 422.
4. S. Mrowec and H. Rickert, *Z. Phy. Chem. NF*, **36** (1963), 22.
5. C. Z. Wagner, *Phys. Chem. B*, **21** (1933), 25.
6. S. Mrowec and T. Werber, *Gas Corrosion of Metals*, Springfield, VA, National Technical Information Service, 1978, p. 383.
7. S. Mrowec and K. Przybylski, *High Temp. Mater.*, **6** (1984), 1.
8. D. J. Young, *Rev. High Temp. Mater.*, **4** (1980), 229.
9. G. Y. Lai, *High Temperature Corrosion of Engineering Alloys*, Materials Park, OH, ASM International, 1990.
10. S. Wegge and H. J. Grabke, *Werkst. u. Korr.*, **43** (1992), 437.
11. W. L. Worrell and H. Kaplan, in *Heterogenous Kinetics at Elevated Temperatures*, eds. W. L. Worrell and G. R. Belton, New York, Plenum Press, 1970, p. 113.
12. K. N. Strafford, The sulfidation of metals and alloys, Metallurgical Review No. 138, *Met. Mater.*, **3** (1969), 409.
13. S. Mrowec, *Oxid. Met.*, **44** (1995), 177.
14. Y. Fueki, Y. Oguri, and T. Mukaibo, *Bull. Chem. Soc. Jpn.*, **41** (1968), 569.
15. S. Mrowec and K. Przybylski, *High Temp. Mater. Process*, **6** (1984), 1.
16. S. Mrowec, *Bull. Acad. Polon. Sci.*, **15** (1967), 517.
17. P. Hancock, *First International Conference on Metallic Corrosion*, London, Butterworth, 1962, p. 193.
18. C. Wagner, *Pittsburgh International Conference on Surface Reactions*, Pittsburgh, PA, Corrosion Publishing Company, 1948, p. 77.
19. C. Wagner, *Atom Movements*, Cleveland, OH, ASM, 1951, p. 153.
20. R. Condit, *Kinetics of High Temperature Processes*, ed. W. D. Kingery, New York, John Wiley, 1959, p. 97.
21. A. Bruckman, *Corr. Sci.*, **7** (1967), 51.
22. M. Cagnet and J. Moreau, *Acta. Met.*, **7** (1959), 427.
23. R. A. Meussner and C. E. Birchenall, *Corrosion*, **13** (1957), 677.
24. S. Mrowec, *Z. Phys. Chem. NF*, **29** (1961), 47.
25. P. Kofstad, P. B. Anderson, and O. J. Krudtaa, *J. Less Comm. Met.*, **3** (1961), 89.
26. T. Rosenqvist, *J. Iron Steel Inst.*, **179** (1954), 37.
27. W. Grosser, D. Meder, W. Auer, and H. Kaesche, *Werkst. u. Korr.*, **43** (1992), 145.
28. M. F. Chen and D. L. Douglass, *Oxid. Met.*, **32** (1989), 185.
29. R. V. Carter, D. L. Douglass, and F. Gesmundo, *Oxid. Met.*, **31** (1989), 341.
30. B. Gleeson, D. L. Douglass, and F. Gesmundo, *Oxid. Met.*, **31** (1989), 209.

31. M. F. Chen, D. L. Douglass, and F. Gesmundo, *Oxid. Met.*, **31** (1989), 237.
32. G. Wang, R. V. Carter, and D. L. Douglass, *Oxid. Met.*, **32** (1989), 273.
33. B. Gleeson, D. L. Douglass, and F. Gesmundo, *Oxid. Met.*, **33** (1990), 425.
34. G. Wang, D. L. Douglass, and F. Gesmundo, *Oxid. Met.*, **35** (1991), 279.
35. G. Wang, D. L. Douglass, and F. Gesmundo, *Oxid Met.*, **35** (1991), 349.
36. C. L. Marshall, *Tin*, New York, Reinhold, 1949.
37. P. L. Daniel and R. A. Rapp, *Advances in Corrosion Science and Technology*, eds. M. G. Fontana and R. W. Staehle, New York, Plenum Press, 1980.
38. C. Ilschner-Gensch and C. Wagner, *J. Electrochem. Soc.*, **105** (1958), 198.
39. S. C. Kuiry, S. K. Roy, and S. K. Bose, *Oxid. Met.*, **46** (1996), 399.
40. C. S. Tedmon, *J. Electrochem Soc.*, **113** (1966), 766.
41. M. J. Maloney and M. J. McNallen, *Met. Trans. B*, **16** (1983), 751.
42. J. M. Oh, M. J. McNallen, G. Y. Lai, and M. F. Rothman, *Met. Trans. A*, **17** (1986), 1087.
43. *Metals Handbook*, 10th edn, Metals Park, OH, ASM International, 1991, vol. 4, p. 542.
44. J. Crank, *Mathematics of Diffusion*, Oxford, UK, Oxford University Press, 1956.
45. H-. J. Grabke, *Härt.-Tech. Mitt.*, **45** (1990), 110.
46. A. Schnaas and H.-J. Grabke, *Oxid. Met.*, **12** (1979), 387.
47. R. F. Hochmann, Catastrophic deterioration of high-temperature alloys in carbonaceous atmospheres. In *Properties of High-Temperature Alloys*, eds. A. Foroulis and F. S. Pettit, Princeton, NJ, Electrochemical Society, 1977, p. 715.
48. H. J. Grabke, *Corrosion*, **51** (1995), 711.
49. G. H. Meier, R. A. Perkins, and W. C. Coons, *Oxid. Met.*, **17** (1982), 235.
50. H. J. Grabke, *Corrosion*, **56** (2000), 801.
51. H. J. Grabke, R. Krajak, and J. C. Nava Paz, *Corrosion Sci.*, **35** (1998), 1141.
52. H. J. Grabke, *Mater. Corr.*, **49** (1998), 303.
53. H. J. Grabke, E. M. Müller-Lorenz, B. Eltester, M. Lucas, and D. Monceau, *Steel Res.*, **68** (1997), 179.
54. H. J. Grabke, E. M. Müller-Lorenz, and S. Strauss, *Oxid. Met.*, **50** (1998), 241.
55. Y. Inokuti, *Trans. Iron Steel Inst. Jpn.*, **15** (1975), 314, 324.
56. H. J. Grabke and E. M. Müller-Lorenz, *Steel Res.*, **66** (1995), 252.
57. A. Schneider, H. Viefhaus, G. Inden, H. J. Grabke, and E. M. Müller-Lorenz, *Mater. Corr.*, **49** (1998), 330.
58. F. S. Pettit, J. A. Goebel, and G. W. Goward, *Corr. Sci.*, **9** (1969), 903.
59. R. A. Perkins, Alloying of chromium to resist nitridation, NASA *Report* NASA-Cr-72892, July, 1971.
60. C. S. Giggins and F. S. Pettit, *Oxid. Met.*, **14** (1980), 363.
61. I. Aydin, H. E. Bühler, and A. Rahmel, *Werkst. u. Korr.*, **31** (1980), 675.
62. U. Jäkel and W. Schwenk, *Werkst. u. Korr.*, **22** (1971), 1.
63. H-. J. Grabke, S. Strauss, and D. Vogel, *Mater. Corr.*, **54** (2003), 895.

7

Reactions of metals in mixed environments

Introduction

Study of the behaviour of pure metals and alloys in single oxidant systems, at high temperature, is vital for developing an understanding of the general mechanisms involved in oxidation and scaling reactions. In almost all cases where alloys are used commercially at high temperature, in gas turbines, in heat exchangers, or structurally in furnaces etc., they are exposed to complex atmospheres that contain a variety of gas species. These situations are complex and demand a thorough understanding of the underlying thermodynamics and kinetic principles to describe the reactions and the mechanisms by which they proceed. In such cases the aim is not simply to explain and understand how the reaction proceeds, but to go a step further and predict how the reactions might be prevented or contained.

In fact, apart from controlled laboratory atmospheres, the gas is always complex in the multi-oxidant sense since even nitrogen in air can form nitrides with some alloy systems in addition to the oxides formed by the oxygen. This is seen particularly in alloys containing metals such as chromium, titanium, and niobium, where the formation of nitrides in air atmospheres interferes with the 'simple' oxidation situation that is observed when using pure oxygen, or oxygen–argon mixtures.[1,2]

Air–H_2O and air–CO_2 atmospheres

The presence of water vapour and carbon dioxide in air and, to a greater extent, in atmospheres derived from the combustion of fossil fuel, has been found to increase the observed reaction rate of steels and other metals noticeably. This was studied and explained by Rahmel and Tobolski[3] who proposed that the gases set up redox systems of H_2–H_2O and CO–CO_2 to transfer oxygen across pores in the oxide structure. When a pore develops in a scale, forming on a metal exposed to air, it is assumed that the total pressure within the pore is the same as that of the external

176

atmosphere. The composition of the atmosphere inside the pore will be decided by the gases that can permeate across the scale, and by some decomposition of the scale. In the case of a metal oxidizing in air, the atmosphere within the pore is expected to be virtually pure nitrogen with a partial pressure of oxygen. The value of the partial pressure of oxygen in the pore will be at, or close to, equilibrium with the oxide at that place in the scale (Equation 7.1). It will be defined by the activity of metal, a_M, as expressed in Equation (7.2),

$$2M\,(s) + O_2\,(g) = 2MO\,(s), \tag{7.1}$$

assuming pure MO:

$$p_{O_2} = \frac{1}{a_M^2}\exp\left(\frac{\Delta G_1^\circ}{RT}\right). \tag{7.2}$$

For the case of iron we have Equation (7.3),

$$2Fe\,(s) + O_2\,(g) = 2FeO\,(s) \quad \Delta G_3^\circ = -528\,352 + 129.33T \text{ J}, \tag{7.3}$$

for which p_{O_2} is given by Equation (7.4):

$$p_{O_2} = \frac{1}{a_{Fe}^2}(1.02 \times 10^{-15}) \text{ atm at 1273 K.} \tag{7.4}$$

Thus, the oxygen partial pressure in pores close to the metal surface, where a_{Fe} is close to unity, will be about 10^{-15} atm, or 10^{-10} Pa. In a growing scale, which has a gradient of oxygen and metal activities across it, oxygen will transfer from the high-oxygen-potential side of the pore to the low-oxygen-potential side of the pore. Such transfer will be very slow due to the low oxygen concentration within the pore. Figure 7.1(a) shows diagrammatically how oxygen is transported across the pore by net decomposition of oxide at the high-oxygen side and net formation of oxide at the low-oxygen side of the pore.

In the presence of water vapour and/or carbon dioxide in the oxidizing atmosphere, the oxygen potentials are not changed but the pore now contains the species H_2, H_2O, CO, and CO_2, from the atmosphere. These species set up H_2–H_2O and CO–CO_2 redox systems in the pore to conform to the oxygen potentials according to Equations (7.5) and (7.6),

$$CO_2\,(g) = CO\,(g) + \frac{1}{2}O_2\,(g) \quad \Delta G_5^\circ = 282\,150 - 86.57T \text{ J};\tag{7.5}$$

$$H_2O\,(g) = H_2\,(g) + \frac{1}{2}O_2\,(g) \quad \Delta G_6^\circ = 245\,993 - 54.84T \text{ J},\tag{7.6}$$

from which we obtain Equation (7.7):

$$p_{O_2} = \left(\frac{p_{CO_2}}{p_{CO}}\right)^2\exp\left(\frac{-2\Delta G_5^\circ}{RT}\right) = \left(\frac{p_{H_2O}}{p_{H_2}}\right)^2\exp\left(\frac{-2\Delta G_6^\circ}{RT}\right). \tag{7.7}$$

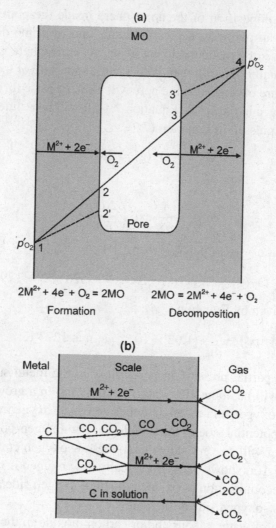

Figure 7.1 (a) Transport of oxygen across a pore in a growing oxide scale. The oxide dissociates at the outer side of the pore releasing oxygen, which migrates across the pore to form oxide at the inner surface. The solid and dashed lines indicate oxygen potential gradients across the scale when transport through the pore is rapid or slow, respectively. (b) Reactions involving CO and CO_2, including transport through the scale and ferrying across pores.

For the system Fe–O, it is instructive to consider the reactions given in Equations (7.8) and (7.9):

$$Fe\,(s) + CO_2\,(g) = FeO\,(s) + CO\,(g)$$
$$\Delta G_8^o = 17\,974 - 21.91T \text{ J}, \tag{7.8}$$
$$Fe\,(s) + H_2O\,(g) = FeO\,(s) + H_2\,(g)$$
$$\Delta G_9^o = -18\,183 + 9.82T \text{ J}. \tag{7.9}$$

Thus the gas compositions in a pore within the oxide scale are given at $1000\,^\circ C$ by Equations (7.10) and (7.11):

$$\frac{p_{CO}}{p_{CO_2}} = a_{Fe} \exp\left(\frac{-\Delta G_8^o}{RT}\right) = 2.56 a_{Fe}, \qquad (7.10)$$

$$\frac{p_{H_2}}{p_{H_2O}} = a_{Fe} \exp\left(\frac{-\Delta G_9^o}{RT}\right) = 1.71 a_{Fe}. \qquad (7.11)$$

Close to the metal surface, where a_{Fe} is about unity, we obtain the values shown in Equation (7.12):

$$\frac{p_{CO}}{p_{CO_2}} = 2.56 \quad \text{and} \quad \frac{p_{H_2}}{p_{H_2O}} = 1.71. \qquad (7.12)$$

If it is assumed that the partial pressures of CO_2 and H_2O in the atmosphere derived from fuel combustion are about 10 kPa, then the partial pressures of CO_2 and H_2O in a pore near the metal surface will be about 2.8 kPa and 3.7 kPa, respectively. These values will be higher in pores that are further from the metal surface, at sites where the activity of iron is lower, eventually tending to 10 kPa at the scale–gas interface.

Thus a pore close to the metal surface is expected to contain the following gas species H_2O (3.7 kPa), H_2 (6.3 kPa), CO_2 (2.8 kPa), CO (7.2 kPa), O_2 (10^{-13} kPa) and N_2 (80 kPa). Clearly, the concentration of O_2 is much less than those of H_2O and CO_2. Consequently, the contribution of the O_2 species to the transfer of oxygen across a pore is negligible compared with the contributions of H_2O and CO_2. Thus, when CO_2 and H_2O are present in the oxidizing atmosphere, higher scaling rates are experienced compared with kinetics measured in air only, for the same pore population. This is illustrated schematically in Figure 7.1(b) for CO_2, and has been demonstrated by Sheasby et al.,[4] who also showed that the transport across pores can only increase the oxidation rate up to the point represented by a compact, pore-free, adherent scale. The effect of H_2O and CO_2 was confined to physical transport across pores and apparently did not involve changes in the defect structure of the scale, which would have shown a change in kinetics for a compact, adherent, scale.

A further problem in the case of a CO–CO_2 atmosphere lies in the effect on the carbon content of the metal. Any CO–CO_2 atmosphere has a carbon potential defined by equilibrium in the reaction shown in Equation (7.13),

$$2CO\,(g) = C\,(s) + CO_2\,(g) \quad \Delta G_{13}^o = -170\,293 + 174.26T \text{ J}, \qquad (7.13)$$

from which a_C is obtained:

$$a_C = \frac{p_{CO}^2}{p_{CO_2}} \exp\left(\frac{-\Delta G_{13}^o}{RT}\right) = \frac{p_{CO}^2}{p_{CO_2}} \left(\frac{2056}{T} - 21.06\right). \qquad (7.14)$$

Thus, in the case of an atmosphere containing carbonaceous gases, a metal may lose or pick up carbon, depending on whether the activity of carbon in the metal is greater or less, respectively, than the value defined by the atmosphere composition. This can be a great nuisance in the case of the decarburization of engineering steels during heat treatment or reheating, as described in Chapter 5. It can also be used with an appropriate atmosphere deliberately to increase the carbon content of the surface of the steel by carburizing.

In the case of water vapour, there exists the possibility of reaction with the metal yielding hydrogen, which can be absorbed by the metal at high temperature, Equation (7.15):

$$M + H_2O\,(g) = MO + 2\underline{H}\,(soln).\qquad(7.15)$$

Carbon dissolving in the metal substrate can also modify the surface properties of the alloy by changing the heat treatment characteristics or causing the carbides of elements such as chromium, titanium, or niobium, etc., to be precipitated. This removes these elements from solution and, in the case of chromium, impairs the oxidation resistance of the alloy. The penetration of oxide scales by CO_2 is an important consideration in these systems.

The behaviour of iron in CO–CO_2 atmospheres was studied by Pettit and Wagner[5] over the temperature range 700–1000 °C, where wustite is stable. They found that the kinetics were controlled by reactions at the scale–gas interface, but carbon pick-up was not observed. In contrast, Surman[6] oxidized iron in CO–CO_2 mixtures in the temperature range 350–600 °C, where wustite was not stable and magnetite exists next to the iron. He observed 'breakaway' oxidation whose onset, after an incubation period, coincided with the deposition of carbon within the scale. This was explained by Gibbs[7] and Rowlands[8] by the penetration of the scale by CO_2 which achieved equilibrium in the scale, dissolved carbon in the metal substrate, and then deposited carbon within the scale, which split open the scale and left the system in a state of rapid 'breakaway' oxidation. The incubation period observed corresponded to the time required to saturate the metal substrate with carbon.

The penetration of carbon through protective scales formed on alloys has been observed[9] in the cases of Fe–15 wt% Cr and Fe–35 wt% Cr alloys. The alloy with the higher chromium content, showing better resistance to carbon penetration, indicating that the Cr_2O_3 scale formed on Fe–35 wt% Cr, is a better barrier than scales containing iron oxides. Nickel–chromium and iron–nickel–chromium alloys are also susceptible to atmospheres rich in carbon monoxide[10–13] between 600 and 1000 °C when formation of internal carbides depletes the alloy of chromium and prevents a protective scale from being maintained. In such atmospheres, the formation of stable protective oxides has been aided by increasing the chromium

(a)

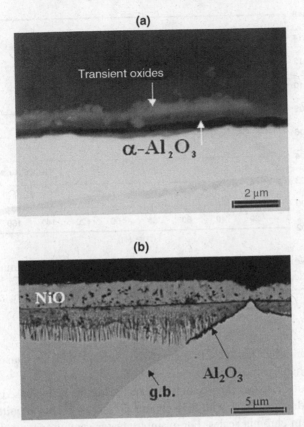

(b)

Figure 7.2 Effect of water vapour on the selective oxidation of Al from a model Ni–8 wt% Cr–6 wt% Al alloy at 1100 °C. (a) Oxidation in dry air, (b) oxidation in air + 10% H₂O; g.b. = grain boundary.

content to 25%,[14] adding silicon and niobium,[15] or by adding 5% of carbon dioxide to the gas phase.[16]

The presence of water vapour has been reported to have a number of additional effects on high-temperature oxidation behaviour, including the following.

- Water vapour may affect the plasticity of oxide scales. Some investigators propose that the plasticity of oxide scales is decreased and, consequently, the spalling of scales is increased.[17] Other investigators have proposed that plasticity of oxide scales is increased,[18] with increased oxidation rates[18] or improved adherence.
- The presence of water vapour adversely affects the selective oxidation of elements such as aluminium and chromium from iron-base[19] and nickel-base[20] alloys. This is illustrated in Figure 7.2 for a Ni–Cr–Al alloy. Exposure in dry air results in a continuous external alumina film while exposure in moist air results in profuse internal oxidation of the Al. It has also been found that water vapour affects the selective oxidation of aluminium from TiAl.[21]

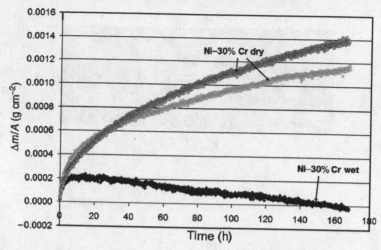

Figure 7.3 Weight change ($\Delta m/A$) versus time data for the isothermal oxidation of Ni–30 wt% Cr at 900 °C in dry air (two runs) and air + 10% H$_2$O (wet). The difference in the two sets of curves is caused by much more rapid oxide volatilization in the moist atmosphere.

- Water vapour can affect transport through oxides such as SiO$_2$ by causing the network structure to be changed.[22]
- Water vapour can cause the vaporization of some oxides to be increased because of the formation of hydrated species with high vapour pressures.[23] This is quite important for alloys which form chromia scales. For example, water-vapour contents in air in excess of 0.1% at 1000 °C cause the partial pressure of CrO$_2$(OH)$_2$ in equilibrium with a Cr$_2$O$_3$ scale to exceed the partial pressure of CrO$_3$, and the partial pressure of CrO$_2$(OH)$_2$ increases with increasing partial pressure of H$_2$O.[24] This means that the reactive evaporation of chromia scales, described in Chapter 4, is more pronounced in atmospheres that contain water vapour.[25–28] This is illustrated in Figure 7.3 which compares the isothermal oxidation rates of a Ni–30 wt% Cr alloy in dry and moist air at 900 °C. The curves for dry air involve small vapour losses of CrO$_3$ but the mass-gain curve mainly reflects chromia growth. The curves in moist air show the results of substantial vapour losses as CrO$_2$(OH)$_2$.
- Water vapour causes the concentration of proton defects to be increased, which can influence defect-dependent properties such as high-temperature creep and diffusion.[29,30]

One of the most serious effects of water vapour on high-temperature oxidation is the increased spalling tendency of Al$_2$O$_3$ and Cr$_2$O$_3$ scales.[20,31] This effect is illustrated for the cyclic oxidation of the alumina-forming superalloy CMSX-4 in Figure 7.4. The water vapour is believed to lower the fracture toughness of the alloy–oxide interface.

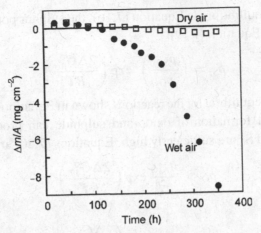

Figure 7.4 Weight change ($\Delta m/A$) versus time data for the cyclic oxidation of CMSX-4 in wet ($p_{H_2O} = 0.1$ atm) and dry air at $1100\,^{\circ}$C.

Metal–sulphur–oxygen systems

Sulphurous gases are present in many commercial atmospheres to which metals are exposed at high temperature. In such cases, the atmospheres contain several oxidants, including sulphur and oxygen. In oxidizing atmospheres, such as the products of combustion of fuels, the sulphur will be present as SO_2 or even SO_3. If the gas has a reducing nature, such as in the reforming of oils, then the sulphur will be present as species such as H_2S and COS, etc. In both types of atmosphere, the simultaneous presence of sulphur and oxygen can lead to the formation of sulphides as well as oxides, seriously compromising the degree of protection afforded to the metal exposed to such atmospheres at high temperatures.

Assume that a metal A, capable of forming oxide A_nO and sulfide A_mS, is exposed to an atmosphere containing sulphur and oxygen, such as air and sulphur dioxide or an inert gas and sulphur dioxide. The basic treatment of this situation is already well established.[1,32] The relevant reactions to consider are those given in Equations (7.16)–(7.18):

$$\frac{1}{2}S_2\,(g) + O_2\,(g) = SO_2\,(g) \quad \Delta G^{\circ}_{16} = -361700 + 76.68T \text{ J}, \quad (7.16)$$

$$nA\,(s) + \frac{1}{2}O_2\,(g) = A_nO\,(s), \quad (7.17)$$

$$mA\,(s) + \frac{1}{2}S_2\,(g) = A_mS\,(s \text{ or } l). \quad (7.18)$$

According to the equilibrium in Equation (7.16), the sulphur potential of the atmosphere is given by Equation (7.19):

$$p_{S_2} = \left(\frac{p_{SO_2}}{p_{O_2}} \right)^2 \exp \left(\frac{2\Delta G_{16}^\circ}{RT} \right). \tag{7.19}$$

By considering the equilibria for the reactions shown in Equations (7.17) and (7.18), it can be seen that formation of oxide and sulphide can proceed if the partial pressures of O_2 and S_2 are sufficiently high, Equations (7.20) and (7.21):

$$p_{O_2} > \frac{a_{A_nO}^2}{a_A^{2n}} \exp \left(\frac{2\Delta G_{17}^\circ}{RT} \right), \tag{7.20}$$

$$p_{S_2} > \frac{a_{A_mS}^2}{a_A^{2m}} \exp \left(\frac{2\Delta G_{18}^\circ}{RT} \right). \tag{7.21}$$

In these equations a_A is the activity of A, and a_{A_nO} and a_{A_mS} are the activities of the oxide and sulphide, respectively. If it is assumed that the oxide and sulphide are immiscible, then the oxide and sulphide activities can be taken to be unity for the A–S–O system. Similarly, at the metal–scale interface the metal activity will be unity but it will fall throughout the scale towards the scale–gas interface.

The value of the metal activity at the scale–gas interface, a_M'', depends on which reaction step is rate determining. The value of a_M'' will be highest if the rate-determining step is a surface reaction or transport through the gas phase to the scale surface. If the reaction rate is determined by diffusion through the scale then the value of a_M'' will be lower and in, or close to, equilibrium with the atmosphere. This was indicated clearly by Rahmel and Gonzales[33] and is shown in Figure 7.5.

From the phase stability diagram of a metal–sulphur–oxygen system, as shown in Figure 7.6, it should be possible to predict which phase will be present in the scale, given the atmosphere composition.

If the atmosphere composition were to lie within a single-phase field and the reaction were to proceed to completion, only a single phase would be produced, either oxide, sulphide, or sulphate. During reaction, however, only the scale–gas interface is in equilibrium with the gas phase. Even so, for atmospheres within single-phase fields, single-phase scales might be expected. The only exception would be for atmospheres lying along boundaries such as the A_nO/A_mS boundary in Figure 7.7, which corresponds to equilibrium in the displacement reaction shown in Equation (7.22):

$$nA_mS + \frac{m}{2}O_2 = mA_nO + \frac{n}{2}S_2. \tag{7.22}$$

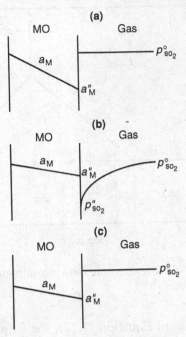

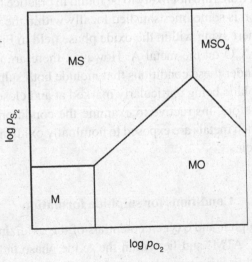

Figure 7.5 Effect of various rate-controlling steps on the activity gradient across the scale and on activities at the scale–gas interface. (a) Diffusion through the scale is rate controlling; (b) Diffusion through the gas phase is rate controlling; (c) Interface reaction is rate limiting. The activity of M at the scale–gas interface increases in the order (a) < (b) < (c).

Figure 7.6 Metal–sulphur–oxygen stability or predominance area diagram.

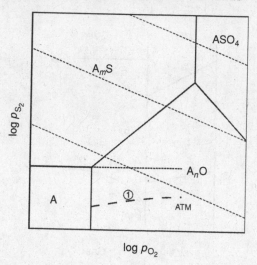

Figure 7.7 The A–S–O stability diagram showing sulphur dioxide isobars and 'reaction path' across scale.

Considering the equilibrium in Equation (7.22), the sulphur and oxygen partial pressures are related by Equation (7.23),

$$\ln p_{S_2} = \frac{m}{n} \ln p_{O_2} - \frac{2}{n} \left(\frac{\Delta G^o_{22}}{RT} \right),$$
(7.23)

and this defines the boundary between the A_nO and A_nS phase fields in Figure 7.7. Such precise gas mixtures are not likely to be found in practice but, as will be shown below, this condition is sometimes satisfied locally within the corrosion product.

Thus an atmosphere lying within the oxide phase field in Figure 7.7 is expected to yield a scale of A_nO on the metal A. However, there are many examples[32,33] of scales growing under these conditions that include both sulphide and oxide, the presence of the sulphide being particularly marked at and close to the metal–scale interface. It is, therefore, instructive to examine the conditions under which sulphides may form when metals are exposed to nominally oxidizing atmospheres that contain sulphur species.

Conditions for sulphide formation

In Figure 7.7, the sulphur and oxygen potentials of the oxidizing gas are indicated at the point marked 'ATM' and lie within the oxide phase field. If a scale forms on metal A, exposed to this atmosphere, and grows at a rate that is controlled by diffusion through the scale, then the scale–gas interface would be in equilibrium with the atmosphere as shown in Figure 7.5(a). The conditions for the scale–gas

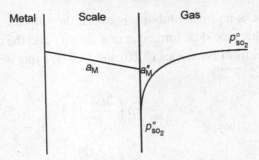

Figure 7.8 Potential gradients in scale and gas where equilibrium with the atmosphere is not established at the scale–gas interface.

interface would correspond to the point marked ATM in Figure 7.7. The dotted line marked 1 would indicate a reaction path across the oxide for this situation and the conditions for the metal–scale interface would correspond to the boundary A/A_nO. This corresponds to Figure 7.5(a) and all of the chemical-potential drop occurs across the scale.

If there is a slow delivery of reactive gas, e.g., to the metal surface, then there will be denudation of SO_2 across a boundary layer in the gas at the metal surface. In this case, the chemical-potential drop is accounted for partly by the fall in SO_2 partial pressure across the boundary layer and partly by the drop in metal activity across the scale, which accounts for the rate of outward transport of ions within the scale. In other words, there are two processes occurring in series, diffusion in the gas and diffusion in the scale. This condition is shown in Figure 7.8, which is an enlargement of Figure 7.5(b).

In all of the situations shown in Figure 7.5 the principal question is where and under what conditions will the sulphide phase form. It is, therefore, important to define the general conditions for sulphide formation. This will occur whenever the local sulphur activity, or chemical potential, is greater than that required to form sulphide with the metal at the activity that obtains at that site. This condition is stated in Equation (7.21).

The sulphur potential in the atmosphere, where the principal source of sulphur is in the form of sulphur dioxide, as in an atmosphere of Ar–SO_2, can be obtained by considering Equation (7.16). The equilibrium constant for such a reaction is given by Equation (7.24),

$$\frac{p_{SO_2}}{p_{O_2} p_{S_2}^{1/2}} = \exp\left(\frac{-\Delta G_{16}^o}{RT}\right), \tag{7.24}$$

from which the sulphur partial pressure is given by Equation (7.19) or (7.25):

$$\ln p_{S_2} = 2 \ln p_{SO_2} - 2 \ln p_{O_2} + \frac{2\Delta G_{16}^o}{RT}. \tag{7.25}$$

Assuming that there is no intersolubility between the sulphide and oxide phases, the oxide and sulphide would be formed at unit activity and the conditions for their formation would be, from Equations (7.20) and (7.21), expressed, respectively, by Equations (7.26) and (7.27):

$$p_{O_2} > \frac{1}{a_A^{2n}} \exp\left(\frac{2\Delta G_{17}^o}{RT}\right), \qquad (7.26)$$

$$p_{S_2} > \frac{1}{a_A^{2m}} \exp\left(\frac{2\Delta G_{18}^o}{RT}\right). \qquad (7.27)$$

Consequently, the condition for sulphide formation at any place is given by combining Equations (7.19) and (7.27) as in Equation (7.28):

$$\left(\frac{p_{SO_2}}{p_{O_2}}\right)^2 \exp\left(\frac{2\Delta G_{16}^o}{RT}\right) > \frac{1}{a_A^{2m}} \exp\left(\frac{2\Delta G_{18}^o}{RT}\right). \qquad (7.28)$$

The oxygen potential at any place within the scale is controlled by the metal activity at that site, as expressed by the identity in Equation (7.26), which may be substituted in Equation (7.28):

$$\left(\frac{p_{SO_2}}{\frac{1}{a_A^{2n}} \exp\left(\frac{2\Delta G_{17}^o}{RT}\right)}\right)^2 \exp\left(\frac{2\Delta G_{16}^o}{RT}\right) > \frac{1}{a_A^{2m}} \exp\left(\frac{2\Delta G_{18}^o}{RT}\right). \qquad (7.29)$$

Rearranging Equation (7.29) indicates that the critical value of sulphur partial pressure at any site is determined by the metal activity at that site as shown in Equation (7.30),

$$p_{SO_2} > \frac{1}{a_A^{2n}} \exp\left(\frac{2\Delta G_{17}^o}{RT}\right) \exp\left(\frac{-\Delta G_{16}^o}{RT}\right) \cdot \frac{1}{a_A^{m}} \exp\left(\frac{\Delta G_{18}^o}{RT}\right), \qquad (7.30)$$

or Equation (7.31),

$$p_{SO_2} > \frac{1}{a_A^{(2n+m)}} \exp\left[\frac{1}{RT}\left(2\Delta G_{17}^o - \Delta G_{16}^o + \Delta G_{18}^o\right)\right]. \qquad (7.31)$$

In using Equation (7.31) it is important to emphasize that its derivation has been obtained assuming the activities of A_mS and A_nO are unity and that the two phases are in equilibrium. Hence, Equation (7.31) is only applicable to the A_mS/A_nO boundary in Figure 7.7. From Equation (7.31) it is seen that the site at which sulphide is most likely to form, for a given value of p_{SO_2}, is at the metal–scale interface where the metal activity is a maximum at unity. This situation is described by

Fe₃O₄ + FeS
duplex

FeO + FeS
duplex

20 µm   Iron

Figure 7.9 Micrograph of duplex iron sulphide–iron oxide scale formed on iron in Ar–1% SO₂ at 900 °C.

Equation (7.32):

$$p_{SO_2} > \exp\left[\frac{1}{RT}(2\Delta G^o_{17} - \Delta G^o_{16} + \Delta G^o_{18})\right]. \tag{7.32}$$

The right hand side of Equation (7.32) corresponds to the triple point A–A$_m$S–A$_n$O in Figure 7.7. Finally, for a given value of p_{SO_2}, the value of metal activity required for sulphide formation is given by Equation (7.33a),

$$a_A^{(2n+m)} > \frac{1}{p_{SO_2}}\exp\left[\frac{1}{RT}(2\Delta G^o_{17} - \Delta G^o_{16} + \Delta G^o_{18})\right], \tag{7.33a}$$

or, alternatively, by Equation (7.33b),

$$a_A > \frac{1}{p_{SO_2}^{1/(2n+m)}}\exp\left[\frac{1}{(2n+m)RT}(2\Delta G^o_{17} - \Delta G^o_{16} + \Delta G^o_{18})\right], \tag{7.33b}$$

where a_A is given by the point at which the SO₂ isobar cuts the A$_m$S–A$_n$O boundary. In the cases of real systems, both duplex scales and predominantly oxide scales have been observed.[34–38] The formation of scales that are duplex mixtures of oxide and sulphide throughout their thickness arises when compound formation is rapid and the supply of reactant from the gas phase is restricted. This situation corresponds to Figure 7.9 for the case of iron reacted with Ar–SO₂ atmospheres at 900 °C.[34] The following mechanism was proposed.

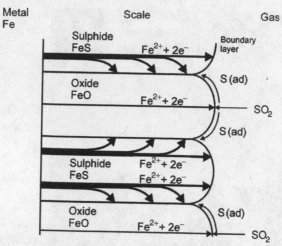

Figure 7.10 Schematic diagram of the mechanism of formation of duplex scale (FeO + FeS) on iron from Ar–SO$_2$ atmospheres, showing how the FeS lamellae provide rapid cation transport and also supply the adjacent FeO lamellae.

Consider the case of an Ar–SO$_2$ atmosphere containing only a small amount of SO$_2$ and whose composition lies in the single-phase FeO field. The inherently rapid reaction quickly denudes a boundary layer of SO$_2$, which must be supplied from the bulk gas by diffusion. With rapid transport properties in the scale, the metal activity is maintained high at the scale–gas interface. The potential gradients are as shown in Figure 7.8. The scale that forms under these conditions has a strong probability of being duplex, with continuous interwoven oxide and sulphide phases, e.g., Figure 7.9. (Note the micrograph in Figure 7.9 was taken under conditions of higher oxygen partial pressure under which Fe$_3$O$_4$ could also form.)

In this situation, the condition for formation of FeO is met but that for formation of FeS directly from the gas phase is not. Consequently, when SO$_2$ impinges on the FeO phase more FeO is formed releasing sulphur. The sulphur is released at an activity that is defined by the sulphur dioxide partial pressure and the iron activity that exist at the FeO–gas interface. The sulphur released causes the nucleation of sulphides adjacent to the FeO nuclei. As more sulphur is released from FeO formation it diffuses to the neighbouring sulphide and, if its activity is high enough, reacts to form more sulphide. The FeO phase does not have sufficiently rapid diffusion for iron ions to sustain the iron activity at the FeO–gas interface at a value high enough to support FeS formation in this way for very long. However, the neighbouring FeS phase is capable of transporting iron ions very rapidly and feeds the FeO phase, thus maintaining a high iron activity at the FeO–gas interface, which is sufficient to promote the formation of sulphide by this mechanism, as indicated in Figure 7.10.

According to Figure 7.8, the sulphur dioxide potential is reduced to p''_{SO_2} at the FeO–gas interface. Thus, according to Equation (7.33b), the iron activity that must be maintained at the FeO–gas interface is given by Equation (7.34):

$$a''_{Fe} = \frac{1}{p''^{1/(2n+m)}_{SO_2}} \exp \left[\frac{1}{(2n+m)RT} (2\Delta G^o_{17} - \Delta G^o_{16} + \Delta G^o_{18}) \right]. \quad (7.34)$$

The data that are available for the Fe–O–S system are shown in Equations (7.35) and (7.36):

$$Fe + \frac{1}{2}O_2 = FeO \quad \Delta G^o_{35} = -264\,176 + 64.7T \text{ J}, \quad (7.35)$$

$$Fe + \frac{1}{2}S_2 = FeS \quad \Delta G^o_{36} = -150\,100 + 51.5T \text{ J}. \quad (7.36)$$

For example, at 800 °C, $\Delta G^o_{35} = -194753$ J and replaces ΔG^o_{17}, and $\Delta G^o_{36} = -94841$ J and replaces ΔG^o_{18}. Also, $\Delta G^o_{16} = -279100$ J. The stoichiometric factors n and m are both unity. Thus, we have an activity expressed by Equation (7.37):

$$a_{Fe} > \frac{1}{p''^{1/3}_{SO_2}} \exp \left[\frac{-204\,925}{3RT} \right]. \quad (7.37)$$

For an experiment using Ar–1% SO_2 the value of p_{SO_2} in the bulk atmospheres is 0.01 atm and will be lower at the scale–gas interface. However, so long as the iron activity is held at a value greater than 2×10^{-3} the possibility exists for the formation of FeO and FeS at the scale–gas interface. Such a structure is shown in Figure 7.9,[34] which corresponds to the diagrammatic Figure 7.10.

The development of a microstructure such as that in Figure 7.9 will now be described in more detail. Consider the exposure of Fe in a gas corresponding to the composition marked 'X' in Figure 7.11. Initially Fe_3O_4 will form. However, the reaction will denude the atmosphere of oxygen in the region of the specimen surface resulting in a local change of atmosphere composition there to lower oxygen partial pressures. In atmospheres containing sulphur dioxide, this results in a local increase in sulphur partial pressure since the sulphur, oxygen, and sulphur dioxide partial pressures are related by the equilibrium constant given in Equation (7.16). Thus the gas composition at the scale–gas interface (or the metal surface in the early stages) follows the reaction path XYZ in Figure 7.11, entering the FeO field at point Y. At this stage the Fe_3O_4, formed earlier, is reduced to FeO and further reaction proceeds as described above with FeO forming and releasing sulphur into the boundary layer where the composition at the surface changes along the path YZ. On reaching point Z, both FeO and FeS may form together as shown in Figure 7.10. Since SO_2 is being consumed, the reaction path must cross the SO_2 isobars which are shown as dashed diagonal lines in Figure 7.11. This represents the establishment of a SO_2

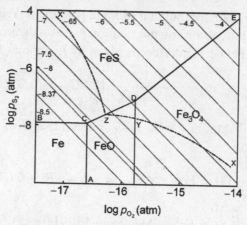

Figure 7.11 The iron–sulphur–oxygen stability diagram at 900 °C showing reaction paths for duplex scale formation. The reaction path starting at X corresponds to the case for which Fe_3O_4 is stable at the bulk gas composition and the one starting at X' corresponds to a bulk gas in which FeS is stable.

partial pressure gradient across the boundary layer in the gas similar to that shown in Figure 7.8.

A similar argument can be made for the formation of duplex scales from atmospheres with high sulphur potentials using the CO–CO_2–SO_2 system[33,35,36] (reaction path $X'–Z$).

As the scale grows, the iron activity at the scale–gas interface must fall and eventually become too low to support sulphide formation. This corresponds to the surface concentration moving from Z in Figure 7.11 up along the FeO–FeS boundary. When it passes point D, Fe_3O_4 becomes the stable oxide and an intermixed FeS–Fe_3O_4 scale begins to grow. This is also observed in Figure 7.9. Eventually the scale becomes so thick that diffusion in the solid begins to control and the surface activities return to point X in Figure 7.11. When this condition arises, oxide only will form giving a scale with an inner duplex zone and an outer oxide zone. This sequence is shown in Figure 7.12, which also shows the metal activity gradient across the scale, and matches the structures to the position on the corresponding graph of weight gain against time. The behaviour of the metal activity in the scale at the outer boundary of the duplex zone is of interest.

Initially, from t_0 the duplex scale grows at a constant rate until time t_2, when the activity of the metal at the scale surface has fallen to the value given by identity in Equation (7.33b). After this point, only oxide is stable and can form on the scale surface. After time t_3, the duplex scale is covered with an outer layer of oxide only, which slows down the reaction rate. Consequently, the metal activity at the duplex-layer–oxide-layer interface rises, reflecting the lower rate of transfer of iron ions

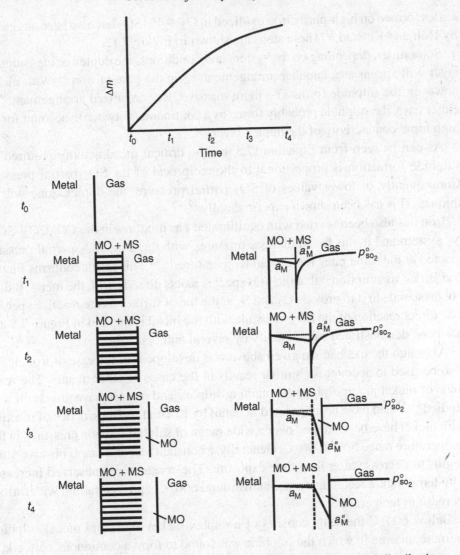

Figure 7.12 Corresponding rates of reaction, structure, and potential distribution for formation of a duplex scale in SO_2-containing atmospheres, eventually forming a surface layer of oxide.

through the scale. This process continues until, at t_4, the metal activity just within the oxide layer has increased and exceeded the critical value for sulphide formation by inward diffusing SO_2, given in Equation (7.33b). Consequently, the formation of sulphide within the oxide layer can be seen as islands of sulphide. It is quite possible that these form an interconnecting network of sulphide within the oxide layer. The formation of a three-dimensional, interconnected sulphide network in scales formed on high-purity iron oxidized in $O_2 + 4\%$ SO_2 has also been suggested by Holt and Kofstad.[37] These steps are shown in Figure 7.12.

Sometimes, depending on the system and conditions, the duplex oxide–sulphide layer will appear as a lamellar arrangement, as in the case of iron.[34] With nickel, however, the sulphide forms in a more massive, less organized arrangement.[38] In either case, the sulphide probably forms as a continuous network to account for the high ionic conductivity of the duplex layer.

As can be seen from Equation (7.33b), the critical metal activity required for sulphide formation is proportional to the reciprocal of the SO_2 partial pressure. Consequently, at lower values of SO_2 partial pressure, the duplex zone will be thinner. This has been shown experimentally.[39,40]

Iron has also been reacted with equilibrated gas mixtures in the $CO–CO_2–SO_2–N_2$ system and it was found that gas mixtures with the same SO_2 partial pressure reacted at the same rates for a given temperature.[41,42] This result confirms Flatley and Birks' assumption[34] that the SO_2 species reacts directly with the metal; it does not dissociate first to provide O_2 and S_2 at the metal surface as the reactive species. The direct reaction of the SO_2 molecule with the metal is implied in Figure 7.8 and has been demonstrated subsequently by several authors.[43–47]

Although the mechanism given above was developed for the case of iron, it can also be used to account for similar results in the cases of other metals. The reactions of nickel in atmospheres containing sulphur and oxygen have also been well studied[48–51] and have been reported in detail by Kofstad.[53] The kinetics of reaction with nickel have been studied over a wide range of sulphur dioxide pressures in the temperature range 500–1000 °C. Generally, a constant reaction rate is observed that begins to decrease after long time exposure. The constant rate observed increases with temperature reaching a maximum around 600 °C, and then falling with further increase in temperature.

Below 600 °C the scale consists of a duplex nickel oxide and nickel sulphide intimate mixture in which the sulphide was found to form a continuous network.[42] As in the case for iron, the continuous nickel sulphide provides a pathway for very rapid nickel-ion diffusion to the scale–gas interface. This results in a high nickel activity there and promotes the duplex formation as described previously. Such a scale is shown in Figure 7.13.

Since the scale grows predominantly by outward transport of nickel, a system of pores and eventually voids form at the metal–scale interface, resulting in the gradual reduction in the reaction rate with time.[51] It has even been observed for this process to proceed to the point where the scale detaches over one side of the specimen.[51]

Above 600 °C, the formation of a eutectic liquid in the Ni–S–O system at the metal–scale interface gives rise to grain-boundary penetration of the metal, as shown

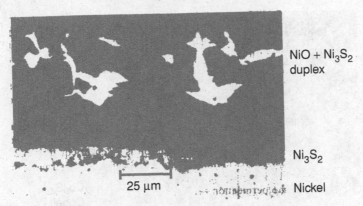

Figure 7.13 Optical micrograph of a duplex scale formed on nickel heated to 500 °C in Ar–1% SO_2.

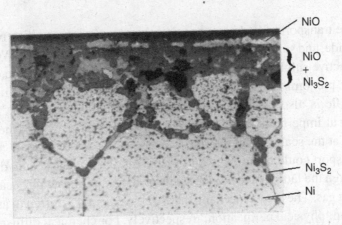

Figure 7.14 Optical micrograph showing an example of grain-boundary penetration by eutectic liquid in the Ni–S–O system following penetration of a preformed oxide layer by SO_2 at 1000 °C.

in Figure 7.14, and the extrusion of eutectic liquid through the scale to the surface, where it begins to convert to oxide.[37,50,51]

The reaction of cobalt with these atmospheres proceeds in a similar manner, producing a fine duplex oxide–sulphide scale as shown in Figure 7.15.[53]

Although iron, nickel, and cobalt behave similarly in Ar–SO_2 atmospheres and they all grow duplex scales, the sulphide and oxide phases are arranged quite differently in the three cases, as can be seen from Figures 7.9, 7.13, and 7.15. The reason for this is not understood and, to the authors' knowledge, has not been addressed.

From the studies of these simple systems, it is also evident that sulphur or sulphur dioxide is capable of transporting across growing oxide scales. Evidence

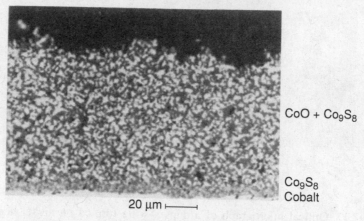

CoO + Co$_9$S$_8$

Co$_9$S$_8$
Cobalt

20 μm ⊢————⊣

Figure 7.15 Optical micrograph showing the fine duplex scale formed on cobalt at 840 °C in Ar–2% SO$_2$.

exists for the transport of hydrogen or water vapour, nitrogen, carbon monoxide or carbon dioxide, and sulphur dioxide across growing scales. Gilewicz-Wolter,[45,54,56] using radioactive sulphur in sulphur dioxide has confirmed in the case of iron that, although the predominant scale growth mechanism is by outward diffusion of iron ions, the scale is also penetrated by sulphur (presumably as SO$_2$). This occurs along physical imperfections in the scale to form sulphide within the scale and, particularly, at the scale–metal interface.

Such transport could occur by two mechanisms. The species could dissolve and diffuse through the lattice of the oxide scale, or it may permeate physically through the scale as a gas. These two mechanisms are conveniently referred to as 'chemical diffusion' and 'physical permeation,' respectively. For chemical diffusion to work, not only must the second oxidant have appreciable solubility in the scale, but its chemical potential in the gas phase must be higher than at the metal–scale interface. For such penetration to result in sulphide formation, the chemical potential in the gas phase must be higher than that required to form sulphide at the metal–scale interface. This condition is given by any atmosphere in the crosshatched area of Figure 7.16. For physical permeation of, say SO$_2$, to result in sulphide formation at the metal–scale interface, the required SO$_2$ partial pressure is given by Equation (7.30). This critical SO$_2$ content of the gas is indicated in Figure 7.16 as the SO$_2$ isobar that passes through the point where the A, AS, and AO phases are in equilibrium. This condition is represented by both the single-hatched and the double-hatched areas in Figure 7.16.

It has been reported[57] that sulphur can diffuse through NiO and CoO. The solubility of sulphur has been measured to be about 0.01%.[58] If this is the case, lattice diffusion should be possible. The permeation of pre-formed NiO and CoO scales has been established[59,60] for all compositions of the hatched areas of Figure 7.16.

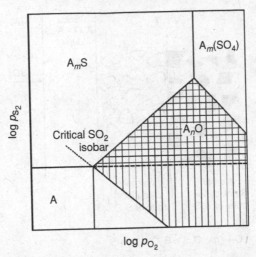

Figure 7.16 An A–S–O stability diagram showing the critical SO_2 isobar and area from which chemical diffusion of S through a scale may proceed (double-hatched) and area from which physical permeation of SO_2 through a scale may proceed (single- and double-hatched) to form sulphide.

Atmospheres corresponding to the areas of Figure 7.16 below the critical SO_2 isobar did not result in sulphide formation at the scale–metal interface. Lobnig, Grabke and coworkers,[61,62] have shown that, in the case of Cr_2O_3, there is no sulphur dissolution in the bulk material, but that the surfaces of pores and cracks in the oxide are lined with adsorbed sulphur. Grabke and coworkers have also reported that tracer studies showed no solubility of S (or C) in oxides such as NiO, FeO, and Fe_3O_4. This implies that the measurements of low values of sulphur solubility in oxides might, at least partially, be accounted for by sulphur adsorbed on internal surfaces of the oxide. This is significant because the determinations quoted above[58] were carried out using oxide powders.

These results confirm that physical permeation of SO_2 gas can definitely occur. The chemical diffusion of sulphur under appropriate conditions cannot be ruled out, but may be unlikely because of the limited solubility of sulphur in most oxides.[61–65] The critical values of SO_2 potentials are generally so low that it is unlikely that gas cleaning can be used to prevent sulphide formation absolutely.

Attack of alloys in complex atmospheres

Consider metal B to be the alloying addition to base-metal A. The additional reactions to be considered are those given in Equations (7.38) and (7.39):

$$pB + \frac{1}{2}O_2 = B_pO, \tag{7.38}$$

$$qB + \frac{1}{2}S_2 = B_qS. \tag{7.39}$$

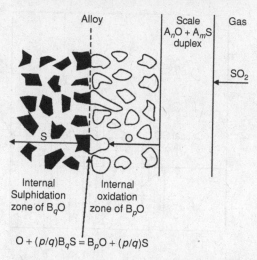

$$O + (p/q)B_qS = B_pO + (p/q)S$$

Figure 7.17 Oxygen diffuses through the internal oxide zone and reacts with sulphides in the front of the internal sulphide zone to form oxide and release sulphur that diffuses deeper into the alloy to form new sulphide.

The metal B is assumed to form more stable compounds than metal A, the oxide of a metal is assumed to be more stable than the corresponding sulphide, and intersolubiity between oxides and sulphides is disregarded.

When the concentration of B is so low that a protective scale of B_pO cannot form, a zone of internal oxidation of B_pO particles in a matrix of A will form. The surface of the alloy, effectively pure A, can now react with the complex atmosphere to form a scale of either A_nO or duplex A_nO and A_mS. Where a duplex scale is formed, the metal–scale interface will be at equilibrium with $A + A_nO + A_mS$; sulphur will dissolve in the metal and diffuse inwards through the internal oxidation zone to form internal B_qS particles. This forms as a second, inner, sulphide-based internal zone of precipitation below the outer internal oxidation zone. Since B_pO is assumed to be substantially more stable than B_qS, sulphide formation is not expected to be seen in the outer internal oxidation zone. As oxygen continues to diffuse inwards it will react with the internal sulphide particles, forming oxide and releasing sulphur to diffuse further into the metal. This is shown in Figure 7.17. Thus, once the internal sulphide zone is established, it can be driven into the alloy by this cascading mechanism, effectively removing the metal B from solution in the alloy.

In the case where B is present at concentrations sufficiently high to form a protective scale of B_pO, low reaction rates would be expected until the scale is penetrated by sulphur, as discussed above. If penetration occurs through a scale of B_pO, which is assumed to be the most stable oxide, then the sulphur species will arrive at a metal–scale interface of low oxygen potential, maybe sufficiently low to generate a sulphur potential high enough to sulphidize A or B from the alloy, or both.

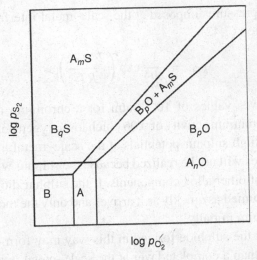

Figure 7.18 Overlay of A–S–O and B–S–O stability diagrams indicating the conditions under which B_pO and A_mS would both be stable.

If the atmosphere has a low oxygen potential which would support formation of B_pO but not A_nO, but has a sulphur potential high enough to form A_mS, then it is possible for a layer of B_pO to form on the alloy. Such a scale may offer temporary protection, but over time sulphur could penetrate the scale to form sulphides at the metal–scale interface. Alloy components could also migrate through the B_pO scale to form sulphides at the scale–gas interface. The sulfide B_qS may or may not form at the scale–gas interface depending on the value of the activity of B at the scale outer surface.

Figure 7.18 shows a simple overlay of the A–S–O and B–S–O stability diagrams. This ignores any solubilities that might occur but indicates clearly the ranges of oxygen and sulphur partial pressures of atmospheres that could support the formation of B_pO and A_mS on the scale surface.

Alloys that are exposed to aggressive atmospheres at high temperatures are usually designed to be 'heat resistant' and one of the most serious problems is how exposure to a 'complex' gas can cause the protective behaviour of scales to change. Usually, such alloys rely on scales based on chromia or alumina formation. Consequently, they rely on sufficiently high concentrations of chromium and aluminium being available in the alloy. This can be disrupted if internal sulphidation removes the chromium or aluminium from solution in the matrix.

Because the oxides chromia and alumina are very stable, see Equations (7.40) and (7.41),

$$\frac{2}{3}Cr + \frac{1}{2}O_2 = \frac{1}{3}Cr_2O_3 \quad \Delta G^{\circ}_{40} = -388\,477 + 74.02T \text{ J}; \qquad (7.40)$$

$$\frac{2}{3}Al + \frac{1}{2}O_2 = \frac{1}{3}Al_2O_3 \quad \Delta G^{\circ}_{41} = -563\,146 + 110.56T \text{ J}, \qquad (7.41)$$

the oxygen partial pressures imposed at the scale–metal interface will be low and are given by Equation (7.42):

$$p_{O_2} = \frac{1}{a_M^{4/3}} \exp\left(\frac{2\Delta G^\circ}{RT}\right). \tag{7.42}$$

At 1100 K, this gives values of 10^{-29} atm for a chromium activity of 0.2 and 10^{-41} atm for an aluminium activity of 0.06. Such low oxygen potentials are capable of generating very high sulphur potentials at the scale–metal interface. Of course equilibrium pressures will not be realized because the sulphur will be consumed by forming sulphides of other alloy components. If the sulphur dioxide penetration is slow, it may be consumed as quickly as it arrives and only the most stable sulphides could be found to form initially.

It is possible that the sulphide formed in this way may form a zone of internal sulphidation rather than a complete layer at the scale–metal interface. This would remove the chromium and aluminium from solution in the matrix and leave the system open for the formation of rapidly growing sulphides of metals such as iron, nickel, or cobalt.

Of course, if the protective scale of chromia or alumina is not penetrated by SO_2, sulphide cannot form at the scale–metal interface. This was found[9] for Ni–20 wt% Cr, Co–35 wt% Cr and Fe–35 wt% Cr alloys exposed to pure SO_2 at 900 °C and emphasizes the resistance of a chromia scale to permeation. On the other hand, alloys in the Fe–Cr–Al, Ni–Cr–Al and Co–Cr–Al systems were exposed to atmospheres in the H_2–H_2S–H_2O system. These atmospheres had compositions that supported the formation of chromia or alumina together with the sulphides of Fe, Ni and Co at the scale–metal interface.[9,66] In these cases, a protective layer of chromia or alumina that formed initially was penetrated by sulphur to form iron, nickel, and cobalt sulphides at the scale–metal interface. Furthermore, iron, nickel, and cobalt ions apparently diffused through the oxide layer to form their sulphides on the outside of the protective scale. Thus the original protective scale was sandwiched between base-metal sulphides.

It was established[40,66–69] that pre-oxidation in a sulphur-free environment does not completely prevent sulphur penetration on subsequent exposure to an atmosphere containing sulphur. This is evident from the edges and corners of the structure, confirming that mechanical factors play a role in the breakdown. Xu et al.[67] found that such penetration of a protective oxide scale by sulphur resulted in its eventual breakdown and the onset of rapid reaction kinetics. These mechanisms appear to be responsible for the deterioration of alloys that form chromia protective layers and that are used in atmospheres of high sulphur potential.[70]

In fact, the permeation of protective scales is not limited to sulphurous species. This has been demonstrated by Zheng and Young,[43] who exposed Fe–28 wt% Cr,

Ni–28 wt% Cr and Co–28 wt% Cr alloys to $CO–CO_2–N_2$ atmospheres at 900 °C. Carbon and nitrogen were found to permeate the Cr_2O_3 scales formed and this was ascribed to molecular transport of gas species through the scales.

The rate at which the protective scale can form is also important as shown by Khanna *et al.*[71] They exposed cast, forged, and single-crystal specimens of a Ni–Cr–Al alloy in air + 1% SO_2 at 1000 °C and found that, when the oxide forms slowly, as on large-grained specimens, there is time for the SO_2 to form sulphides with the alloy constituents, and to prevent a fully protective scale from forming. The smaller-grain-size alloys, having a larger grain-boundary diffusion component, develop the protective scale more rapidly. The stability of Cr_2O_3 layers formed on Fe–25 wt% Cr alloys was found[72,73] to be improved by the addition of 4 wt% Ti and 7 wt% Nb. This promoted the formation of $Cr_2Ti_2O_2$ and $Nb_{0.6}Cr_{0.4}O_2$, which suppressed sulphide formation and delayed breakaway to longer times. Permeation of alumina scales by sulphur species to form sulphides as an inner zone at the scale–metal interface has also been observed.[74]

The above discussion has addressed primarily the behaviour of sulphur dioxide as the source of the two oxidants sulphur and oxygen. The thermodynamic and kinetic principles used are valid for all atmospheres that contain multiple oxidants. Zheng and Rapp[76] have studied the attack of Fe–Cr and Ni–Cr alloys in atmospheres of $H_2–H_2O–HCl$ at 800 °C. They found that, in atmospheres of low-HCl and high-H_2O contents, where a stable Cr_2O_3 scale could form, the alloys showed a resistance to corrosion, which increased with the chromium content. As the HCl was increased, the alloys became less resistant and eventually evaporative losses became dominant.

Many atmospheres that contain SO_2 are derived from the combustion of sulphur-bearing fuels and, therefore, contain several percent of 'excess' oxygen to ensure full combustion. Several workers have addressed the situation of the oxidation of metals in $SO_2–O_2$ atmospheres.[9,48,76–78] In this case the possibility of forming sulphur trioxide must be taken into account:

$$SO_2(g) + \frac{1}{2}O_2(g) = SO_3(g) \quad \Delta G^{o}_{43} = -99\,000 + 93.95T \text{ J.} \quad (7.43)$$

The presence of oxygen promotes the formation of SO_3, but the reaction is slow and needs a catalyst.[76] The higher oxygen potential and, thereby, higher sulphur trioxide potential also increases the possibility of forming the metal sulphate, as can be seen readily from the stability diagrams. A second effect of having oxygen in the atmosphere is that it holds the sulphur potential to very low values in the atmosphere, according to the equilibrium in Equation (7.19).

When using $SO_2–O_2$ atmospheres, a catalyst is almost essential in order to obtain reproducible results.[76] In the cases of copper[79] and nickel[80] an abrupt increase in reaction rate is seen when the atmosphere composition is such as to support formation of the sulphate phase, i.e., when the SO_3 partial pressure exceeds that at which sulphate can form. Under these conditions, with nickel, a mixture of NiO and Ni_3S_2 forms, whereas only NiO forms below this critical value.

The reaction rate of nickel in SO_2–O_2 atmospheres was found to be constant and thought to be controlled by the adsorption of SO_3 on to the scale surface from the gas.[80] The adsorbed SO_3 then forms $NiSO_4$ with the NiO of the scale. The final step in the process is proposed to be the reaction of the $NiSO_4$ with Ni ions and electrons to form NiO and Ni_3S_2. This mechanism has the doubtful sequence of forming a compound that is then immediately consumed. The assumption of $NiSO_4$ as an intermediate product has the advantage of offering an explanation of the increase in kinetics at SO_3 partial pressures above the value required for sulphate formation. The detailed mechanism remains unresolved, however. Cobalt has also been found to behave similarly to nickel in SO_2–O_2 atmospheres.[81]

In all of the work quoted above, the salient feature is that, however an initial protective layer is formed, eventually the second oxidizing species involving sulphur, carbon, nitrogen, or chlorine will penetrate the oxide layer and initiate a breakaway situation. The only thing to be determined is how long it will take. This will vary depending on the alloy, atmosphere, and temperature, especially temperature cycling. Certainly pre-oxidation is not a cure, although it may buy a little time.

It is generally thought that sulphide formation is always deleterious for high-temperature situations. While this is generally the case, one exception is the behaviour of Fe–35 wt% Ni–20 wt% Cr at 540 °C in a coal gasifier (syngas) atmosphere. This alloy was used in a heat exchanger to cool the syngas, which can start out at 1200 °C. The syngas is usually described as 'reducing' and contains some H_2O, CO_2, HCl, and H_2S impurities. Somewhat surprisingly, it was found[82] that, in atmospheres with high-sulphur and low-oxygen potentials, although Cr_2O_3 did not form, somewhat protective behaviour was shown because of the formation of a scale of $FeCr_2S_4$ with internal oxidation and sulphidation. More rapid reaction was found in atmospheres of lower sulphur potential when an external scale of Fe(NiCr)S was formed. This is a case when a sulphide layer can be somewhat protective.

References

1. F. S. Pettit, J. A. Goebel, and G. W. Goward, *Corr. Sci.*, **9** (1969), 903.
2. R. A. Perkins and S. J. Vonk, Materials problems in fluidized bed combustion systems, EPRI-FP-1280, Palo Alto CA, Electric Power Research Institute, 1979.
3. A. Rahmel and J. Tobolski, *Werkst. u. Korr.*, **16** (1965), 662.
4. J. Sheasby, W. E. Boggs, and E. T. Turkdogan, *Met. Sci.*, **18** (1984), 127.
5. F. S. Pettit and J. B. Wagner, *Acta met.*, **12** (1964), 35.

6. P. L. Surman, *Corr. Sci.*, **13** (1973), 825.
7. G. B. Gibbs, *Oxid. Met.*, **7** (1973), 173.
8. P. C. Rowlands, in *Metal–Slag–Gas Reactions and Processes*, eds. Z. A. Foroulis and W. W. Smeltzer, Toronto, Electrochemical Society, 1975, p. 409.
9. C. S. Giggins and F. S. Pettit, *Oxid. Met.*, **14** (1980), 363.
10. F. A. Prange, *Corrosion*, **15** (1959), 619.
11. F. Eberle and R. D. Wylie, *Corrosion*, **15** (1959), 622.
12. W. B. Hoyt and R. H. Caughey, *Corrosion*, **15** (1959), 627.
13. P. A. Lefrancois and W. B. Hoyt, *Corrosion*, **19** (1963), 360.
14. H. Lewis, *Br. Corr. J.*, **3** (1968), 166.
15. B. E. Hopkinson and G. R. Copson, *Corrosion*, **16** (1960), 608.
16. R. A. Perkins and S. J. Vonk, Corrosion chemistry in low oxygen activity atmospheres. Annual report, EPRI Report FP1280 Palo Alto, CA, Electric Power Research Institute.
17. R. L. McCarron and. J. W. Shulz, The effects of water vapour on the oxidation behavior of some heat resistant alloys. Proceedings *Symposium on High Temperature Gas–Metal Reactions in Mixed Environments*, New York, AIME, 1973, p. 360.
18. C. W. Tuck, M. Odgers, and K. Sachs, *Corr. Sci.*, **9** (1969), 271.
19. I. Kvernes, M. Oliveira, and P. Kofstad, *Corr. Sci.*, **17** (1977), 237.
20. M. C. Maris-Sida, G. H. Meier, and F. S. Pettit, *Metall. Mater. Trans.*, **34A** (2003), 2609.
21. R. Kremer and W. Auer, *Mater. Corr.*, **48** (1997), 35.
22. E. A. Irene, *J. Electrochem. Soc.*, **121** (1974), 1613.
23. J. F. Cullinan, 'The oxidation of carbon–carbon composites between 300 °C and 900 °C in oxygen and oxygen/water vapor atmospheres', M.S. Thesis, University of Pittsburgh, Pittsburgh, PA, 1989.
24. K. Hilpert, D. Das, M. Miller, D. H. Peck, and R. Weiss, *J. Electrochem. Soc.*, **143** (1996), 3642.
25. H. Asteman, J.-E. Svensson, M. Norell, and L.-G. Johansson, *Oxid. Met.*, **54** (2000), 11.
26. H. Asteman, J.-E. Svensson, and L.-G. Johansson, *Corr. Sci.*, **44** (2002), 2635.
27. H. Asteman, J.-E. Svensson, and L.-G. Johansson, *Oxid. Met.*, **57** (2002), 193.
28. J. E. Segerdahl, J.-E. Svensson, and L.-G. Johansson, *Mater. Corr.*, **53** (2002), 247.
29. A. J. Sedriks, *Corrosion of Stainless Steel*, 2nd edn, New York, NY, John Wiley and Sons, Inc., 1996.
30. P. Kofstad, in *Microscopy of Oxidation*, eds. M. J. Bennett and G. W. Lorimer, London, The Institute of Metals, 1991, p. 2.
31. R. Janakiraman, G. H. Meier, and F. S. Pettit, *Metall. Mater. Trans.*, **30A** (1999), 2905.
32. N. Birks, *High Temperature Gas-Metal Reactions in Mixed Environment*, ed. S. A. Jansson and Z. A. Foroulis, New York, NY, AIME, 1973, p. 322.
33. A. Rahmel and J. A. Gonzales, *Corr. Sci.*, **13** (1973), 433.
34. T. Flatley and N. Birks, *J. Iron Steel Inst.*, **209** (1971), 523.
35. A. Rahmel, *Werkst. u. Korr.*, **23** (1972), 272.
36. A. Rahmel, *Oxid. Met.*, **9** (1975), 401.
37. A. Holt and P. Kofstad, *Mater. Sci. Eng.*, **A120** (1989), 101.
38. M. R. Wootton and N. Birks, *Corr. Sci.*, **12** (1972), 829.
39. F. Gesmundo, D. J. Young, and S. K. Roy, *High Temp. Mater. Proc.*, **8** (1989), 149.
40. D. J. Young and S. Watson, *Oxid. Met.*, **44** (1995), 239.
41. G. McAdam and D. J. Young, *Oxid. Met.*, **37** (1992), 281.

42. G. McAdam and D. J. Young, *Oxid. Met.*, **37** (1992), 301.
43. X. G. Zheng and D. J. Young, *Oxid. Met.*, **42** (1994), 163.
44. W. J. Quadakkers, A. S. Khanna, H. Schuster, and H. Nickel, *Mater. Sci., Eng.*, **A120** (1989), 117.
45. J. Gilewicz-Wolter, *Oxid. Met.*, **46** (1996), 129.
46. J. Gilewicz-Wolter and Z. Zurek, *Oxid. Met.*, **45** (1996), 469.
47. B. Gillot and M. Radid, *Oxid. Met.*, **33** (1990), 279.
48. K. L. Luthra and W. L. Worrell, *Met. Trans.*, **9A** (1978), 1055.
49. K. L. Luthra and W. L. Worrell, *Met. Trans.*, **A10** (1979), 621.
50. P. Kofstad and G. Akesson, *Oxid. Met.*, **12** (1978), 503.
51. M. Seiersten and P. Kofstad, *Corr. Sci.*, **22** (1982), 487.
52. P. Kofstad, *High Temperature Corrosion*, New York, NY, Elsevier (1988).
53. P. Singh and N. Birks, *Oxid. Met.*, **12** (1978), 23.
54. J. Gilewicz-Woer, *Oxid. Met.*, **11** (1977), 81.
55. J. Gilewicz-Woer, *Oxid. Met.*, **29** (1988), 225.
56. J. Gilewicz-Woer, *Oxid. Met.*, **34** (1990), 151.
57. R. H. Chang, W. Stewart, and J. B. Wagner, Proceedings of the 7th International Conference on Reactivity of Solids, Bristol, July 1972, p. 231.
58. M. C. Pope and N. Birks, *Oxid. Met.*, **12** (1978), 173.
59. M. C. Pope and N. Birks, *Oxid. Met.*, **12** (1978), 191.
60. P. Singh and N. Birks, *Oxid. Met.*, **19** (1983), 37.
61. R. E. Lobnig, H. J. Grabke, H. P. Schmidt, and K. Henessen, *Oxid. Met.*, **39** (1993), 353.
62. R. E. Lobnig, H. P. Schmidt, and H. J. Grabke, *Mater. Sci. Eng.*, **A120** (1989), 123.
63. I. Wolf and H. J. Grabke, *Solid State Comm.*, **54** (1985), 5.
64. H. J. Grabke and I. Wolf, *Mater. Sci Eng.*, **87** (1987), 23.
65. I. Wolf, H. J. Grabke, and H. P. Schmidt, *Oxid. Met.*, **29** (1988), 289.
66. W. F. Chu and A. Rahmel, *Rev. High Temp. Mater.*, **4** (1979), 139.
67. H. Xu, M. G. Hocking, and P. S. Sidky, *Oxid. Met.*, **41** (1994), 81.
68. F. H. Stott and M. J. Chang, in *Corrosion Resistant Materials for Coal Conversion Systems*, eds. M. J. Meadowcroft and M. J. Manning, Amsterdam, Elsevier, 1983, p. 491.
69. F. H. Stott, F. M. Chang, and C. A. Sterling, in *High Temperature Corrosion in Energy Systems*, ed. M. J. Rothman, Warrendale, PA, AIME, 1985, p. 253.
70. R. A. Perkins, 'Corrosion in high temperature gasification environments', Third Annual Conference on Coal Conversion and Utilization, Germantown, MD, DOE, October, 1978.
71. A. S. Khanna, W. J. Quadakkers, X. Yang, and H. Schuster, *Oxid. Met.*, **40** (1993), 275.
72. C. R. Wang, W. Q. Zhang, and R. Z. Zhu, *Oxid. Met.*, **33** (1990), 55.
73. D. J. Baxter and K. Natesan, *Oxid. Met.*, **31** (1989), 305.
74. W. Kai and R. T. Huang, *Oxid. Met.*, **48** (1997), 59.
75. X. Zheng and R. A. Rapp, *Oxid. Met.*, **48** (1997), 527.
76. C. B. Alcock, M. G. Hocking, and S. Zador, *Corr. Sci.*, **9** (1969), 111.
77. M. G. Hocking and V. Vasantasree, *Corr. Sci.*, **16** (1976), 279.
78. K. P. Lillerud, B. Haflan, and P. Kofstad, *Oxid. Met.*, **21** (1984), 119.
79. N. Tattam and N. Birks, *Corr. Sci.*, **10** (1970), 857.
80. B. Haflan and P. Kofstad, *Corr. Sci.*, **23** (1983), 1333.
81. T. Froyland, Masters Thesis, University of Oslo, Norway, 1984.
82. W. T. Bakker, *Oxid. Met.*, **45** (1996), 487.

8

Hot corrosion

Introduction

In addition to attack by reactive gases, alloys used in practical environments, particularly those involving the combustion products of fossil fuels, undergo an aggressive mode of attack associated with the formation of a salt deposit, usually a sulphate, on the metal or oxide surface. This deposit-induced accelerated oxidation is called hot corrosion. The severity of this type of attack, which can be catastrophic, has been shown to be sensitive to a number of variables including deposit composition, and amount, gas composition, temperature and temperature cycling, erosion, alloy composition, and alloy microstructure.[1] A number of comprehensive reviews on hot-corrosion have been prepared.[1-3] The purpose of this chapter is to introduce the reader to the mechanisms by which hot corrosion occurs. The examples used will be those associated with Na_2SO_4 deposits which are often encountered in practice. However, the effects of some other deposits will be briefly described at the end of this chapter.

Once a deposit has formed on an alloy surface the extent to which it affects the corrosion resistance of the alloy will depend on whether or not the deposit melts, how adherent it is and the extent to which it wets the surface, and the status of equilibrium conditions at the interfaces. A liquid deposit is generally necessary for severe hot corrosion to occur although some examples exist where dense, thick, solid deposits have, apparently, resulted in considerable corrosion.[4]

A major problem in studying hot corrosion, as well as describing it, arises because the hot corrosion process is affected by test methods, or field-use conditions. For example, the attack of a Ni–8 wt% Cr–6 wt% Al alloy, as indicated by the weight change per unit area versus time, Figure 8.1, is dependent upon the amount of Na_2SO_4. Significant differences occur when using immersion versus thin deposits obtained by spraying with an aqueous solution of Na_2SO_4 prior to oxidation. Another example is presented in Figure 8.2 which shows that the gas flow rate

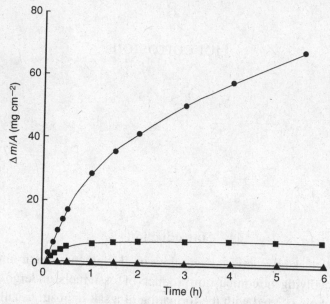

Figure 8.1 Weight change versus time data for the hot corrosion attack of Ni–8% Cr–6% Al specimens with different amounts of Na_2SO_4 at 1000 °C. The amount of degradation increases as the amount of the deposit is increased: ▲, 0.5 mg cm^{-2} Na_2SO_4; •, immersion in crucible with 1g Na_2SO_4; ■, 5 mg cm^{-2} Na_2SO_4.

can affect the hot corrosion attack of a Ni–8 wt% Cr–6 wt% Al–6 wt% Mo alloy. When hot corrosion first became a problem in gas turbines, it developed in aircraft gas turbines. It was found that testing using Na_2SO_4 deposits applied to alloys, followed by exposures at elevated temperatures in air, resulted in the development of degradation microstructures similar to those observed in practice. Consequently, much of the initial hot corrosion research involved testing in air, or using burner rigs where the SO_2 and SO_3 pressures were low and similar to those in aircraft gas turbines burning fuel with low sulphur contents. As gas turbines began to be used to generate electric power, and especially to propel marine vessels, it became clear that testing had to be performed in gases containing SO_2 and SO_3 in order to develop degradation microstructures typical of those observed in these latter applications. In this chapter it will be attempted to describe the various hot corrosion processes and to show the effects of test conditions. Of course an important aspect of such an endeavour is a description of the hot-corrosion test procedures. There are a great variety of test procedures.[5–7] In this chapter the hot corrosion data are generally applicable to coupon specimens (1 cm × 1 cm × 0.3 cm) coated with thin layers (0.5–5 mg cm^{-2}) of Na_2SO_4 and exposed to air, pure oxygen, or air containing SO_2 and SO_3 at temperatures from 700 to 1000 °C.

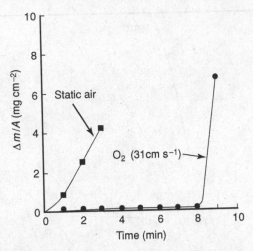

Figure 8.2 Weight change versus time data obtained for the isothermal hot corrosion of a Ni–8% Cr–6% Al–6% Mo alloy in static air at 1000 °C and in oxygen having a linear flow rate of 31 cm s^{-1}.

The degradation sequence of hot corrosion

Once an alloy surface has been partially or completely wetted by a molten salt deposit, conditions for severe corrosion may develop. The hot corrosion of virtually all susceptible alloys is observed to occur in two stages: an initiation stage during which the rate of corrosion is slow and similar to that in the absence of the deposit, and a propagation stage in which rapid, sometimes catastrophic, corrosion occurs. This sequence is illustrated for isothermal oxidation of the commercial alloy IN-738, in Figure 8.3 and for the cyclic oxidation of some M–Cr–Al–Y coating alloys in Figure 8.4. During the initiation stage, the alloy and the deposit are being altered to make the alloy susceptible to rapid attack. This alteration may include depletion of the element responsible for forming the protective scale on the alloy, incorporation of a component from the deposit (e.g., sulphur) into the alloy, dissolution of oxides into the salt, and development of cracks or channels in the scale. This alteration usually results in shifts in deposit composition toward more corrosive conditions. The length of the initiation stage varies from seconds to thousands of hours, Figures 8.3 and 8.4, and depends on a large number of variables including alloy composition, alloy microstructure, deposit composition, gas composition and flow rate, temperature, extent of thermal cycling, deposit thickness, equilibrium conditions at the deposit interfaces, specimen geometry, and the presence or absence of erosive conditions. In many cases, the end of the initiation stage follows the local penetration of the liquid deposit through the scale and subsequent spreading along the scale–alloy interface. This situation, in which the deposit reaches

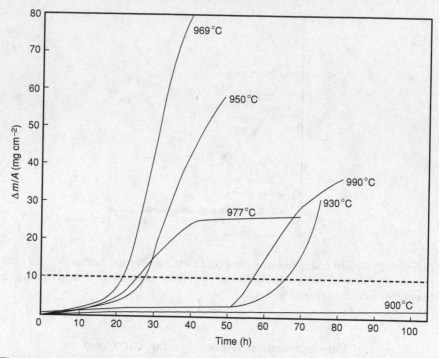

Figure 8.3 Isothermal mass change versus time for IN-738 coated with $1\,mg\,cm^{-2}$ Na_2SO_4 in 1 atm O_2. These data consist of an initiation stage with small weight changes and a propagation stage with larger weight changes. (The dashed line gives an arbitrary measure of the end of the initiation stage).

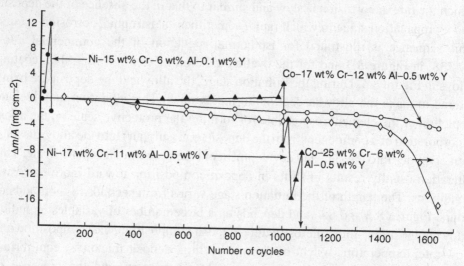

Figure 8.4 Weight change versus time for the cyclic hot corrosion (1 h cycles) of Na_2SO_4-coated ($\sim 1\,mg\,cm^{-2}$ Na_2SO_4 applied every 20 h) alloys in air. Severe hot-corrosion attack was evident at times where abrupt weight increases or decreases occurred.

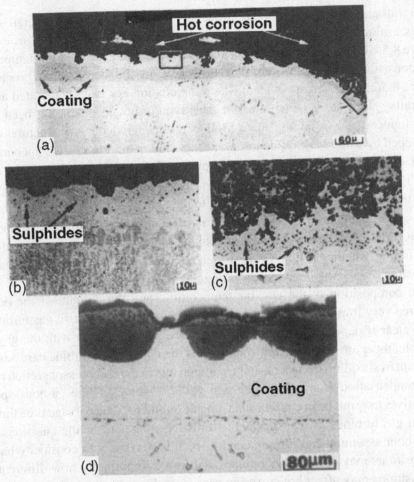

Figure 8.5 Photomicrographs showing the microstructural features developed during the hot corrosion of gas-turbine materials in aircraft and marine service. (a) High-temperature (aircraft) hot corrosion of an aluminide-coated nickel-base superalloy (B-1900), Type I. The characteristic features developed in the coating and superalloy substrate where the coating was penetrated are shown in (b) and (c), respectively. (d) Low-temperature (marine) hot corrosion of a Co–Cr–Al–Y coating after 4200 h of service, Type II.

sites of low oxygen activity and is in contact with an alloy depleted in Al or Cr,[*] generally leads to the rapid propagation stage. The propagation stage can proceed by several modes depending on the alloy and exposure conditions. The propagation modes are characterized by rather definitive microstructures as shown in Figure 8.5 where features resulting from hot corrosion attack in gas turbines used

---

[*] The alloys to be considered in this chapter will be nickel or cobalt base containing Cr and Al. Nickel- and cobalt-base alloys are used at elevated temperatures (700–1100 °C) where chromium and/or aluminium is present in order to form protective scales of Cr_2O_3 or α-Al_2O_3.

in aircraft and ship propulsion are presented. The specimen from aircraft service shows an aluminide-coated superalloy, Figure 8.5(a), in which sulphides are evident, Figure 8.5(b). This form of hot corrosion has been called Type I or high-temperature hot corrosion, and the sulphides play a role in the degradation process. In Figure 8.5(c), features are evident where the coating has been penetrated and the superalloy substrate is being attacked. This type of hot corrosion has been called alloy-induced self-sustaining attack. Both of these degraded microstructures can be developed by using Na_2SO_4 deposits and exposure in air. Finally, the hot corrosion attack developed in marine service is shown in Figure 8.5(d) and has been called Type II or low-temperature hot corrosion. In this type of hot corrosion sulphides are often not present in the alloy adjacent to the corrosion products; however, some alloys, especially those rich in nickel, can have substantial sulphide formation.[8] The SO_3 partial pressure in the gas and temperature are also important factors in determining whether sulphides are formed. To develop the Type II degradation microstructure testing in gases containing SO_3 is necessary.

The composition of the gas, the temperature, and the amount of the deposit are three very important parameters in hot corrosion processes. In gas turbines it is not clear if deposits on turbine hardware are in equilibrium with the gas. For example, there are data available that suggest the deposits on turbine hardware are the result of shedding from deposits on components in the compressor section rather than condensation from the combustion gases.[9] Moreover, the gaseous species themselves may not be in equilibrium with one another.[10] The passage time through aircraft gas turbines is approximately 10 ms. Nevertheless, while questions may exist about assuming equilibrium conditions, it is necessary to consider what gas compositions may exist in gas turbines, and more importantly how different gas compositions may affect hot corrosion processes. For a liquid fuel containing 0.5 wt% sulphur, the SO_3 partial pressures at 1400 K (1127 °C), 1100 K (827 °C) and 700 K (427 °C) are 9.8×10^{-5}, 3.8×10^{-4} and 7.5×10^{-4} atm, respectively.[10] For a fixed sulphur content the SO_3 pressure decreases as the temperature is increased. On the other hand, the SO_3 pressure required to form sulphates increases as the temperature is increased. For example, the equilibrium SO_3 pressure required to form Na_2SO_4 at 1000 K (727 °C) is 1.4×10^{-24} atm compared to 1.3×10^{-15} atm at 1300 K (1027 °C). In this chapter hot corrosion in gas turbines will be emphasized, but the treatment will be sufficiently broad so as to provide the fundamentals for all types of liquid deposit-induced accelerated corrosion.

Initiation stage of hot corrosion

During the initiation stage elements in the alloy are oxidized and electrons are transferred from metallic atoms to reducible species in the deposit. Initially the

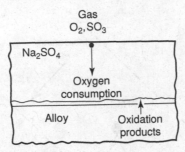

Figure 8.6 Schematic diagram illustrating oxygen consumption by the alloy during the initiation stage of hot corrosion.

reducible species is oxygen which comes from the Na_2SO_4 and the gas environment, Figure 8.6. Consequently, the reaction product barrier that forms beneath the deposit on the alloy surface often exhibits features similar to those for the gas–alloy reaction in the absence of the deposit. Nevertheless, there are differences because sulphur also enters the alloy from the deposit. As a result of such reactions with the alloy, the Na_2SO_4, especially adjacent to the alloy, begins to change in composition. The exact changes depend upon the composition of the alloy, the exposure time, the thickness of the deposit, and the composition and flow rate of the gas. The composition of the gas is critical. The solubilities of O_2 and SO_2 in Na_2SO_4 are extremely low[11] and, as proposed by Luthra,[12] transport of oxygen in Na_2SO_4 may consist of an $S_2O_7^{2-} - SO_4^{2-}$ exchange reaction. Consequently, in a gas with oxygen, but no SO_3, transport of oxygen through the Na_2SO_4 is probably negligible. The important point is that the deposit can become more basic or more acidic compared to the as-deposited condition as indicated schematically in the thermodynamic stability diagram[13] in Figure 8.7. These changes in composition of the deposit begin to affect the reaction-product barrier. This barrier can be compromised whereby the liquid deposit has access to the alloy and the degradation process proceeds to the propagation stage.

In discussing the composition of Na_2SO_4 (melting point in air, 884°C), as is evident in Figure 8.7, the composition of Na_2SO_4 at a given temperature is fixed by the oxygen partial pressure, and by either the activity of Na_2O in the melt, a_{Na_2O}, or the SO_3 partial pressure, since in view of the reaction shown in Equation (8.1),

$$Na_2SO_4 = SO_3 + Na_2O, \qquad (8.1)$$

the product $a_{Na_2O} \cdot p_{SO_3}$ is equal to a constant (K_1) at a fixed temperature. This relationship can be used to define the basicity of the Na_2SO_4. When Na_2SO_4 is in equilibrium with pure Na_2O (the Na_2O–Na_2SO_4 boundary in Figure 8.7), a_{Na_2O} can be taken as 1, and $(p_{SO_3})_{eq}$ can be determined. As the SO_3 partial pressure in equilibrium with the melt is increased beyond $(p_{SO_3})_{eq}$, a_{Na_2O} decreases and the

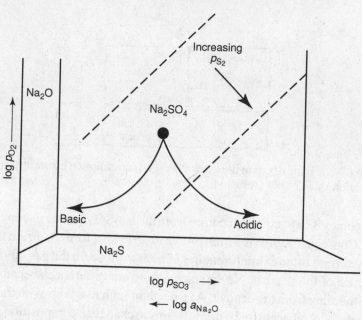

Figure 8.7 A thermodynamic stability diagram for the Na–O–S system at constant temperature in which some possible composition changes of the Na$_2$SO$_4$ phase are indicated.

melt becomes less basic. In many of the reactions used to describe hot corrosion, ionic species are used. This is the case because it is necessary to identify ions that may be formed or consumed. However, the activities of individual charged species cannot be measured and care must be used in applying the law of mass action to such equations. For example, in pure Na$_2$SO$_4$ the reaction shown in Equation (8.2) may be written in place of Equation (8.1),

$$SO_4^{2-} = SO_3 + O^{2-},$$ (8.2)

which shows that sulphate ions dissociate into SO$_3$ and oxide ions. Upon applying the law of mass action to this reaction and taking the activity of sulphate ions as unity one obtains Equation (8.3),

$$p_{SO_3} a_{O^{2-}} = K_2.$$ (8.3)

If it is assumed that $a_{O^{2-}} = a_{Na_2O}$, it can be shown that $K_1 = K_2$. Consequently the basicity of the Na$_2$SO$_4$ can be defined in terms of oxide ions or Na$_2$O. In this chapter the activities of charged species will not be used.

Propagation modes of hot corrosion

The propagation modes of hot corrosion are determined by the mechanisms by which the molten deposits cause the protectiveness of the reaction-product oxide scales to be destroyed. Two possible propagation modes are basic and acidic fluxing

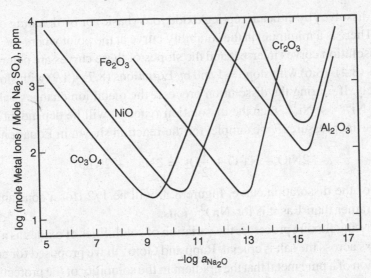

Figure 8.8 Measured oxide solubilities in fused Na_2SO_4 at 927 °C (1200 k) and 1 atm O_2.[17] (This figure originally appeared in Y. Zhang, and R. A. Rapp, *Corrosion*, **43** (1987), 348, and is reproduced with the kind permission of NACE International, © NACE International, 2005.)

of a protective oxide scale. Basic fluxing occurs because oxide ions in the melt react with the oxide to form soluble species, and acidic fluxing involves dissolution of the oxide by donating its oxide ions to the melt. Rapp and coworkers[14–17] and Stern and Dearnhardt[18] have determined solubility curves for a number of oxides in Na_2SO_4 as a function of the activity of Na_2O in the melt. Some typical curves developed by Rapp[17] are presented in Figure 8.8. Inspection of these curves shows that the solubility of most oxides in molten Na_2SO_4 is a function of the activity of Na_2O in the Na_2SO_4 (or the SO_3 pressure), and dissolution may occur via basic or acidic reactions. For example, considering NiO, equivalent basic fluxing reactions are shown in Equations (8.4)–(8.6), where the ion in solution is assumed to be NiO_2^{2-},

$$NiO + Na_2O = 2Na^+ + NiO_2^{2-}; \quad (8.4)$$

$$NiO + O^{2-} = NiO_2^{2-}; \quad (8.5)$$

$$NiO + Na_2SO_4 = 2Na^+ + NiO_2^{2-} + SO_3; \quad (8.6)$$

whereas equivalent acidic fluxing reactions for NiO with Ni^{2+} ions in solution are given in Equations (8.7)–(8.9):

$$NiO + SO_3 = Ni^{2+} + SO_4^{2-}, \quad (8.7)$$

$$NiO = Ni^{2+} + O^{2-}, \quad (8.8)$$

$$NiO + Na_2SO_4 = Ni^{2+} + SO_4^{2-} + Na_2O. \quad (8.9)$$

The NiO dissolves by either accepting oxide ions (basic) or by donating oxide ions (acidic). There is a minimum in the solubility curve at the point where the basic and acidic dissolution curves intersect and the slopes of these curves are determined by Equations (8.4)–(8.6) with slope $= 1$ and by Equations (8.7)–(8.9) with slope $= -1$, respectively. If, during the dissolution process, the metal ion changes valence, for example, $Ni^{2+} \rightarrow Ni^{3+}$, then the dissolution reaction will be dependent upon the oxygen partial pressure. For example, for the reaction shown in Equation (8.10),

$$2NiO + Na_2O + \frac{1}{2}O_2 = 2Na^+ + 2NiO_2^-, \qquad (8.10)$$

the slope of the dissolution curve, Figure 8.8, will be $1/2$ (for a constant oxygen pressure) rather than 1 as it is for NiO_2^{2-} ions.

In discussing fluxing processes the variation of solubilities of oxides as a function of thickness across the salt is crucial. Rapp and Goto[19] have proposed for continued hot corrosion of a pure metal that the gradient in the solubility of the protective oxide (C_{Oxide}) in the salt film is negative at the oxide–salt interface, Equation (8.11),

$$\left(\frac{dC_{Oxide}}{dx} \right)_{x=0} < 0, \qquad (8.11)$$

where x is the thickness of the deposit. This condition allows the oxide to dissolve at the oxide–salt ($x = 0$) interface and precipitate out in the salt as discontinuous particles ($x > 0$). When this solubility gradient is positive the salt becomes saturated with oxide and protective oxide will eventually form at the metal surface. Shores[20] has examined the Rapp–Goto criterion for fluxing and has shown that for some cases this criterion cannot be sustained indefinitely as a function of exposure time. Nevertheless, as will be shown subsequently, even when transitory, this criterion is an important aspect of hot corrosion processes. In fact the transitory condition may be close to practice because in many cases, especially gas turbines, deposits are formed intermittently. Shores[20] has also considered the form of the salt deposit on the metal surface and has shown that in many cases the salt distributes itself in pores of the oxide scale. Surface tension effects play a significant role in determining the distribution of the salt. It is reasonable to consider the salt deposit as a continuous layer on the surface of the metal or alloy during the initiation stage of hot corrosion, but as the propagation stage becomes dominant the salt is more likely to be distributed in the pores of the non-protective oxide and consequently the attack may not be uniform over all of the specimen's surface.

There are other propagation modes in addition to the fluxing propagation modes. For example, as will be shown in more detail subsequently, in the case of Na_2SO_4-induced hot corrosion a significant amount of sulphur can be transferred from the Na_2SO_4 into the metal or alloy. In some alloys the oxidation of these sulphides is a major factor in the degradation. Hence, oxidation of sulphides is a propagation

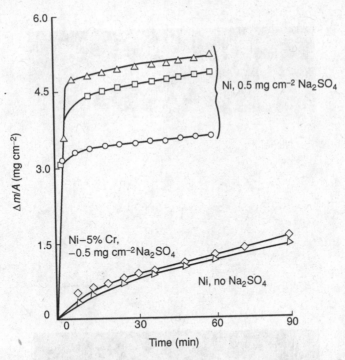

Figure 8.9 Comparison of the oxidation behaviour of pure nickel with and without a coating of Na_2SO_4. The Na_2SO_4-coated specimens undergo accelerated oxidation (three separate runs are shown) but the attack is not self sustaining and accelerated oxidation stops after approximately 6 min for specimens coated with 0.5 mg cm^{-2} Na_2SO_4. A Ni–5 wt% Cr alloy does not undergo accelerated oxidation when coated with 0.5 mg cm^{-2} Na_2SO_4.

mode. In the late 1950s and early 1960s, many investigators believed that hot corrosion occurred due to sulphidation. Bornstein and DeCrescente showed that hot corrosion could occur due to basic fluxing.[21,22] It is important to note that sulphidation is a form of hot corrosion, but all hot corrosion does not occur via the sulphidation propagation mode.

Basic fluxing

In order to show the development of basic-fluxing conditions, the hot corrosion of nickel[23] in pure oxygen will be considered. Weight changes versus time measurements at 1000 °C are compared for nickel with and without Na_2SO_4 deposits in Figure 8.9. The Na_2SO_4 causes increased oxidation but the attack is not self sustaining and eventually subsides. The amount of attack, however, does increase as the amount of the Na_2SO_4 deposit is increased. The NiO scales that form upon the nickel with and without a deposit are compared in Figure 8.10. The NiO which has

(a)

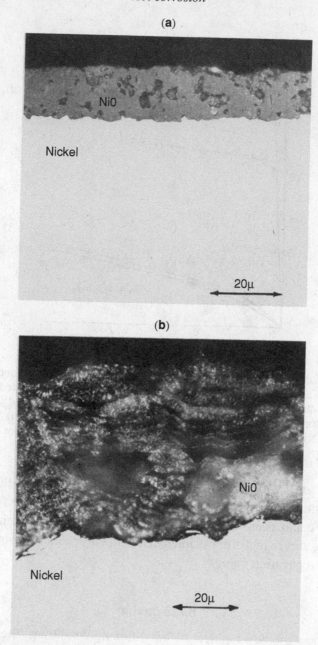

(b)

Figure 8.10 Optical micrographs showing transverse sections through specimens of nickel after oxidation in air at 1000 °C. (a) Dense layer of NiO formed after 3 h of oxidation in the absence of Na_2SO_4. (b) Porous scale of NiO formed after 1 min of oxidation with specimen coated with 0.5 mg cm^{-2} Na_2SO_4.

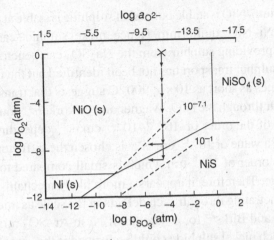

Figure 8.11 Stability diagram showing the phases of nickel that are stable in the Na_2SO_4 region of Fig. 8.7 at 1000°C. The dashed lines are sulphur isobars. The arrows show how the composition of Na_2SO_4 can change because of removal of oxygen and sulphur from the Na_2SO_4. The 'X' indicates the starting composition of the Na_2SO_4.

developed in the presence of the Na_2SO_4 is not dense and it is not protective. In Figure 8.11 a thermodynamic stability diagram is presented to show the phases of nickel which exist in the Na_2SO_4 region of the Na–S–O stability diagram as shown in Figure 8.7 for 1000°C. In Figure 8.11 the composition of the as-deposited Na_2SO_4[†] is indicated, and it can be seen that NiO is stable in this Na_2SO_4. As oxidation begins the oxygen potential in the Na_2SO_4 decreases, since NiO begins to form on the nickel and oxygen cannot move rapidly enough through the Na_2SO_4 from the gas. This is especially the case when SO_3 is not present in the gas. As can be seen from the sulphur isobars in Figure 8.11, as the oxygen potential in the Na_2SO_4 decreases the sulphur potential concomitantly increases. Eventually sulphur potentials can be achieved which are sufficient to form nickel sulphide. Inspection of Figure 8.11 shows that there are two mechanisms by which nickel sulphide may be formed. When the sulphur partial pressure is greater than $10^{-7.1}$, but less than that for the

---

[†] As discussed previously the Na_2SO_4 deposits in gas turbines may not be in equilibrium with the gas. If they are in equilibrium with the gas the SO_3 pressures will depend upon the amount of sulphur in the fuel, and the temperature. Consequently a rather wide range of SO_3 pressures is possible. Based upon degradation microstructures obtained from field experience, Figure 8.5, it is reasonable to assume that SO_3 pressures may range from 10^{-5} to 10^{-3} atm. The equilibrium pressures of oxygen and SO_3 in reagent grade Na_2SO_4 used in laboratory tests are usually not known. Reagent grade Na_2SO_4 does not react with an α-Al_2O_3 crucible exposed at 1000°C in air. The equilibrium SO_3 pressure for this reaction is 10^{-7} atm. In Figure 8.11 the pressures of oxygen and SO_3 in Na_2SO_4 prior to reaction with nickel have been taken as 1 and 3×10^{-5}, respectively. In this chapter the compositions of gases containing oxygen and sulphur will be defined in terms of the oxygen and SO_3 pressures. This is done because it is SO_3 that reacts with Na_2SO_4 deposits. Moreover, the pressures of other species can be determined when necessary from these pressures. For example at a constant temperature, $p_{S_2} = p_{SO_3}^2/Kp_{O_2}^3$; and $p_{SO_2} = p_{SO_3}/K'p_{O_2}^{1/2}$.

NiO–NiS equilibrium, NiO is stable but nickel sulphide is stable at oxygen potentials below that for the Ni–NiO equilibrium. Hence, nickel sulphide can be formed at the Ni–NiO interface providing sulphur from the Na_2SO_4 can penetrate the NiO layer. The nature of the sulphur transport has not been identified but the fact that sulphides are observed in times as short as 10 s at 900 °C suggests that transport is not lattice diffusion of sulphur through the NiO. Wagner and coworkers[24] found the diffusivity of S in NiO to be of the order of 10^{-14}–10^{-12} $cm^2\,s^{-1}$ depending on the oxygen partial pressure. If a value of 10^{-13} $cm^2\,s^{-1}$ is chosen the diffusion distance in 10 s at 900 °C is of the order of 10^{-6} μm which is small compared to the thickness of the initial NiO scale. Therefore, it appears a more likely mechanism is transport by SO_2 molecules penetrating through such defects in the scale as microcracks, as was found by Wootton and Birks[25] for oxidation of Ni in Ar–SO_2 mixtures. The other mechanism by which nickel sulphide could be formed is the case where the oxygen potential in the Na_2SO_4 is reduced to values whereby the sulphur potential in the Na_2SO_4 equals or exceeds that required for the NiO–NiS equilibrium, Figure 8.11. Such high sulphur partial pressures could be developed in cracks containing SO_2 as discussed previously. To verify that nickel sulphide could be formed beneath NiO scales and that this condition could develop by using Na_2SO_4, an experiment was performed[23] whereby nickel was placed in an evacuated quartz tube at 1000 °C with Na_2SO_4, but the nickel was not in contact with the Na_2SO_4. As shown in Figure 8.12(a), nickel sulphide was observed to be present beneath NiO. However, when the same experiment was performed where the quartz tube was backfilled with oxygen to give 0.9 atm at 1000 °C, no nickel sulphide was formed, Figure 8.12(b).

The net effect of the oxygen and sulphur removal from the Na_2SO_4 deposit on nickel oxidized in air is to increase the oxide-ion concentration, or the activity of Na_2O in the Na_2SO_4 as indicated in the equivalent reactions, shown in Equations (8.12) and (8.13),

$$SO_4^{2-} = O^{2-}\text{(in }Na_2SO_4) + \frac{3}{2}O_2 \text{ (used to form NiO)}$$

$$+ \frac{1}{2}S_2 \text{ (used to form NiS);} \tag{8.12}$$

$$Na_2SO_4 = Na_2O \text{ (in }Na_2SO_4) + \frac{3}{2}O_2 \text{ (used to form NiO)}$$

$$+ \frac{1}{2}S_2 \text{ (used to form NiS);} \tag{8.13}$$

and as shown schematically in Figure 8.11. At the beginning of the oxidation reaction some NiO is formed on the nickel surface as shown schematically in Figure 8.13(a). Some of the oxygen for this reaction is removed from the Na_2SO_4 deposit, adjacent to the nickel, which causes the sulphur potential in the Na_2SO_4

(a)

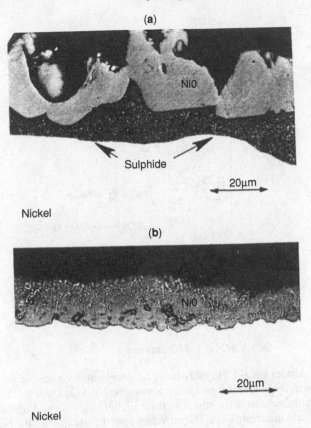

Figure 8.12 Optical micrographs showing a comparison of nickel specimens heated 24 h at 1000 °C in a sealed quartz tube which contained Na_2SO_4 in an Al_2O_3 crucible. The nickel specimens did not touch the Na_2SO_4. (a) The quartz tube was evacuated and sealed. A layer of nickel sulphide (arrows) has formed on the metal beneath a dense continuous layer of NiO. The nickel sulphide has been etched electrolytically with $NaNO_3$. (b) The quartz tube was backfilled with oxygen to give a pressure of 0.9 atm at 1000 °C. A dense layer of NiO has been formed. No evidence of sulphide formation was observed.

to increase. Sulphur from the Na_2SO_4 penetrates the NiO to form nickel sulphide beneath the NiO scale and the oxide-ion activity of the sulphate increases to levels at which the oxide ions begin to react with NiO to form nickelate ions, namely NiO_2^{2-}, Equations (8.4)–(8.6) or NiO_2^-, Equation (8.10). The increased oxide-ion activity is restricted to the Na_2SO_4 adjacent to the nickel and, therefore, as the NiO_2^{2-} ions diffuse away from the nickel they decompose into NiO particles and O^{2-} ions, Figure 8.13(b). The Rapp–Goto criterion is satisfied. The NiO particles are not protective and consequently the initially protective NiO is destroyed. It is important to note that as the oxygen potential in the Na_2SO_4 adjacent to the nickel comes to some steady state value, Figure 8.11, the production of oxide ions will cause the sulphur potential to decrease and a condition can be reached where sulphur

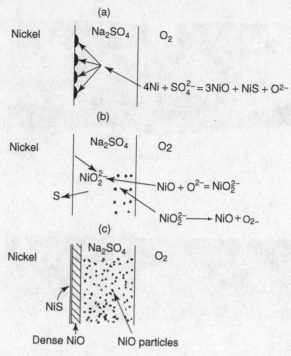

Figure 8.13 Model for the Na_2SO_4-induced accelerated oxidation of nickel. (a) An oxygen activity gradient is produced across the Na_2SO_4 layer by the formation of NiO. (b) Sulphur enters the metal to form nickel sulphide and oxide ions react with NiO to form nickelate ions. The nickelate ions diffuse toward the Na_2SO_4–gas interface where they decompose to NiO particles and oxide ions. NiO is not stable on the metal but forms away from the metal as a non-protective scale. Accelerated oxidation takes place. (c) Sulphur stops entering the alloy and oxide ions are no longer produced. The Na_2SO_4 becomes saturated with nickel and NiO forms as a continuous layer on the metal surface. Accelerated oxidation no longer occurs.

can no longer react with nickel. Then the concentration of oxide ions in the Na_2SO_4 becomes stabilized at some value higher than the oxide-ion concentration of the initially deposited Na_2SO_4. The Rapp–Goto criterion is no longer satisfied. The Na_2SO_4 is saturated with nickel oxide and is at some point on the basic side of the NiO solubility curve shown in Figure 8.8. As oxidation continues a protective layer of NiO is formed on the nickel and the hot corrosion attack ceases, Figure 8.13(c) and 8.14. The hot corrosion is not self sustaining. It is important to emphasize that this type of attack prevails when SO_3 is not present in the gas, or if present, the pressure is less than about 10^{-3} atm, or equilibrium is not achieved at the deposit–gas interface.

Otsuka and Rapp[26] have studied the hot corrosion of nickel at 900 °C in O_2–0.1% SO_2 ($p_{SO_3} = 3 \times 10^{-4}$ atm). These investigators devised a potentiometric measurement technique to determine simultaneously the basicity and the oxygen

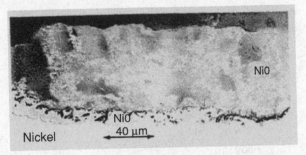

Figure 8.14 Optical micrograph showing transverse section of a nickel specimen coated with 0.5 mg cm^{-2} Na$_2$SO$_4$ and then oxidized 20 h in 1 atm oxygen at 1000 °C. The oxide scale is NiO and is composed of a porous and a dense zone. The dense portion forms beneath the porous zone after the accelerated period of oxidation.

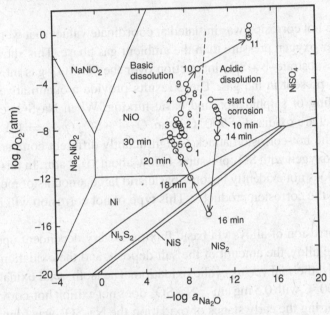

Figure 8.15 Trace of basicity and oxygen activity measured on a pre-oxidized 99% Ni coupon with a Na$_2$SO$_4$ film at 1173 K in O$_2$−0.1 mol% SO$_2$ gas. The central dashed line in the NiO stable field indicates the minimum in NiO solubility. Numbers in the figure designate the reaction time in hours, except as indicated.[26] (This figure originally appeared in: N. Otsuka and R. A., Rapp, *J. Electrochem. Soc.*, **137** (1990), 46, and is reproduced by permission of The Electrochemical Society, Inc.)

pressure at the substrate–Na$_2$SO$_4$ interface during the hot corrosion of nickel induced by a thin film of Na$_2$SO$_4$. The results from this experiment are plotted as a function of time onto the Ni–Na–S–O stability diagram presented in Figure 8.15. The dashed line in this figure indicates the minimum in the NiO solubility (which depends on p_{O_2}) with basic dissolution on the left and acidic dissolution on

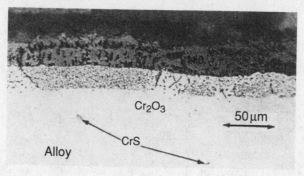

Figure 8.16 Optical micrograph showing transverse section of a Ni–5 wt% Cr alloy that has been coated with 0.5 mg cm^{-2} Na$_2$SO$_4$ and then oxidized 20 h at 1000 °C in 1 atm of oxygen. A dense protective scale of NiO has been formed on the alloy surface. The subscale is predominantly Cr$_2$O$_3$ with some CrS particles.

the right. The hot corrosion was initiated at coordinate values that were more basic and at a lower oxygen pressure than the ambient gas phase. This shows the dominance of the substrate–Na$_2$SO$_4$ interaction over the Na$_2$SO$_4$–gas interaction even when SO$_3$ is present in the gas. These results provide a quantitative verification for the coupling of sulphidation and basic fluxing. When Na$_2$SO$_4$-coated specimens of nickel are oxidized at 900 °C in an O$_2$-4% SO$_2$ ($p_{SO_3} = 1.2 \times 10^{-2}$ atm) gas mixture the hot-corrosion attack is significantly different compared to that in oxygen or in oxygen with SO$_3$ pressures below about 10^{-3} atm. In this former case the Na$_2$SO$_4$ does not evidently become basic and large amounts of nickel sulphide are present in the corrosion product.[27] This type of hot corrosion will be discussed subsequently.

The hot corrosion of alloys via basic fluxing is very dependent upon the composition of the alloy, the amount of the salt deposit, and the oxidation conditions, for example, isothermal versus cyclic oxidation. The isothermal oxidation of Ni–5 wt% Cr at 1000 °C with 0.5 mg cm^{-2} Na$_2$SO$_4$ does not exhibit hot-corrosion attack, Figure 8.9. During the early stages of oxidation the Na$_2$SO$_4$ was found to contain water-soluble chromium, which shows chromate ions had been formed. A transverse section through this oxidized specimen is presented in Figure 8.16, where a relatively dense layer of NiO has formed above a subscale of Cr$_2$O$_3$ with some CrS particles. Evidently some Cr$_2$O$_3$ reacts with the Na$_2$SO$_4$ to form chromate ions and the Na$_2$SO$_4$ melt cannot become sufficiently basic to permit nickelate ions to be formed. A possible reaction is shown in Equation (8.14):

$$Cr_2O_3 + 2O^{2-} + \frac{3}{2}O_2 = 2CrO_4^{2-}. \tag{8.14}$$

It is also important to note that the formation of CrS prevents nickel sulphide formation. Furthermore, the solubility of Cr$_2$O$_3$ in Na$_2$SO$_4$ increases with oxygen

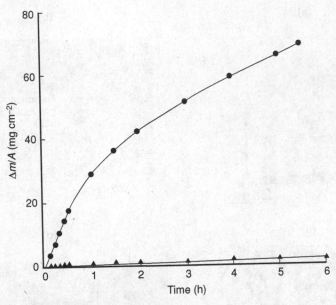

Figure 8.17 Weight-change versus time data showing that increased chromium concentration in Ni–Cr–Al alloys (•, Ni–8 wt% Cr–6 wt% Al; ▲, Ni–15 wt% Cr–6 wt% Al) extends the initiation stage for hot corrosion attack induced by immersion in a crucible with 1 g Na_2SO_4 in static air isothermally at 1000 °C.

partial pressure.[28] Consequently, as proposed by Rapp,[29] any dissolved chromate in the salt will tend to precipitate back onto the reducing metal substrate (e.g., at cracks or pores in the NiO scale), and thereby maintain the continuity of the scale. As the amount of the salt deposit is increased, or if cyclic oxidation conditions are used, substantial hot corrosion attack can occur. For example all of the Ni–Cr–Al–Y and Co–Cr–Al–Y alloys for which cyclic hot corrosion data are presented in Figure 8.4 are resistant to hot corrosion attack under isothermal conditions and with a small 0.5 mg cm^{-2} deposit of Na_2SO_4. During cyclic testing, however, with Na_2SO_4 applied at 20 h time intervals, hot corrosion occurs due to basic fluxing and sulphidation.

A similar basic fluxing mode to that of nickel is also observed in the case of some Ni–Cr–Al alloys where Cr_2O_3 and Al_2O_3 are involved in addition to NiO. The hot corrosion process is dependent upon the composition of the alloy and the amount of Na_2SO_4. In the case of Ni–8 wt% Cr–6 wt% Al, substantial basic hot corrosion attack is observed at 1000 °C in air under isothermal conditions with 5 mg cm^{-2} of Na_2SO_4 or immersion in Na_2SO_4 (Figure 8.1) but Ni–15 wt% Cr–6 wt% Al is not attacked, Figure 8.17. Evidently the higher chromium concentration inhibits the hot corrosion. Microstructural features of a Ni–8 wt% Cr–6 wt% Al specimen after immersion in Na_2SO_4 at 900 °C for 10 min are presented in

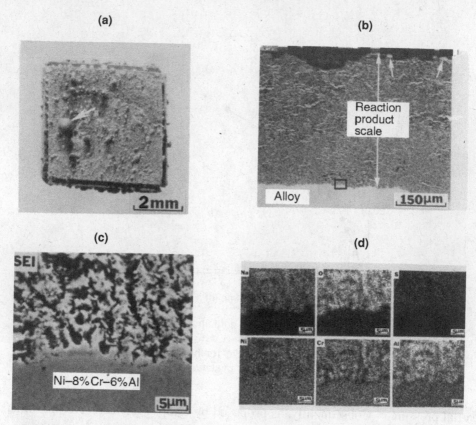

Figure 8.18 (a) Surface and (b) and (c) microstructural features developed in
a Ni–8 wt% Cr–6 wt% Al specimen after exposure at 1000 °C in air to thick
melts (1 g Na$_2$SO$_4$) for 10 min; (a) and (b) are optical micrographs, (c) is a
scanning electron micrograph. These features are typical of hot corrosion via the
basic fluxing propagation mode. Nickel sulphide is evident in the outer part of
the reaction product, arrows, (a) and (b), and the X-ray images (d) show that the
reaction-product scale adjacent to the alloy (c) consists of nickel stringers and
Na$_2$SO$_4$ containing aluminium and chromium.

Figure 8.18. The Na$_2$SO$_4$ has penetrated into the alloy preferentially removing
chromium and aluminium. Since the Na$_2$SO$_4$ layer is thick, virtually no oxygen
is supplied by the gas phase. Due to reaction with chromium and aluminium the
oxygen pressure is reduced below that to oxidize nickel, and nickel sulphide particles are formed at the surface of the specimen, Figure 8.18(a) and 8.18(b). Within
the reaction-product scale sulphate ions provide oxygen to oxidize aluminium and
chromium and sulphur to form nickel sulphide. Consequently, the oxide-ion concentration is increased to levels at which the Cr$_2$O$_3$ and Al$_2$O$_3$ can react with oxide
ions as described in Equations (8.14) and (8.15):

$$Al_2O_3 + O^{2-} = 2AlO_2^-. \tag{8.15}$$

Depending upon the amount of Na_2SO_4 that is present, the chromate and aluminate ions migrate out through the salt layer to sites of higher oxygen potential but lower oxide-ion activities close to the salt–gas interface, and Cr_2O_3 and Al_2O_3 precipitate out releasing oxide ions according to the reverse reactions of Equations (8.14) and (8.15). As in the case of nickel, the oxide-ion concentration in the melt becomes stabilized at a value higher than that of the initial deposit and the whole melt becomes saturated with both Cr_2O_3 and Al_2O_3, consistent with the solubility curves for these two oxides, Figure 8.8; the negative solubility gradient is eliminated. This is another example of where the Rapp–Goto criterion is satisfied initially, but the hot corrosion attack subsides. It is not self sustaining in the absence of additional salt deposition and a protective scale begins to develop upon the alloy, as shown for the hot corrosion of Ni–8 wt% Cr–6 wt% Al with a deposit of 5 mg cm^{-2} of Na_2SO_4, Figure 8.19. The nickel sulphide that was formed is converted to NiO and some of the sulphur that is released forms chromium sulphides in the alloy beneath the oxide scale.

The basic fluxing that has been described has a number of distinct features. Metal sulphides are usually found in the alloy substrate or the corrosion product since sulphur is removed from the Na_2SO_4. Furthermore, the amount of attack depends on the production of oxide ions in the melt; the oxide-ion concentration in the melt eventually reaches a constant value, greater than that in the initial deposit, and the melt becomes saturated with the oxide or oxides that participated in the basic fluxing process. The Rapp–Goto criterion is satisfied initially but not for the longer exposure times. Finally, this form of hot corrosion is usually restricted to high temperatures (above about 900 °C, 1170 k) and in gases that do not contain a substantial amount of an acidic component (e.g., $p_{SO_3} < 10^{-3}$ atm) or where the deposit is not in equilibrium with the gas when a substantial amount of an acidic component is present. This is a form of Type I hot corrosion.

Acidic fluxing

Acidic fluxing may be further subdivided into alloy-induced acidic fluxing, in which the acid conditions in the salt are established by dissolution of species from the alloy, which react strongly with Na_2O, or the O^{2-} ions, and gas-induced acidic fluxing in which the acid conditions are established by interaction with the gas phase.

Alloy-induced acidic fluxing

In alloy-induced acidic fluxing, elements such as molybdenum, tungsten, or vanadium in the alloy cause deposits to become acidic as oxides of these elements are incorporated into the deposits. Typical reactions considering molybdenum are given

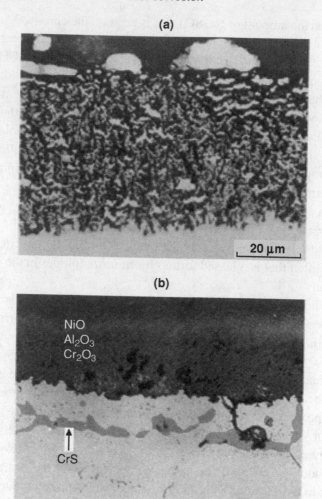

Figure 8.19 Microstructural photomicrographs showing features of Ni–8% Cr–6% Al specimens after exposure at 1000 °C in air to 5 mg cm^{-2} Na$_2$SO$_4$ for (a) 2 min, and (b) 1 h. Degradation via basic fluxing is evident after 2 min, but the Na$_2$SO$_4$ becomes consumed after 1 h hence the rapid attack ceases and the microstructure no longer exhibits the basic fluxing features (b).

in Equations (8.16)–(8.20):

$$Mo + \frac{3}{2}O_2 = MoO_3, \tag{8.16}$$

$$MoO_3 + Na_2SO_4 = Na_2MoO_4 + SO_3, \tag{8.17}$$

$$Al_2O_3 + 3MoO_3 \ (\text{in } Na_2SO_4) = 2Al^{3+} + 3MoO_4^{2-}, \tag{8.18}$$

$$Cr_2O_3 + 3MoO_3 \ (\text{in } Na_2SO_4) = 2Cr^{3+} + 3MoO_4^{2-}, \tag{8.19}$$

$$3NiO + 3MoO_3 \ (\text{in } Na_2SO_4) = 3Ni^{2+} + 3MoO_4^{2-}. \tag{8.20}$$

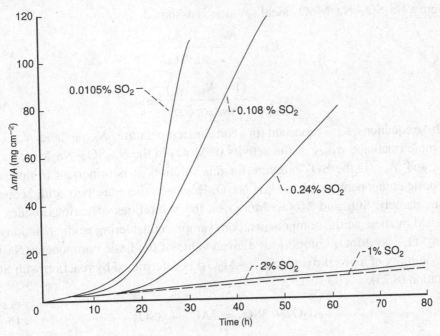

Figure 8.20 Effect of O_2–SO_2 gas mixtures on the corrosion kinetics for U-700, coated with 0.7 mg cm^{-2} Na$_2$SO$_4$, and exposed at 950 °C. As the SO_2 concentration is increased, p_{SO_3} is increased.[30] (This figure originally appeared in: A. K., Mishra, *J. Electrochem. Soc.*, **133** (1986), 1038, and is reproduced by permission of The Electrochemical Society, Inc.)

The Al$_2$O$_3$, Cr$_2$O$_3$, and NiO dissolve into the melt in regions where the MoO$_3$ activity is high and precipitate out wherever the activity of MoO$_3$ is low. This could occur in the region of the deposit near the gas phase where MoO$_3$ is being lost to the gas phase via vaporisation. As has been shown by Misra,[30,31] alloy-induced acidic fluxing is affected by the gas composition. As can be seen in Figure 8.20, catastrophic corrosion of Udimet 700 (Ni–14.8 wt% Cr–17.5 wt% Co–4.4 wt% Al–2.95 wt% Ti–5.03 wt% Mo) occurs in O_2–SO_2 gas mixtures with SO_2 concentrations up to 0.24%, but not in gas mixtures with 1 or 2% SO_2. When the SO_3 pressure is increased ($SO_2 + \frac{1}{2}O_2 = SO_3$) the incorporation of MoO$_3$ into the Na$_2$SO$_4$, Equation (8.17), is evidently restricted. The attack is still more severe than simple oxidation but involves sulphidation–oxidation attack, which will be discussed subsequently.

Both Misra[30,31] and Fryburg *et al.*,[32] who studied the hot corrosion of B-1900 (Ni–8 wt% Cr–6 wt% Al–6 wt% Mo–10 wt% Co–1.0 wt% Ti–4.3 wt% Ta–0.11 wt% C–0.15 wt% B–0.072 wt% Zr), emphasize that in order to have catastrophic hot corrosion of Udimet 700 or B-1900, the Na$_2$SO$_4$ deposit must be converted to a Na$_2$MoO$_4$–MoO$_3$ melt. Melts more complex than Na$_2$SO$_4$ will be discussed in more detail subsequently, but when MoO$_3$ reacts with Na$_2$SO$_4$, Equation (8.17), the equilibrium condition shown in Equations (8.21), or (8.22), must be considered, where a $Na_2SO_4-Na_2MoO_4$ ideal solution is assumed

$$K_{17} = \frac{N_{Na_2MoO_4} p_{SO_3}}{N_{Na_2SO_4} a_{MoO_3}} \qquad (8.21)$$

$$a_{MoO_3} = \frac{(1 - N_{Na_2SO_4})}{N_{Na_2SO_4} K_{17}} p_{SO_3} \qquad (8.22)$$

In these equations, K_{17} is constant for a constant temperature, $N_{Na_2SO_4}$ and $N_{Na_2MoO_4}$ are mole fractions, a_{MoO_3} is the activity of MoO_3 in the $Na_2SO_4-Na_2MoO_4$ solution, and p_{SO_3} is the SO_3 pressure for this solution. It is important to note that the basic component in this melt is Na_2O. However, there are two acidic components, namely, SO_3 and MoO_3. Moreover, the solubilities of various oxides are affected by these acidic components. For example, considering acidic reactions, in a $Na_2SO_4-Na_2MoO_4$ solution for a given value of the basic component (Na_2O), the solubility of a given oxide (e.g., α-Al_2O_3) is determined by reaction with SO_3, Equation (8.23),

$$Al_2O_3 + 3SO_3 = 2Al^{3+} + 3SO_4^{2-} \qquad (8.23)$$

or by reaction with MoO_3 as described in Equation (8.18), whichever is the larger. The amount of oxide dissolved is determined by the values of a_{MoO_3} and p_{SO_3}. These two values are interrelated via Equation (8.22). Inspection of this equation shows that the value of a_{MoO_3} is affected by the composition of the liquid mixture. In particular, for a given pressure of SO_3, a_{MoO_3} increases as $N_{Na_2SO_4}$ decreases. Such remarks are consistent with the observations of Misra and Fryburg *et al.*, and it appears in order to have fluxing due to MoO_3; $N_{Na_2SO_4}$ must approach zero and a_{MoO_3} approaches unity. When the SO_3 pressure is increased, a_{MoO_3} is expected to increase according to Equation (8.22), but evidently the amount of Na_2SO_4 is also increased and consequently a_{MoO_3} is actually decreased. Therefore the hot corrosion attack is decreased as can be seen in Figure 8.20.

The oxidation of B-1900 in oxygen with Na_2SO_4 deposits is a particularly good example of alloy-induced acidic fluxing. Fryburg *et al.*[32] have studied the initiation of hot corrosion in B-1900 and some of their results are presented in Figure 8.21. The application of a Na_2SO_4 deposit (3 mg cm^{-2}) produces a substantial increase in the oxidation rate, Figure 8.21(a). (The rate of simple oxidation of this alloy is too low to be observable on the scale of Figure 8.21.) Fryburg *et al.* have measured the evolution of SO_2 from the deposit, Figure 8.21(b), as well as the amounts of water-soluble species in the deposit, Figure 8.21(c), after various times of hot corrosion. Using these results it appears that during the initiation stage, Cr_2O_3, Al_2O_3, and MoO_3 are formed. The Cr_2O_3 and MoO_3 react with the Na_2SO_4 to form Na_2CrO_4 and Na_2MoO_4. Sulphur trioxide is evolved but decomposes to sulphur dioxide

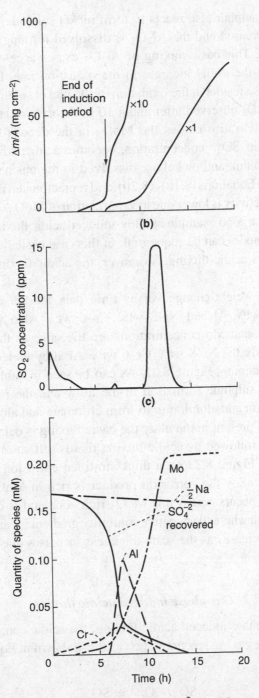

Figure 8.21 Preoxidized B-1900 with 3 mg cm^{-2} Na$_2$SO$_4$, tested at 900 °C in air (sample area = 8 cm^{-2}). (a) Typical weight change curve shown at two scale factors, (b) concentration of SO$_2$ (g) evolved as a function of time, and (c) quantity of water-soluble species found after various times of hot corrosion testing. (This figure originally appeared in: G. C., Fryburg, F. J., Kohl, C. A. Stearns and W. L., Fielder, *J. Electrochem. Soc.*, **129** (1982), 571, and is reproduced by permission of The Electrochemical Society, Inc.)

and oxygen, and sulphur also reacts to form nickel sulphide, whereby the oxide-ion activity is increased and the Al_2O_3 is dissolved forming $NaAlO_2$, Equations (8.13) and (8.15). This basic fluxing of Al_2O_3 causes accelerated oxidation and is responsible for the initial increase in the oxidation rate, Figure 8.21(a). This fluxing process also results in the oxidation of the nickel sulphide, which produces the evolution of SO_2 observed after about 10 h of corrosion, Figure 8.21(b). This accelerated oxidation also causes the MoO_3 in the deposit to increase and the deposit, now low in SO_4^{2-} concentration, becomes acidic; whereby the oxides of aluminium, chromium, and nickel are dissolved in regions near the alloy via the reactions shown in Equations (8.18)–(8.20), and reprecipitated near the gas interface where the MoO_3 activity is lower due to vaporization of MoO_3. The hot corrosion of B-1900 is not only a good example of alloy-induced acidic fluxing, but it also shows that propagation modes can be sequential. In this case basic fluxing with sulphide formation precedes acidic fluxing. Moreover, the acidic fluxing is self sustaining and catastrophic.

In Figure 8.22 weight-change versus time data for hot corrosion in air of Ni–8 wt% Cr–6 wt% Al and Ni–8 wt% Cr–6 wt% Al–6 wt% Mo are compared. The Cr, Al, and Mo concentrations are the same as those in B-1900. As discussed previously the Ni–8 wt% Cr–6 wt% Al alloy undergoes basic fluxing that is not self sustaining, Figure 8.22. As can be seen in Table 8.1 sulphate ions are consumed via sulphide formation in the alloy and the resulting oxide ions react with chromium and aluminium to form chromate and aluminate ions in solution. When Mo is present in the alloy the basic fluxing is delayed but eventually occurs and is then followed by acidic fluxing that is self sustaining, Figure 8.22. As can be seen in Figure 8.23(a), a thick stratified corrosion product is formed and just above the alloy the corrosion product is rich in MoO_3, Figure 8.23(c). The acidic fluxing occurs across this MoO_3-rich zone and is consistent with the Rapp–Goto criterion where the negative solubility gradient is caused by the activity of MoO_3 which decreases as the scale thickness increases due to vaporization of MoO_3.

Gas-phase induced acidic fluxing

In the case of gas-phase induced acidic fluxing, the acidic component is supplied to the deposit by the gas via reactions such as those shown in Equations (8.24) and (8.25):

$$SO_3 + O^{2-} = SO_4^{2-}, \tag{8.24}$$
$$SO_3 + SO_4^{2-} = S_2O_7^{2-}. \tag{8.25}$$

Table 8.1 *Analyses of wash water from Ni–8% Cr–6% Al specimens with Na$_2$SO$_4$ deposits after exposure at 1000 °C in air for different times*

Time (min)	pH*	%Remaining		Cr (μg)	Al (μg)	Ni (μg)
		Na	SO$_4^{2-}$			
1	6.4	100	100	<20	40	<5
2	7.9	100	71	50	260	<5
10	8.1	74	29	420	260	<5
30	8.0	72	19	1310	200	<5

* The pH of water prior to use was 5.4.

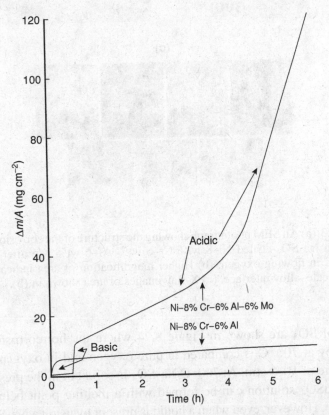

Figure 8.22 Comparison of the isothermal hot corrosion of Na$_2$SO$_4$-coated Ni–8 wt% Cr–6 wt% Al and Ni–8 wt% Cr–6 wt% Al–6 wt% Mo. Both alloys undergo basic fluxing, but Ni–8 wt% Cr–6 wt% Al–6 wt% Mo eventually undergoes alloy-induced acidic fluxing.

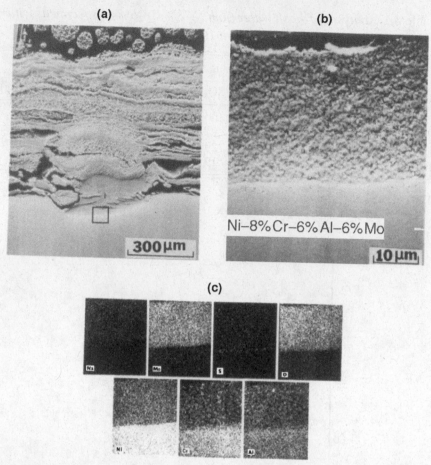

Figure 8.23 (a) An SEM photograph showing the structure of the corrosion product formed on Na_2SO_4-coated Ni–8 wt% Cr–6 wt% Al–6 wt% Mo after exposure at 1000 °C in flowing oxygen. (b) Higher magnification of area indicated in (a) showing scale–alloy interface. (c) X-Ray images of area shown in (b).

The effects of SO_3 are shown in Figure 8.24 where the hot corrosion of a Co–Cr–Al–Y alloy at 700 °C is compared in pure oxygen and in oxygen with a SO_3 partial pressure of 10^{-4} atm. At 700 °C Na_2SO_4 is solid but in the presence of SO_3 a Na_2SO_4–$CoSO_4$ solution can be formed with a melting point below 700 °C,[33] Figure 8.25(a). However, even when a liquid is present by using a Na_2SO_4–$MgSO_4$ mixture which is liquid at 700 °C in oxygen, Figure 8.25(b), much more attack is observed when SO_3 is present in the gas phase. Transport through the salt deposits is affected by the presence of SO_3. This type of hot corrosion is more severe at 700 °C than 1000 °C. For example, as shown in Figure 8.26, the attack of Co–Cr–Al–Y is greater at 700 °C than 1000 °C when the pressure of SO_3 is the

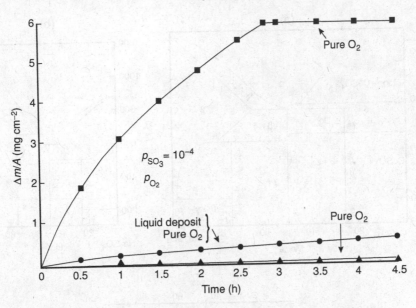

Figure 8.24 Weight-change versus time data obtained for the isothermal hot corrosion of Co–Cr–Al–Y coated IN-792 at 700 °C. Hot corrosion was induced by using Na_2SO_4 deposits ($\sim$1 mg cm^{-2}). In one experiment a Na_2SO_4–40 mol% $MgSO_4$ deposit was used to obtain a liquid deposit at the test temperature. The gas was flowing oxygen except in one experiment where an SO_2–O_2 gas mixture was passed over a platinum catalyst to develop an SO_3 pressure of 10^{-4} atm during the first 2.9 h of the experiment.

same at both temperatures, and the attack appears to decrease as the SO_3 pressure is decreased.

A complete description of gas-phase acidic fluxing is difficult because the effects are different for different metals and alloys. Also, as the SO_3 pressure is increased sulphide formation in the alloy begins to play more of a role in the degradation process. One distinctive feature of this type of hot corrosion is that the degradation rates are greater at low temperatures (e.g., 650–750 °C) compared to higher temperatures (e.g., 950–1000 °C). Hence it is often called 'low-temperature' or Type II hot corrosion. This low-temperature characteristic results from the need to form sulphates such as $CoSO_4$ and $NiSO_4$ (not necessarily at unit activity), which are necessary to form liquid sulphate solutions and require higher SO_3 partial pressures as temperature is increased. Furthermore, as mentioned previously, the equilibrium SO_3 partial pressures in most combustion environments, for a fixed amount of sulphur, decrease as temperature is increased,[10] and therefore this type of hot-corrosion attack tends to fall off as temperature is increased.

Gas-phase-induced hot corrosion exhibits certain microstructural characteristics that depend upon alloy composition. Typical degradation microstructures for

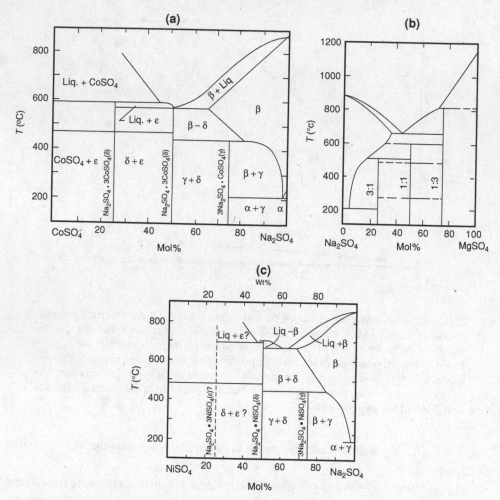

Figure 8.25 Pseudo-binary phase diagrams of the (a) Na₂SO₄–CoSO₄, (b) Na₂SO₄–MgSO₄, and (c) Na₂SO₄–NiSO₄ systems.

Co–Cr–Al–Y alloys are presented in Figures 8.5(d), 8.27(a) and 8.28(a). The attack is often localized and pit shaped, Figure 8.5(d) and 8.27(a), but is more uniform when specimens are vapour honed to remove pre-formed oxide scales. The salt deposit is present at the corrosion front, Figure 8.27(b), and virtually no depleted zone is evident in the alloy adjacent to the corrosion front, Figure 8.28(a). Finally, the chromium and aluminium distributions in the corrosion product are the same as those in the alloy, Figure 8.28(a) and (b), which indicates that very little diffusion of these two elements occurs during the corrosion process. On the other hand, the cobalt is present in the outer part of the corrosion product, Figure 8.27(a), which shows that it has diffused through the corrosion product that is permeated with liquid salt. Several models have been developed to account for the hot corrosion

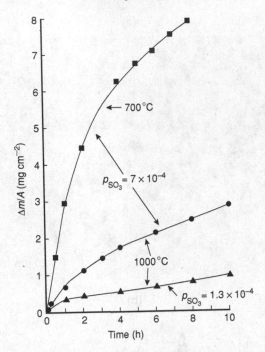

Figure 8.26 Weight-change versus time data for the hot corrosion attack of a Co–Cr–Al–Y coating on IN-738 using 1 mg cm^{-2} Na$_2$SO$_4$ deposits and SO$_2$–O$_2$ gas mixtures. When the SO$_2$–O$_2$ ratios were adjusted to give the same SO$_3$ pressure, more attack occurred at the lower temperature. When the same ratio was used at both temperatures, a lower SO$_3$ pressure was developed at the higher temperature.

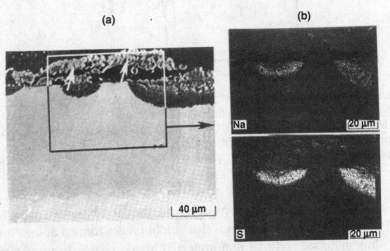

Figure 8.27 Degradational features developed during the hot corrosion of Co–Cr–Al–Y coatings as a result of exposure at 704 °C to Na$_2$SO$_4$ deposit ($\sim$ 1 mg cm^{-2}) and oxygen containing SO$_3$ at 7×10^{-4} atm: (a) SEM image shows localized attack is evident and the outer zone of the corrosion product (arrows) is rich in cobalt; (b) X-ray images show that sodium and sulphur are present in the corrosion product.

(a)

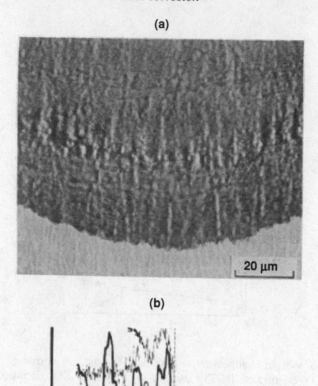

(b)

Figure 8.28 (a) Microstructural features developed in a Co–Cr–Al–Y alloy during 17.3 h of exposure to a Na_2SO_4 deposit (2.5 mg cm^{-2}) and oxygen containing SO_3 (7×10^{-4} atm) at 704 °C. Ghost images of the corrosion front and the α-cobalt phase in the alloy are evident. (b) Results from microprobe analyses of the corrosion product and the alloy adjacent to the corrosion front show similar Al and Cr distributions.

of Co–Cr–Al–Y alloys. All of these models have similarities, but there are also distinct differences.[12,34–36] One proposes that sulphides are formed at the corrosion–product alloy interface and that non-protective oxide phases form as a result of their oxidation.[36] The effects of sulphide formation certainly become important at higher temperatures and higher SO_3 pressures, as will be discussed subsequently, but sulphide formation may not be important in Type II hot corrosion of

Co–Cr–Al–Y. Another model[35] proposes that, due to the low oxygen pressure at the corrosion-front–alloy interface, sulphate ions can be converted to sulphite ions, but the presence of significant amounts of sulphite ions in the corrosion product has not been substantiated. A reasonable model for the Type II hot corrosion of Co–Cr–Al–Y alloys and other cobalt-base alloys has been proposed by Luthra.[12,34] In this model, after a Na_2SO_4–$CoSO_4$ solution has formed, cobalt diffuses outward through the Na_2SO_4–$CoSO_4$ melt via Co^{2+} ions. These ions react with either SO_3, Equation (8.26), or O_2, Equation (8.27), at, or near, the salt–gas interface, depending on the SO_3 partial pressure, to form $CoSO_4$ or Co_3O_4 and Co^{3+} ions:

$$3Co^{2+} + SO_3 + \frac{1}{2}O_2 = CoSO_4 \text{ (s)} + 2Co^{3+}, \qquad (8.26)$$

$$3Co^{2+} + \frac{2}{3}O_2 = \frac{1}{3}Co_3O_4 \text{ (s)} + 2Co^{3+}. \qquad (8.27)$$

The Co^{3+} ions then diffuse inward to the scale–alloy interface. In this mechanism the rate is controlled by the counter fluxes of Co^{2+} and Co^{3+} ions:

$$\underbrace{Co^{2+} + 2e^-}_{\text{Scale–salt interface}} + \underbrace{2Co^{3+}}_{\text{diffusing in through the liquid salt}}$$

$$\rightarrow \underbrace{3Co^{2+}}_{\text{diffusing out through the liquid salt}}. \qquad (8.28)$$

The Al and Cr in the Co–Cr–Al–Y are converted to oxides *in situ* but cannot form protective scales because of the rapid removal of cobalt from the alloy surface.

This degradation is self sustaining. Metal ions go into solution at the alloy–salt interface and precipitate as a non-protective solid in the salt, but the metal that dissolves and reprecipitates is the more noble metal rather than the elements that would form the protective scale in the absence of hot corrosion.

Sulphur-induced hot corrosion (sulphidation)

As mentioned previously, hot corrosion can occur due to the accumulation of sulphides in certain alloys. Bornstein and DeCrescente showed that hot corrosion induced by Na_2SO_4 could occur via basic fluxing.[21, 22] In their experiments they sulphidized the alloy B-1900 to introduce sulphur in a quantity equal to that in a Na_2SO_4 deposit that had caused hot corrosion attack of this alloy when exposed in air at 900 °C, and showed that hot-corrosion attack did not occur. On the other hand, hot corrosion of this alloy did occur when a deposit of $NaNO_3$ was used. These investigators proposed that it was the Na_2O, or oxide ions, in Na_2SO_4 and $NaNO_3$ that caused the hot-corrosion attack. As has been discussed previously for basic fluxing induced by Na_2SO_4, the basic-fluxing process involves sulphide formation in the alloys that undergo hot-corrosion attack. Sulphide formation in the

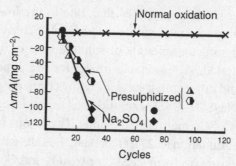

Figure 8.29 Comparison of the cyclic oxidation data obtained for Ni–25 wt% Cr–6 wt% Al specimens that were coated with Na₂SO₄ to those pre-sulphidized in an H₂S–H₂ gas mixture. Approximately 5 mg cm⁻² of Na₂SO₄ was added to one specimen after every 5 h of exposure up to 20 h and then after every 10 h interval beyond 20 h. The pre-sulphidation was performed at the same time intervals that the Na₂SO₄ was applied and the sulphur picked up was equivalent to the sulphur in a 5 mg cm⁻² Na₂SO₄ deposit. The specimens were oxidized in air at 1000 °C.

alloy is one of the reasons that the Na₂SO₄ becomes basic. Continued applications of Na₂SO₄ can result in severe hot corrosion via sulphidation. In Figure 8.29 results are presented for the cyclic oxidation in air of a Ni–25 wt% Cr– 6 wt% Al alloy with repeated applications of Na₂SO₄ and for pre-sulphidation of a specimen at the same intervals where the sulphur picked up by the specimen was equivalent to the sulphur in the Na₂SO₄ deposits. As can be seen this alloy is initially resistant to both the Na₂SO₄ applications and the pre-sulphidation, but eventually both treatments caused severe degradation, Figure 8.29. The microstructures of degraded samples which have been exposed to Na₂SO₄ and pre-sulphidation are similar as can be seen in Figure 8.30. Nickel-base alloys are especially susceptible to this type of degradation. This attack involves the presence of nickel sulphide, Figure 8.31(a), containing elements such as chromium. Protective oxides are not formed as these sulphides are oxidized, Figure 8.31(b). Sulphide formation in alloys has been shown to affect the subsequent oxidation resistance.[37] The situation depends upon the alloy composition and the amount and type of sulphides that are developed in the alloy.

Basic fluxing and sulphidation are forms of Type I hot corrosion. These two types of hot corrosion are interrelated. At high temperatures (i.e., >850 °C) and in gases with low SO₃ pressures (i.e., <10⁻⁴ atm) basic-fluxing conditions develop because of sulphur removal from the Na₂SO₄ by the metal or alloy. Consequently as alloys undergo hot corrosion attack via basic fluxing, in some cases it is possible to have the hot corrosion mechanism change to sulphidation. This can occur when large amounts of sulphur are introduced into the alloy, or the Na₂O component of the Na₂SO₄ is reduced. The Co–Cr–Al–Y alloys are especially resistant to Type I hot

(a)

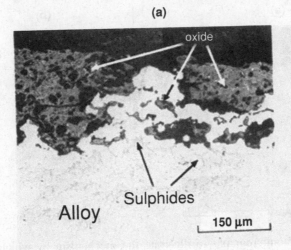

(b)

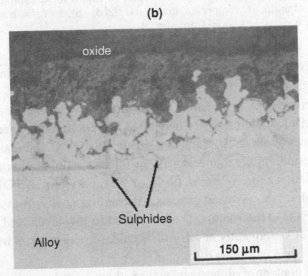

Figure 8.30 Optical micrographs showing comparison of the microstructure that developed in Ni–25 wt% Cr–6 wt% Al specimens after isothermal oxidation at 1000 °C (a) where the specimens were coated with 5 mg cm^{-2} Na$_2$SO$_4$ or (b) pre-sulphidized, as described in Fig. 8.29.

corrosion, Figure 8.4. These alloys are also resistant to pre-sulphidation followed by oxidation, Figure 8.32.

The discussion in the two preceding paragraphs has involved oxidation in air or oxygen. It is important to emphasize that the corrosion characteristics can change significantly when SO$_2$ and SO$_3$ are also present in the gas phase. This raises the question: what should the composition of the gas be for valid testing? The answer

(a)   (b)

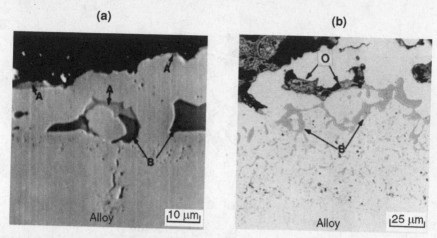

Figure 8.31 Photomicrographs to illustrate the mechanism by which sulphide phases in alloys can result in the formation of non-protective oxide scales during oxidation. (a) Sulphide phases composed of liquid nickel sulphide, A, and chromium sulphide, B, are preferentially oxidized, (b), to form non-protective oxides O, as well as additional sulphides, B.

is: it depends upon the application of interest. In the case of gas turbines, SO_3 partial pressures as high as 10^{-3} atm seem reasonable.

As mentioned previously, Misra[30,31] has observed very significant changes in hot corrosion attack of Udimet 700 when the SO_3 pressure is changed, Figure 8.20. For SO_3 pressures of about 10^{-3} atm and greater at 950 °C, the scale morphology showed considerable internal attack with an internal sulphidation zone ahead of an internal oxidation zone. Whereas, for lower SO_3 pressures, MoO_3 played a role in the catastrophic hot corrosion process. At the higher SO_3 pressures the mole fraction of Na_2SO_4 in the solution is not reduced to low levels and a_{MoO_3} is low, Equation (8.22). Consequently numerous sulphides are developed in the alloy and the degradation is via oxidation of sulphides.

As the nickel content of alloys is increased, there is an increasing tendency for sulphides to be formed during Type II hot corrosion. It is difficult to determine how much of the attack is due to oxidation of these sulphides and how much is due to a fluxing process similar to that described for Co–Cr–Al–Y alloys. Sulphide formation is prevalent in the hot corrosion of nickel at temperatures from 700 to 1000 °C providing the SO_3 pressure in the gas is sufficiently high. Kofstad and coworkers[27,38] have studied the Na_2SO_4-induced hot corrosion of nickel in $O_2 - 4\%$ SO_2 gas mixtures over this temperature range. At 700 °C ($p_{SO_3} = 3.5 \times 10^{-2}$) $NiSO_4$ must be formed in order to have a liquid Na_2SO_4–$NiSO_4$ solution develop, Figure 8.25(c). Consequently, there is an initiation stage prior to the hot corrosion attack. After the liquid solution is formed, the propagation stage occurs as a two-zoned corrosion

(a)

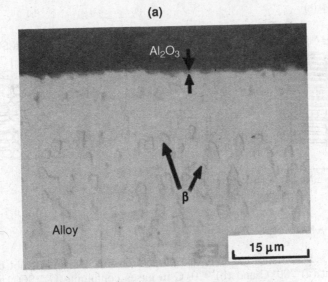

(b)

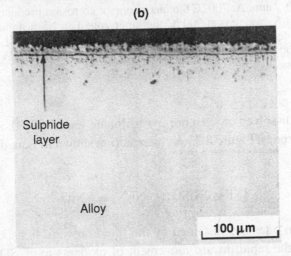

Figure 8.32 Comparison of the microstructure of Co–Cr–Al–Y specimens after isothermal oxidation at 1000 °C in 1 atm of oxygen where (a) the specimens were coated with 5 mg cm^{-2} of Na_2SO_4, or (b) pre-sulphidized for 20 s in an H_2S–H_2 gas mixture with $H_2S/H_2 = 0.2$ prior to oxidation. Protective oxide scales have been formed on both specimens.

product is developed, as shown schematically in Figure 8.33(a), consisting of an inner zone of NiO and Ni_3S_2, and an outer zone of NiO within a liquid solution of Na_2SO_4–$NiSO_4$. When the SO_3 pressure is sufficiently high, solid $NiSO_4$ may also be present at the scale–gas interface. It is proposed that nickel diffuses[27] from the metal through the Ni_3S_2, and SO_3 and oxygen diffuse inward through the liquid-sulphate solution. The diffusing species are not known but calculations[27] suggest

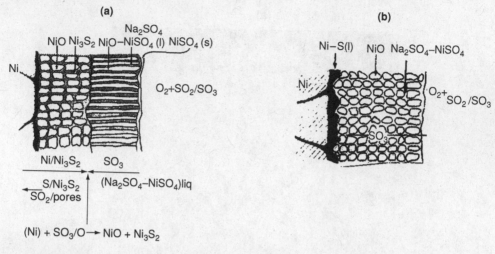

Figure 8.33 Schematic diagrams of the reaction mechanisms of hot corrosion of nickel at (a) 700 °C and (b) 900 °C in gases containing O_2, SO_2, and SO_3; $p_{SO_3} = 4 \times 10^{-3}$ atm. At 700 °C the inner zone of corrosion product is a mixture of NiO and Ni_3S_2 whereas at 900 °C the inner zone is a Ni–S liquid. The outer zones at both temperatures are a mixture of NiO and a liquid Na_2SO_4–$NiSO_4$ solution. At 700 °C when the SO_3 pressure is high enough, $NiSO_4$ is also present in the outer zone.

that they are not dissolved oxygen nor pyrosulphate ions ($S_2O_7^{2-}$). At the interface between the two zones, Figure 8.33(a), the reaction shown in Equation (8.29) takes place:

$$9Ni + 2NiSO_4 = Ni_3S_2 + 8NiO. \tag{8.29}$$

The rapid reaction rates are sustained by the rapid outward transport of nickel in the sulphide and the rapid inward movement of oxidants (e.g., SO_3) through the liquid sulphate from the gas.

In the case of Na_2SO_4-induced hot corrosion of nickel in gas mixtures with $O_2 - 4\% SO_2$ at temperatures between 900° and 1000 °C, where the SO_3 pressure is $\sim 1 \times 10^{-2}$ atm or greater, corrosion products similar to those shown in Figure 8.33(a) are developed but the sulphide layer is continuous with no NiO, as shown schematically in Figure 8.33(b). Again nickel diffuses outward through sulphide and SO_3 diffuses inward through the liquid Na_2SO_4–$NiSO_4$ solution. At the sulphide–sulphate interface reaction between nickel and $NiSO_4$ occurs as described by Equation (8.29). Under these acidic conditions the molten Na_2SO_4 serves as a solvent for $NiSO_4$, the Na_2SO_4 is not consumed, and consequently the hot corrosion attack is self sustaining.

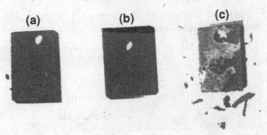

Figure 8.34 Optical micrographs showing effects of Na_2SO_4 and NaCl on scale spalling for the cobalt-base superalloy FSX-414: (a) no salt, (b) Na_2SO_4, and (c) Na_2SO_4 and NaCl.

The effects of other deposits

In some cases NaCl can be present along with deposits of Na_2SO_4. For example, sea water can be ingested in gas turbines used in marine environments. Chlorine in deposits may affect the hot corrosion of alloys in at least two ways. Firstly, chlorine concentrations in the ppm range have been shown to increase the propensity of oxide scales (e.g., Al_2O_3, Cr_2O_3) on alloys to crack and spall,[39] Figure 8.34. The mechanism by which chlorine causes spalling is not well understood but it may be similar to that of sulphur (Chapter 5, refs. 101–103). The increased tendency for spalling of protective oxides will cause alloys to progress to the propagation stage of hot corrosion after shorter exposure times.

It has also been observed that large concentrations of chloride cause aluminium and chromium to be rapidly removed from M–Cr–Al–Y alloys as shown in Figure 8.35 where it can be seen that as the amount of NaCl in Na_2SO_4 is increased a Co–Cr–Al–Y coating is much more rapidly degraded. During this process a rather unique microstructure is developed as shown in Figure 8.36 for the removal of aluminium from Co–Cr–Al–Y. The external scale is rich in aluminium oxides, Figure 8.36(a)–(c), and an internal zone is developed in the alloy, Figure 8.36(d). This internal zone contains porosity, Figure 8.36(d), but the pores contain alumina out near the external scale, Figure 8.36(e), and aluminium chloride in this zone adjacent to the unaffected alloy, Figure 8.36(e)–(f). The attack occurs preferentially at the β-phase in the alloy, Figure 8.37(a), but the pores do not have the exact shape of this phase, Figure 8.37(b). It is proposed that the salt becomes deficient in oxygen and sulphur due to reaction with aluminium and chromium in the alloy. Consequently, at points along the salt–alloy interface gaseous metal chlorides begin to form. This reaction takes place first with elements for which the thermodynamic conditions are most favourable. For example, reaction with aluminium is observed before chromium but chromium does react when the aluminium concentration has been reduced. As the gaseous metallic chlorides move outward through the melt, oxygen pressures are encountered for which the metallic chlorides are converted to

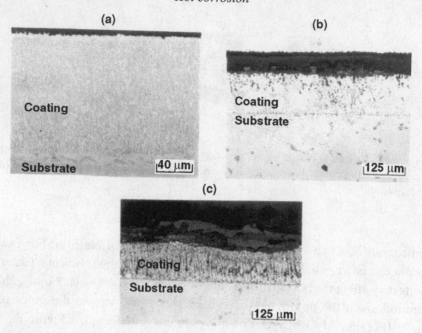

Figure 8.35 Photomicrographs to compare the degradation of Co–Cr–Al–Y coatings on IN-738 after exposure at 899 °C in air with Na_2SO_4 deposits containing different amounts of NaCl: (a) 500 h with Na_2SO_4; (b) 500 h with Na_2SO_4–5 wt% NaCl; (c) 40 h with Na_2SO_4–90 wt% NaCl.

non-protective metallic oxides and the chlorine is recycled to react with elements in the alloy. Continuation of this process results in the development of pores that are covered with discontinuous oxide particles. The pores have been observed to form at temperatures as low as 650 °C. Strikingly similar structures have been observed at much lower temperatures as a result of aqueous corrosion.[40] In these cases, pore growth via preferential removal of an element is accounted for by surface diffusion of those elements not reacting with the liquid, or by their solution into the liquid followed by subsequent precipitation on the sides of the pore. Such effects would account for the development of pores associated with phases rich in aluminium but not having their exact shape. As the process of pore development continues chloride is gradually lost to the gas phase and eventually, depending upon temperature, salt composition, gas composition, and alloy composition, the chloride concentration becomes insufficient to react with the alloy. When such a condition is reached the innermost portions of the pores begin to react with sulphur and the degradation proceeds via Na_2SO_4-induced hot corrosion. The presence of chloride, however, has caused the alloy to become depleted of aluminium and/or chromium, and the surface area of the alloy available for reaction with Na_2SO_4 has been increased due to the formation of pores.

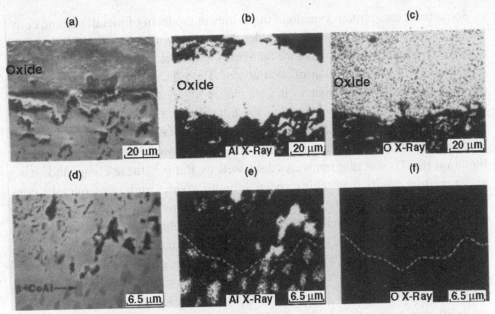

Figure 8.36 Microstructural photomicrographs and X-ray images of a Co–25 wt% Cr–6 wt% Al–0.5 wt% Y specimen after 100 h of cyclic hot corrosion testing at 900 °C where 1 mg cm^{-2} Na$_2$SO$_4$–90 wt% NaCl was applied after every 20 h. The structural features at the external scale–alloy porous-zone interface and at the alloy porous-zone–unaffected alloy interface are shown in (a) and (d), respectively; (b) and (c) are X-ray images of the area shown in (a), and (e) and (f) are X-ray images of the area defined in (d). Note lack of oxygen in particles in porous zone adjacent to the unaffected alloy (f).

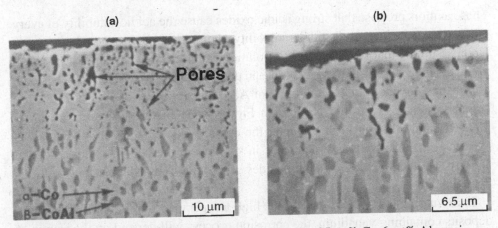

Figure 8.37 Scanning electron micrographs of a Co–25 wt% Cr–6 wt% Al specimen after exposure to Na$_2$SO$_4$–90% NaCl deposit in air at 900 °C. (a) A network of pores is evident and these pores appear to coincide with the β-CoAl phase of the alloy, but (b) the pores do not have the exact shape of the β-phase.

Some fuels can contain vanadium impurities in the form of metal–organic complexes and consequently vanadium oxides can be present in deposits. Some superalloys may contain vanadium but the number is small because of the adverse effects of vanadium on the corrosion of such alloys. The effects produced by vanadium oxides in deposits are similar to the effects of MoO_3 and WO_3, but the former usually comes from the fuel whereas the latter come from alloying elements in alloys used for high-temperature applications. All three of these oxides have the tendency to complex with oxide ions in solution to form vanadates, molybdates, Equation (8.17), and tungstates. As discussed by Rapp,[29] these compounds have lower melting points than their respective binary oxides and, consequently, contribute to a reduction of the liquidus for their solutions with Na_2SO_4. Moreover, the complexing with oxide ions affects the acid–base chemistry of the salt deposit, Equations (8.21) and (8.22). Zhang and Rapp[41] have measured the solubilities of CeO_2, HfO_2, and Y_2O_3 at 1173 K as a function of salt basicity in a solution with mole fractions of $0.7Na_2SO_4$–$0.3NaVO_3$ and, for comparison, the solubility of CeO_2 in pure Na_2SO_4. The results obtained for CeO_2 are presented in Figure 8.38. The solubility minima for the Na_2SO_4–$NaVO_3$ solution occurred at a higher solubility and at a higher basicity compared to the results obtained with pure Na_2SO_4. Similar shaped curves were obtained for HfO_2 and Y_2O_3 in Na_2SO_4–$NaVO_3$. Zhang and Rapp[41] propose that the metavanadate solution complexes with the oxide ions provided by the acidic dissolution of CeO_2 to form the orthovanadate ion via the reaction shown in Equation (8.30):

$$3CeO_2 + 4NaVO_3 = 2Na_2O + Ce_3(VO_4)_4. \qquad (8.30)$$

These authors propose that strong acidic oxides cause the acidic solubility of every other oxide to be increased with its solubility minimum shifted to more basic melts. Upon assuming that Na_2SO_4–$NaVO_3$ solutions are ideal, Hwang and Rapp[42] have predicted the solubilities of some oxides in these solutions as a function of basicity. The results obtained for the solubility of Al_2O_3 in a Na_2SO_4–$0.3NaVO_3$ melt are compared to those in pure Na_2SO_4 in Figure 8.39. In the discussion of alloy-induced acidic fluxing in this chapter the acidic oxide, i.e., MoO_3 in Equations (8.16)–(8.20), has been assumed to be in solution in the Na_2SO_4 with a defined activity. The complexing of acidic oxides with oxide ions is consistent with this approach.

In discussing the specific case of the high-temperature corrosion of alloys with deposits containing vanadium, the corrosion process will depend on the source of the vanadium. If the vanadium comes from the alloy, then the corrosion process will be similar to that described previously for molybdenum under alloy-induced acidic fluxing. When the vanadium arrives via deposition from the combustion of the fuel, the vanadium component of the fuel is oxidized in the combustion

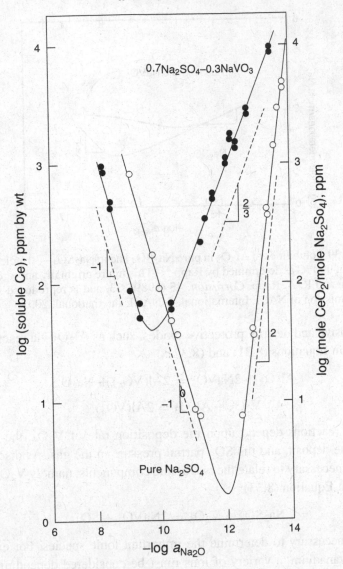

Figure 8.38 Measured solubilities of pure CeO₂ in pure Na₂SO₄ and 0.7 Na₂SO₄−0.3NaVO₃ at 1173 K.[41] (This figure originally appeared in: R. A. Rapp, *Corr. Sci.*, **44** (2002), 209, and is reproduced by permission of Elsevier.)

chamber and carried as solid V_2O_4 particles to deposit on turbine blades and other hardware.[43] With the influx of secondary air, the solid V_2O_4 particles are oxidized on hardware surfaces to form V_2O_5 with the low melting temperature of 691 °C (964 K). When Na_2SO_4 is also present in the deposit, liquid solutions are formed with melting points below 700 °C (973 K). These solutions, whether comprised of fused V_2O_5, a sodium vanadate, Na_2SO_4, or a solution of several of these salts, are

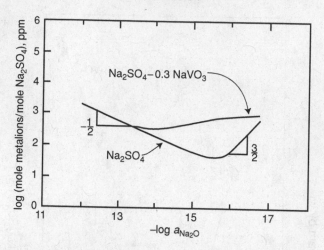

Figure 8.39 Solubilities of Al_2O_3 in pure Na_2SO_4 and in Na_2SO_4-30 mole percent $NaVO_3$ at 900 °C as determined by Rapp.[42] (This figure originally appeared in: Y. S. Huang and R. A. Rapp, *Corrosion*, **45** (1989), 33, and is reproduced with the kind permission of NACE International, © NACE International, 2005.)

highly corrosive and destroy protective oxides, such as Al_2O_3 via reactions of the type shown in Equations (8.31) and (8.32):

$$Al_2O_3 + 2NaVO_3 = 2Al(VO_4) + Na_2O, \tag{8.31}$$

$$V_2O_5 + Al_2O_3 = 2Al(VO_4). \tag{8.32}$$

The precise reactions depend upon the deposition rate of V_2O_4, the amount of sodium in the deposit, and the SO_3 partial pressure in the gas. As discussed previously, it is necessary to relate the two acidic components, namely V_2O_5 and SO_3, to each other, Equation (8.33):

$$Na_2SO_4 + V_2O_5 = 2NaVO_3 + SO_3. \tag{8.33}$$

It is also necessary to determine the important ionic species. For example, in the case of vanadium, a variety of ions must be considered depending upon the composition of the salt deposit. This includes orthovanadate, VO_4^{3-}, pyrovanadate, $V_2O_7^{4-}$, and metavanadate, VO_3^- ions. It is then possible to formulate reactions for relevant processes. For example, in the case of Al_2O_3 scales, the alumina could be dissolved by reaction with VO_3^- ions, Equation (8.34),

$$Al_2O_3 + 3VO_3^- = 2Al^{3+} + 3VO_4^{3-}, \tag{8.34}$$

with the possible reactions occurring at the melt–gas interface, Equations (8.35) and (8.36).

$$VO_4^{3-} + V_2O_5 = 3VO_3^- \tag{8.35}$$

$$VO_4^{3-} + SO_3 = VO_3^- + SO_4^{2-} \tag{8.36}$$

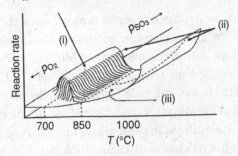

Figure 8.40 Schematic diagram to show the relationship of the different hot corrosion mechanisms as a function of temperature and SO_3 pressure. (i) Type II, gas-phase induced acidic fluxing. (ii) At high SO_3 pressures ($PSO_3 > 10^{-3}$ atm) pronounced sulphide formation accompanied by oxidation of sulphides and fluxing reactions. (iii) Type I (alloy-induced acidic fluxing; basic fluxing, sulphidation).

In such cases the Al_2O_3 stays in solution and the salt deposit is replenished with VO_3^{1-} ions by reaction with V_2O_5 or SO_3 that is supplied by the gas.

The hot-corrosion process is more complex with deposits containing a number of components because a greater number of ionic species must be considered and different compounds may be formed, some of which can be solid. For example, in the case of hot corrosion of nickel- and cobalt-base alloys at $700\,°C$ in Na_2SO_4–$NaVO_3$, solid deposits of $Co_2V_2O_7$ and $Ni_3V_2O_8$ affect the low-temperature hot corrosion process.[44] Nevertheless, the general phenomena are essentially the same as have been described for Na_2SO_4 deposits.

The effect of temperature and gas composition on Na_2SO_4-induced hot corrosion

It is of value to attempt to describe how the mechanisms of Na_2SO_4-induced hot corrosion vary with gas composition (O_2–SO_2–SO_3 mixtures) and temperature. In Figure 8.40 a schematic diagram is presented showing the possible Na_2SO_4-induced hot-corrosion mechanisms as a function of gas composition and temperature. It is important to emphasize that this diagram is applicable to nickel- and cobalt-base alloys and that the boundaries between the different mechanism regimes are diffuse. For example, in regions where sulphur plays a significant role, nickel-base alloys are more susceptible than cobalt-base alloys. Also oxygen is the major component of the gas with the pressures of SO_2 and SO_3 never totaling more than 0.05 atm.

Inspection of Figure 8.40 at $700\,°C$ shows that at low SO_3 pressures, or in pure oxygen, there is no hot corrosion attack because the Na_2SO_4 is solid. As the SO_3 pressure at $700\,°C$ is increased the deposit becomes liquid due to the formation of sulphate solutions, or if the SO_3 pressure is high enough, $Na_2S_2O_7$. Hot corrosion (Type II) then occurs via gas-phase-induced acidic fluxing. As the temperature is increased a transition from Type II to Type I hot corrosion begins to take place. When the SO_3 pressure is low, or in pure oxygen at temperatures above the melting point of Na_2SO_4 (884 °C) Type I hot corrosion occurs via basic fluxing or sulphidation. In the case of superalloys with elements such as Mo, W, or V alloy-induced acidic fluxing is also prevalent. At all temperatures between 700 and 1000 °C as the SO_3 pressure is increased heavy sulphide formation in the alloy begins to become a dominant feature in all of the alloys. Nickel-base alloys are more susceptible to such conditions than cobalt-base alloys. In some alloys the observed increase in corrosion appears to result from the oxidation of these sulphides. Udimet is an example of this type of degradation. For others, the sulphides permit rapid transport of metal out through the corrosion product, which is accompanied by inward movement of SO_3 through the outer sulphate layer. Nickel is an example of this latter type of hot corrosion attack.

Test methods

Finally, it is important to discuss the type of experiments that should be performed to determine the hot corrosion properties of metals and alloys. The answer, of course, is that the test should duplicate the conditions for the application of interest. However, this is not always possible, because the operating conditions may not be well defined and they may be difficult to simulate experimentally. When this is the case, it is important to document the degradation microstructures from service hardware, and then attempt to develop similar microstructures using selected conditions in laboratory experiments. Care must be exercised in using this approach, however, because a full understanding of the effects of all the experimental variables may not be ascertained. This is especially the case where different combinations of the important experimental variables are considered. For example, in the case of gas turbines the early work was done by coating specimens with Na_2SO_4 and oxidizing in air at 900 °C, or by using burner rigs, with the objectives of developing degradation microstructures similar to those in service, where the service experience came primarily from aircraft gas turbines. As gas turbines began to be used in marine environments it was necessary to use SO_3 at temperatures lower than 900 °C in the laboratory experiments, and to duct the burner rigs in order to develop the degradation microstructures observed in marine service. The importance of incorporating SO_3 into the test conditions was only appreciated after low-temperature hot corrosion became a problem.

References

1. F. S. Pettit and C. S. Giggins, 'Hot corrosion'. In *Superalloys II*, eds. C. T. Sims, N. S. Stoloff, and W. C. Hagel, New York, NY, John Wiley & Sons, 1987.
2. J. Stringer, *Ann. Rev. Mater. Sci.*, **7** (1976), 477.
3. Y. S. Zhang and R. A. Rapp, *J. Met.*, **46** (December) (1994), 47.

4. J. Stringer, in *High Temperature Corrosion*, NACE-6, ed. R. A. Rapp, Houston, TX, National Association of Corrosion Engineers, 1983, p. 389.
5. S. R. J. Saunders, 'Corrosion in the presence of melts and solids'. In *Guidelines for Methods of Testing and Research in High Temperature Corrosion*, eds. H. J. Grabke and D. B. Meadowcroft, London, The Institute of Materials, 1995, p. 85.
6. F. S. Pettit, 'Molten Salts'. In *Corrosion Tests and Standards*, ed. R. Baboian, ASTM Manual Series: Manual 20 Philadelphia, PA, American Society for Testing and Materials.
7. D. A. Shifler, 'High temperature gaseous corrosion testing', *ASM Handbook*, Materials Park, OH, ASM International, 2003, vol. 13A, p. 650.
8. R. H. Barkalow and G. W. Goward, in *High Temperature Corrosion*, NACE-6, ed. R. A. Rapp, Houston, Texas, National Association of Corrosion Engineers Houston, TX, (1983), p. 502.
9. N. S. Bornstein and W. P. Allen, *Mater. Sci. Forum*, **251–254** (1997), 127.
10. J. G. Tschinkel, *Corrosion*, **28** (1972), 161.
11. R. E. Andersen, *J. Electrochem. Soc.*, **128** (1979), 328.
12. K. L. Luthra, *Met. Trans.*, **13A** (1982), 1853.
13. J. M. Quets and W. H. Dresher, *J. Mater.*, **4** (1969), 583.
14. D. K. Gupta and R. A. Rapp, *J. Electrochem. Soc.*, **127** (1980), 2194.
15. P. D. Jose, D. K. Gupta, and R. A. Rapp, *J. Electrochem. Soc.*, **132** (1985), 73.
16. Z. S. Zhang and R. A. Rapp, *J. Electrochem. Soc.*, **132** (1985), 734; 2498.
17. R. A. Rapp, *Corrosion*, **42** (1986), 568.
18. M. L. Dearnhardt and K. H. Stern, *J. Electrochem. Soc.*, **129** (1982), 2228.
19. R. A. Rapp and K. S. Goto, 'The hot corrosion of metals by molten salts'. In *Molten Salts*, eds. J. Braunstein and J. R. Selman, Pennington, New Jersey, Electrochemical Society, 1981, p. 81.
20. D. A. Shores, in *High Temperature Corrosion*, NACE-6, ed. R. A. Rapp, Houston, Texas, National Association of Corrosion Engineers, 1983, p. 493.
21. N. S. Bornstein and M. A. DeCrescente, *Trans. Met. Soc. AIME*, **245** (1969), 1947.
22. N. S. Bornstein and M. A. DeCrescente, *Met. Trans.*, **2** (1971), 2875.
23. J. A. Goebel and F. S. Pettit, *Met. Trans.*, **1** (1970), 1943.
24. D. R. Chang, R. Nemoto, and J. B. Wagner, Jr., *Met. Trans.*, **7A** (1976), 803.
25. M. R. Wootton and N. Birks, *Corr. Sci.*, **12**, (1972), 829.
26. N. Otsuka and R. A. Rapp, *J. Electrochem. Soc.*, **137** (1990), 46.
27. K. P. Lillerud and P. Kofstad, *Oxid. Met.*, **21** (1984), 233.
28. Y. Zhang, *J. Electrochem. Soc.*, **137** (1990), 53.
29. R. A. Rapp, *Corr. Sci.*, **44** (2002), 209.
30. A. K. Misra, *Oxid. Met.*, **25** (1986), 129.
31. A. K. Misra, *J. Electrochem. Soc.*, **133** (1986), 1038.
32. G. C. Fryburg, F. J. Kohl, C. A. Stearns, and W. L. Fielder, *J. Electrochem. Soc.*, **129** (1982), 571.
33. K. L. Luthra and D. A. Shores, *J. Electrochem. Soc.*, **127** (1980), 2202.
34. K. L. Luthra, *Met. Trans.*, **13A** (1982), 1647; 1843.
35. R. H. Barkalow and F. S. Pettit, 'On Oxidation Mechanisms for Hot Corrosion of CoCrAlY Coatings in Marine Gas Turbines,' Proceedings of the 14th Conference on Gas Turbine Materials in a Marine Environment, Naval Sea Systems Command, Annapolis, MD, 1979, p. 493.
36. K. T. Chiang, F. S. Pettit, and G. H. Meier, 'Low temperature hot corrosion.' *High Temperature Corrosion*, NACE-6, ed. R. A. Rapp, Houston, Texas, National Association of Corrosion Engineers, 1983, p. 519.

37. J. A. Goebel and F. S. Pettit, *Met. Trans.*, **1** (1970), 3421.
38. P. Kofstad and G. Akesson, *Oxid. Met.*, **14** (1980), 301.
39. J. B. Johnson, J. R. Nicholls, R. C. Hurst, and P. Hancock, *Corr. Sci.*, **18** (1978), 543.
40. M. G. Fontana and N. D. Greene, *Corrosion Engineering*, 2nd edn, New York, McGraw Hill, 1978, p. 67.
41. Y. Zhang and R. A. Rapp, *Corrosion*, **43** (1987), 348.
42. Y. S. Huang and R. A. Rapp, *Corrosion*, **45** (1989), 33.
43. C. G. Stearns and D. Tidy, *J. Inst. Energy*, **56** (1983), 12.
44. B. M. Warnes, The influence of vanadium on the sodium sulfate induced hot corrosion of thermal barrier coating materials, Ph.D. Dissertation, University of Pittsburgh, PA, 1990.

9

Erosion–corrosion of metals in oxidizing atmospheres

Introduction

Under many service situations, materials are exposed to high-temperature gas streams. A gas stream can entrain solid particles, picking up larger particles at higher speeds. In fact it is probable that a particle-free gas flow can only be provided under the most carefully contrived laboratory conditions. In most practical cases, gas streams will contain solid particles. Good examples of this are found in gas turbines, oil refining and reforming, as well as in power generation. This chapter will address the ways in which the oxidation and solid-particle erosion of metals proceed simultaneously and interact.

Erosion of materials

To understand how erosion and oxidation interact, it is necessary to understand how metals and oxides erode. Previous work into the erosion of metals and ceramics at room temperature[1-4] has shown that the erosion of ductile material involves plastic deformation, cutting, plowing, and fatigue. These mechanisms can occur by the impact of a single particle, as in the case of cutting, or may require several subsequent impacts as in the cases of plowing and fatigue. Brittle materials, on the other hand, erode by the introduction of intersecting brittle crack systems.[5] Most of the metals in use respond in a basically ductile manner at high temperatures. The oxides that arise from oxidation can be either quite ductile (wustite, nickel oxide, etc.) or quite brittle (alumina, chromia, etc.). Depending on the oxide and the circumstances, the oxide scale can be deformed plastically by the erosive impacts; alternatively the oxide could spall under impact.

Ductile and brittle materials respond quite differently to variations in the angle of attack of the erodent on the substrate. The rate of erosion of a brittle material increases with the angle of attack, reaching a maximum at normal incidence, as

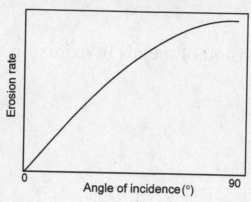

Figure 9.1 Variation in rate of erosion of a brittle material with angle of incidence of erosive stream.

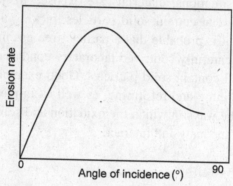

Figure 9.2 Variation of rate of erosion of a ductile material with angle of incidence.

shown in Figure 9.1. In the case of ductile materials, the rate of erosion increases to a maximum at about 25–30° incidence and then falls with further increase to reach a low, finite, value at normal incidence, as shown in Figure 9.2. Finnie[1] described the erosion of ductile materials as a cutting, or machining, process (Figure 9.3) and was successful in representing the process for angles up to 60° incidence, but predicted zero erosion at normal incidence. The observation of finite rates of erosion at normal incidence was explained by Bitter,[2,3] who proposed that the metal surface work hardened under more normal impacts, inducing metal removal by fatigue, etc.

Tilly[4] proposed that the removal of ductile metal could also take place in two or more stages, in which the initial impact(s) produced lips by plastic deformation and subsequent impacts removed metal from the lips by cutting. The extent of plastic deformation of the metal substrate can be increased by adiabatic heating, which is generated, to form features such as shear lips.[6,7] An extensive review of the erosion of ductile materials has been given by Sundararajan.[8]

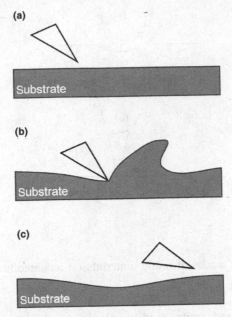

Figure 9.3 Modes of removal of ductile material by erosion; (a) before impact; (b) during cutting; (c) rebound after cutting.

Erosion at room temperature has also been found to include processes such as work hardening and removal of platelets[9] as well as low cycle fatigue.[10,11] It should be remembered that processes that involve work hardening will become less significant as the temperature increases.

The erosion of brittle materials has been described in terms of interacting Hertzian crack systems, or by the formation of lateral and median vents for sharp indentors.[5] This is indicated in Figure 9.4. It has also been observed that brittle materials are capable of undergoing plastic deformation[12] particularly when under erosion by very small particles.[10,11] This is important because, at high temperature, some oxides formed as scales may show a ductile reaction to erosive impact.

Although the damage of materials by erosion at room temperature can be very serious, in terms of reducing the life of a component, the damage that occurs at high temperature under conditions of erosion in an oxidizing atmosphere is found to be several times more severe. In fact, the severity of this form of attack, due to simultaneous erosion and corrosion, indicates that the two individual processes combine to give a strongly synergistic result.

Study of erosion–corrosion processes at high temperature

In studying erosion–corrosion interactions at high temperature, reaction rates must be measured and the structure of the reaction zone must be observed. The

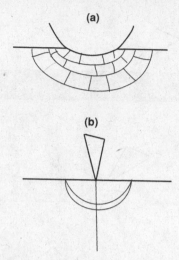

Figure 9.4 Fracture modes of brittle materials on indentation by (a) round and (b) sharp indentors.

morphology is obtained quite readily using standard techniques of optical microscopy, scanning electron microscopy, and X-ray diffraction. Transmission electron microscopy is also becoming popular, although there are considerable problems associated with accurate placing of the initial perforation of the foil.

To define the conditions of erosion–corrosion exposures, it is necessary to measure and control several parameters such as the following.

- Specimen temperature.
- Atmosphere composition.
- Particle size, shape, and loading.
- Particle speed.
- Angle of incidence of the particle stream to the specimen.
- Time of exposure.

Some of these parameters are difficult to fix and measure, which makes fundamental work more difficult. However, to compare materials in a testing sense it is only necessary to have constant conditions, which is not too difficult to achieve.

Two broad types of equipment have been used. In the first, shown diagrammatically in Figure 9.5(a), the erosive particles are accelerated down a tube to impact on a specimen at pre-determined angle, speed, and temperature. In the second, indicated in Figure 9.5(b), the specimen takes the shape of, or is attached to, an arm, which is attached perpendicular to a shaft that rotates about its axis. Such a specimen may be rotated through a fluidized bed of erosive particles, or it may be rotated just above such a bed in the zone where particles are ejected into the gas space by the fluidizing action of the gas. In Figure 9.5(c), the specimen is shown

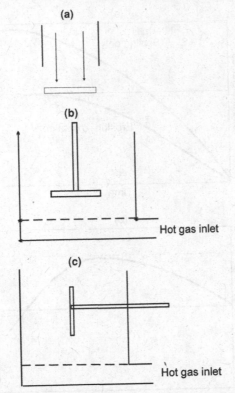

Figure 9.5 Diagrammatic representation of acceleration tube equipment and rotating specimen equipment for the study of erosion–oxidation. (a) Vertical impact of hot gas; (b) horizontally or (c) rotating specimen in hot fluid bed.

to rotate in the vertical plane and could be arranged to spend half of its cycle in the fluid bed and half in the gas space above the bed. All types of equipment have been used successfully and have yielded consistent results.

The erosion–corrosion attack of a specimen can be followed by measuring the change in weight as the reaction proceeds. Alternatively, dimensional changes can be measured. Specimens exposed using acceleration tubes are usually flat discs (surfaces other than the reaction surface can be protected by aluminizing). Specimens used in the rotating mode may be either round or flat in section.

Reaction kinetics can be obtained from measurements of the weight, or thickness, of the specimen at time intervals. This can be done using one specimen, which is reinserted for further exposure, or using several specimens, each of which yields one data point only. In the latter case a specimen is available for examination for each measurement taken, which allows the sequence of development of surface features to be established. Whichever technique is used it is convenient and sensible to consider the attack of pure metals before considering the more complex situations

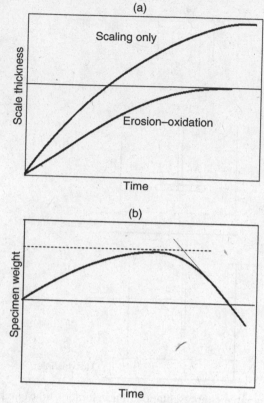

Figure 9.6 Plots of scale thickness (a) and weight change (b) as functions of time for erosion–oxidation. The diagrams correspond on a common time base.

of alloys. Corresponding plots of scale thickness and specimen weight as a function of time are shown in Figure 9.6.

Erosion–corrosion of pure metals at high temperature

Erosion–corrosion can be regarded as the combined effect of two competing processes. Corrosion, or oxidation, forms a surface film that more or less passivates the metal, causing the reaction to proceed more slowly as the film thickens. Erosion, on the other hand, removes the surface of the material and would, therefore, remove or thin down the passivating film that is being formed by oxidation. Thus, in the presence of both processes, the erosive action prevents the oxide film from developing its full protection. Indeed the oxide film may be removed completely; allowing oxide formation to proceed at its highest rate as the product is continually removed by erosion. This can give rise to very high rates of degradation. The intensity of erosion that is required to cause complete scale removal continuously depends on the oxide that is being formed. A higher erosive intensity would be required for

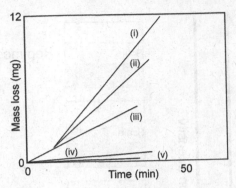

Figure 9.7 Effect of oxide formation and temperature on the rate of degrada-tion of nickel under erosion–corrosion conditions (i) 140 m s^{-1}, 800 °C, air, (ii) 140 m s^{-1}, 650 °C, air, (iii) 90 m $^{-1}$, 800 °C, air, (iv) 140 m s^{-1}, 800 °C, nitrogen; (v) 73 m s^{-1}, 25°C, air.

oxide scales that grow more quickly. Consequently, the extent and type of erosive attack depends on the relative intensities with which the two processes, erosion and oxidation, proceed. A further, important, observation[6] is that, if the oxidation component can be removed altogether, by eroding at room temperature or at high temperature in an inert atmosphere, the degradation occurs only very slowly. This is illustrated in Figure 9.7 for the case of nickel.[13] Similar results are also obtained for cobalt.

The interactions between erosion and oxidation can be clarified by considering several separate regimes of interaction.[8,13–16] As mentioned above, if the process of oxidation can be excluded, then erosion alone causes substantially slower degra-dation than when the two processes are operating together, synergistically. The results of this synergism and the mechanisms by which it occurs are divided into two regimes of primary interaction. Regime 1 refers to conditions under which the erosion component affects only the scale. This refers to the situation where the oxide growth is vigorous and the erosion is relatively mild. The scale thickness and specimen weight change as functions of time, which are presented in Figure 9.6, are applicable to Regime 1. In Regime 2, the erosion is strong and the oxide growth is slow. The transition from Regime 1 to Regime 2, as the erosion and corrosion intensities vary, is shown in Figure 9.8.

In Regime 1, the oxide scale is not removed completely by erosion but is only thinned down and the erosive impacts do not affect the metal substrate. Consequently the surface of the specimen remains flat, but contains particle cutting features as shown in Figure 9.9. In Regime 2, the scale is removed and the metal substrate undergoes plastic deformation by the impact of the erosive particles. The plastic deformation of the metal surface results in the formation of surface features

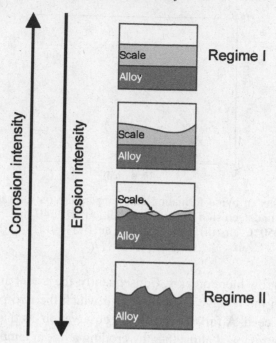

Figure 9.8 Regimes of interaction of erosion and corrosion as erosive and corrosive intensities are varied.

Figure 9.9 Scanning electron micrograph of the surface of cobalt impacted for 30 min at 90° at 800 °C and 90°m s^{-1} by 20 °μm alumina erosive particles.

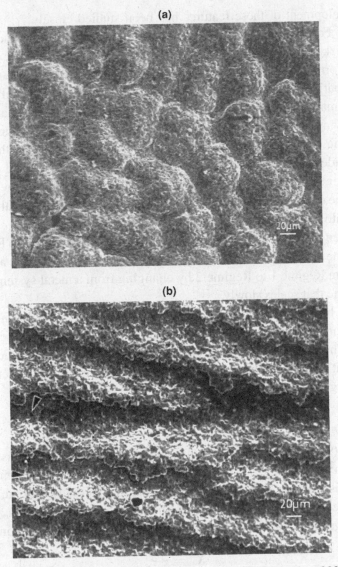

Figure 9.10 Surface topography developed on eroding nickel in air at 800 °C using 20 μm alumina particles: (a) impacting at 140 m s^{-1} at 90° for 30 min; and (b) impacting at 90 m s^{-1} at 30° for 30 min.

in relief that can be substantially larger than the erosive particles themselves. These surface features can resemble hills and valleys at normal incidence of the erosive flow, as shown in Figure 9.10(a), or systems of ripples at oblique incidence, Figure 9.10(b).[13] It is believed that the component of the erosive flow that is parallel to the surface is responsible for the alignment of the ripples and for their slow progression that is observed to occur across the metal surface.[17]

The development of these features has been studied closely[13-18] using nickel and cobalt. For the reasons explained in Chapter 3, the nickel oxide scale that forms on nickel grows slowly whereas the less tightly stoichiometric cobalt oxide scale that forms on cobalt grows rapidly in the temperature range 600–800 °C. Upon comparing cobalt and nickel under similar conditions, namely an erosive stream of small (20 µm) alumina particles loaded at between 500 and 1000 mg min^{-1}, flowing at 90–140 m s^{-1}, and impacted at 90° to the specimen surface, the following differences are evident.[13,17] In the case of cobalt, with the rapidly growing oxide scale, a surface scale formed and grew in spite of the erosive flux, and the outer scale surface remained flat, Figure 9.9. The more slowly oxidizing nickel, on the other hand, developed no continuous scale and the metal surface was deformed into a system of hills and valleys, Figure 9.10(a).

For a given metal system, it is possible to move from Regime 1 to Regime 2 by increasing the erosion rate. For a given set of erosion conditions, it is also possible to move from Regime 1 to Regime 2 by changing from a metal system that scales rapidly to one that scales slowly.

Regime 1 has been described as 'erosion-enhanced oxidation,' because the scale is thinner than it would have been in the absence of erosion, thus the scaling rate is enhanced. The scale grows under diffusion control but its outer surface is eroded at a constant rate by the erosive flux. This situation can be represented by the relationship given in Equation (9.1):

$$\frac{dX}{dt} = \frac{k}{X} - k_{eo}. \tag{9.1}$$

In Equation (9.1), X is the instantaneous scale thickness, t is time, k is the scaling constant (cm^2 s^{-1}) and k_{eo} is the erosion constant for the oxide (cm s^{-1}). This relationship is valid, so long as the erosive impacts affect only the outer surface of the scale. The scale will continue to grow, so long as the term $k/X > k_{eo}$. However, as the scale grows it reaches a critical thickness, X^*, given by Equation (9.2),

$$X^* = \frac{k}{k_{eo}}; \tag{9.2}$$

X^* remains constant because the rate of growth of the scale is then zero. This steady-state thickness, X^*, becomes smaller as k_{eo} increases with more aggressive erosion. This behaviour has been demonstrated using cobalt[19] and is illustrated in Figure 9.11. This paralinear behaviour is similar to that described in Chapter 4 for the oxidation of chromium under conditions where volatile CrO_3 evaporates from the scale–gas interface.[20]

The above considerations are illustrated in Figure 9.6. On exposure to erosion–oxidation, the cobalt specimen begins to gain weight as the oxide scale forms faster than it can be removed by erosion. As the scale thickens it grows more slowly

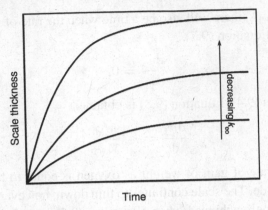

Figure 9.11 Development of a constant-thickness scale on cobalt under erosion–oxidation in Regime 1 at 800 °C for varying k_{eo}.

and results in a lower rate of weight gain. Eventually, the scale reaches its critical thickness where the rate of scale growth by oxidation is equal to the rate of scale removal by erosion. From this point, the specimen loses weight at a constant rate that represents the rate at which metal is being removed as oxide from the outer surface of the scale.

This situation can be treated as follows, measuring weight change per unit area. The oxide formed is MO, w_s is the specimen weight per unit area, w° is the original weight of the metal specimen per unit area, w_O is the weight of oxygen in the scale per unit area, M_M is the atomic weight of the metal, V_{MO} is the molar volume of the oxide MO and t is time. At any time we have the w_s given in Equation (9.3):

$$w_s = w^{\circ} + w_O - k_{eo}\frac{M_M}{V_{MO}}t.\tag{9.3}$$

The rate of change of specimen weight is given by differentiating Equation (9.3) with respect to time:

$$\frac{dw_s}{dt} = \frac{dw_O}{dt} - k_{eo}\frac{M_M}{V_{MO}}.\tag{9.4}$$

When a scale of constant thickness has been achieved, there is no further net gain of oxygen, Equation (9.5):

$$\frac{dw_O}{dt} = 0.\tag{9.5}$$

Consequently, at this stage the rate of change of specimen weight can be expressed as in Equation (9.6):

$$\frac{dw_s}{dt} = -k_{eo}\frac{M_M}{V_{MO}}.\tag{9.6}$$

As the scale develops, there will also be a time when the rate of weight gain of the specimen is zero, Equation (9.7),

$$\frac{dw_s}{dt} = 0,$$ (9.7)

and, from Equation (9.4), Equation (9.8) is obtained:

$$\frac{dw_O}{dt} = k_{eo}\frac{M_M}{V_{MO}}.$$ (9.8)

At this point, the rate of gain of weight as oxygen is equal to the rate of weight loss of metal as oxide. The scale continues to thin down, however, and the constant scale thickness is only achieved when Equations (9.5) and (9.6) are satisfied, as shown in Figure 9.6.

Regime 1 has been studied by several groups of workers[13–17,19] and an interpretation of the mechanism by which the reactions in this regime proceed has been developed.[18,21] The basic concept in this interpretation is that the erosive impacts remove a portion of the surface of the protective scale over a 'footprint' caused by an individual particle impact.[18] During the time between successive impacts at the same site the scale can form more oxide and 'heal.' If the amount of scale formed during healing is less than that which is removed at each impact, then the scale will thin down.

The value of the erosion constant will vary with the erosion conditions, thus under strong erosion the actual rate of scale formation will increase as the scale thickness is held to low values. The rate of scale formation is also the rate at which the metal degrades by conversion to oxide.

From Equation (9.1) it can be seen that if the rates of oxide formation and erosion are equal, then the scale will remain at constant thickness, X^*, given by Equation (9.2). Clearly, the higher the value of k_{eo}, the thinner the scale will be and the faster the metal will degrade. Eventually, the erosion process can be made so severe that the behaviour pattern moves from Regime 1 to Regime 2, when the erosion begins to deform the metal surface.

A semi-quantitative analysis of the model for Regime 1 has been given,[21] as follows. It is assumed that all particles are identical, of volume v and density ρ. They are impinged on the specimen of area A at a loading rate of M (g s^{-1}), each impact produces a footprint of area a and removes material to a depth d. The time between impacts, t^*, for any given site can be written as in Equation (9.9):

$$t^* = \frac{v\rho A}{Ma}.$$ (9.9)

The volume of scale removed at each impact is ad. Since the impact rate of particles on the surface is $M/v\rho$ s^{-1}, the rate of material removal is $Mad/v\rho$ cm^3 s^{-1}. A macroscopic measurement of the erosion rate gives the rate of material removal as $k_{eo}A$. Thus, the depth removed per impact, d, can be obtained, Equation (9.10):

$$d = k_{eo}\frac{Av\rho}{Ma} = k_{eo}t.$$ (9.10)

The time between impacts can be calculated from Equation (9.9) and the depth removed per impact, d, can be calculated from k_{eo}, previously measured for the oxide. The value obtained for d can now be compared with the thickness of oxide that would be expected to form in the time between impacts. If d is greater than the oxide growth, then the scale will become thinner, by the action of erosion. An expression for the scale growth can be derived as follows. For the growth of the oxide we have Equations (9.11) or (9.12):

$$\frac{dX}{dt} = \frac{k}{X};$$ (9.11)

$$X\,dX = k\,dt;$$ (9.12)

the latter may be integrated indefinitely to give Equation (9.13):

$$X^2 = 2kt^* + C.$$ (9.13)

At $t^* = 0$, $X = X^*$ so that Equation (9.14) is obtained,

$$X^2 = X^{*2} + 2kt^*,$$ (9.14)

where k is the parabolic scaling constant, X is the scale thickness after time t^* (time between impacts) from an initial steady-state thickness, X^*, i.e., $X = X^* + d$ under steady state because d is formed and removed every t^* seconds. Thus, substituting into Equation (9.14), the steady-state thickness, X^*, may be written as in Equation (9.15):

$$X^* = \frac{2kt^* - d^2}{2d}.$$ (9.15)

If this value is positive, then the reaction produces a limiting scale of finite thickness and the system is reacting in Regime 1. If the value calculated for X^* were to be negative then this would mean that the scale is not capable of growing quickly enough to overcome the effects of erosion and would, therefore, be reacting in Regime 2. Such values have been calculated for the cases of cobalt and nickel under erosion–oxidation at 700 °C.[13]

The erosion rate constants for nickel oxide, NiO, and cobalt oxide, CoO, at 700 °C in a stream of 20 μm alumina particles flowing at 463 mg min^{-1}, were determined to be 9×10^{-7} and 3.5×10^{-7} cm s^{-1}, respectively. The time between impacts for this erosive flow was 10.5 s yielding values of d, the thickness removed

Table 9.1 *Comparison of calculated with measured steady-state thicknesses of scales formed during erosion–oxidation exposures at 700 °C using 20 μm alumina particles flowing at 463 mg min^{-1}. The time between impacts was 10.55 S*

System	Erosion constant, k_{eo} (cm s^{-1})	Time between impacts, t^*(s)	$d = k_{eo}t^*$(cm)	X^* (cm) Calculated from Equation (9.12)	X^* (cm) measured (ref. 13)
NiO	9.0×10^{-7}	10.5	9.5×10^{-6}	-4.5×10^{-6}	No scale
CoO	3.5×10^{-7}	10.5	3.7×10^{-6}	1.9×10^4	$3–4 \times 10^{-4}$

after each impact of 9.5×10^{-6} and 3.7×10^{-6} cm for NiO and CoO, respectively. The parabolic scaling constants for these oxides were taken to be 2.04×10^{-13} and 6.73×10^{-11} cm^2 s^{-1} yielding values for X^*, from Equation (9.15), of -4.5×10^{-6} and 1.9×10^{-4} cm for NiO and CoO, respectively. These calculations are shown in Table 9.1. A comparison of calculated and measured values for X^* are also presented in Table 9.1.

From these results, the erosion–oxidation of nickel under the above conditions would place the system in Regime 2, whereas cobalt would be in Regime 1. This was observed and confirms this simple semi-quantitative analysis.

The above treatment works quite well for scales formed of oxides such as CoO, which is not brittle at high temperature. Nickel oxide, while showing primarily ductile response to erosive impact, has occasionally shown signs that some scale removal occurred by spalling. It is to be expected that scales formed of more strongly stoichiometric oxides, such as chromia and alumina, would be more likely to fail by spallation.

Region 2 is the least understood region of erosion–corrosion interaction. It is possible to describe some of the mechanisms that are responsible for the degradation in this regime, but it is extremely difficult to describe them quantitatively. In this regime, the rapid degradation of the substrate leaves little trace of oxide on the surface. This implies that the oxide is being removed completely as it forms. Consequently, oxide formation, and its removal, must be proceeding at the maximum possible rate that applies to an initially clean metal surface; in which case the rate of heating due to the release of oxidation enthalpy must be high. In addition, the erosive impacts deform the substrate into a convoluted surface with increased surface area. Furthermore, the energy of the impacts is expended in plastic deformation of the substrate surface, so rapidly that substantial heating must be expected.[5,6,22–24] It is difficult to account quantitatively for all of these phenomena.

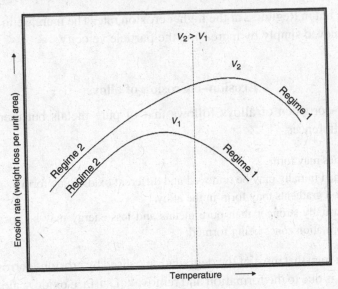

Figure 9.12 Kinetic features of erosion–oxidation in Regime 1 and in Regime 2.

It is interesting to consider the erosion–corrosion process as a function of temperature and several authors have expressed their results in this way obtaining a 'bell curve' plot of rate of attack against temperature for two different velocities.[25–27] This behaviour, shown in Figure 9.12, can also be interpreted in terms of the mechanisms discussed above. At low temperatures, corrosion or oxidation proceeds very slowly, the erosion–corrosion interaction is dominated by the erosion term. Thus, the overall rate of degradation is low. As the temperature is increased, the oxidation component becomes more active and converts the surface metal to oxide at a higher rate. This oxide is removed by erosion and the overall degradation rate increases. Initially, at low temperatures the system starts out in Regime 2. As the temperature is increased, the system enters Regime 1 at a temperature where the erosion is incapable of removing all of the oxide that forms. At this point the rate of degradation is at its maximum value for these conditions. A further increase in temperature results in the formation of a steady-state oxide layer and a corresponding reduction in degradation kinetics. As the temperature is increased further, the substrate and oxide become more plastic allowing more of the incident energy to be absorbed by plastic deformation of the target surface. Under these conditions, the weight loss due to erosion is reduced and the weight gain due to oxidation is increased, giving a net reduction in degradation rate.

Figure 9.12 shows that increasing the erosive intensity will move the entire curve to the right, giving a peak rate at a higher temperature. It can also be seen that, at a temperature indicated by the dotted line, the system is in Regime 1 at the lower erosion rate, but in Regime 2 at the higher erosion rate. The increase in erosion rate could be achieved simply by increasing the particle velocity.

Erosion–corrosion of alloys

The erosion–corrosion of alloys follows that of pure metals but there are some important differences.

- Several oxides may form.
- Oxides formed initially may be removed and different oxides may form.
- Concentration gradients may form in the alloy.
- Alloys are usually stronger than pure metals and less energy may be absorbed due to thinner deformation zones being formed.

It has been seen that most of the degradation caused by erosion–corrosion at high temperatures is due to the formation and removal of surface oxide scale. The direct removal of metal by erosion plays a relatively minor role. Thus, alloys that form slow-growing, protective, oxide scales would be expected to degrade more slowly under erosion–corrosion than alloys whose scales form more rapidly. This has been reported[28] for cases where the erosive intensity was very low. However, there are two circumstances that are capable of defeating this expectation. The first is that, as an oxide scale is removed and reforms on an alloy, it draws the metal forming the most protective oxide scale from the alloy, and forms a diffusion gradient in the alloy surface. In this case the concentration of the protective metal at the surface will fall with time and will eventually no longer be able to support formation of this protective oxide. This is similar to the situation with persistent spalling. The second circumstance is that, under erosion, the protective oxide that forms on the alloy surface is unable to grow laterally to form complete coverage of the surface. In this case, the alloy surface is held by the erosion in a state of extended transient oxidation, when all oxides that are thermodynamically capable of forming do so. Since the major alloy partner is usually a metal such as iron or nickel that forms a rapidly growing oxide, this situation is expected to result in rapid degradation rates, in spite of the presence in the alloy of elements that normally provide a passivating oxide layer.

This type of behaviour has been observed[29] for the nickel–chromium alloy system. In this work, pure nickel, Ni–20 wt% Cr and Ni–30 wt% Cr alloys were exposed to erosion–corrosion at 700 and 800 °C. An erosive stream, loaded at 400 mg min^{-1} of 20 μm alumina particles and flowing at 75 and 125 m s^{-1}, was used to impact normally on the specimens. Under simple oxidation in air at these temperatures, the alloys both developed protective scales of chromia and showed very low rates of

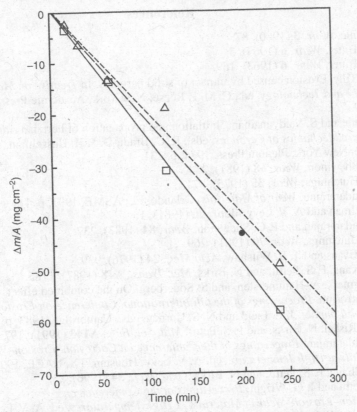

Figure 9.13 Effect of alloying with chromium on the erosion–oxidation kinetics of nickel exposed to an erosive stream at 90° incidence, 700 °C and 75 m s^{-1} Ni (---); Ni–30 wt% Cr (– –); Ni–20 wt% Cr(—).

attack. Under erosion–corrosion conditions however, both alloys degraded at high rates that were comparable, or even identical, to the rate at which the pure nickel was attacked. The results are shown in Figure 9.13. Both nickel oxide and chromia were found to exist on the eroded surface but there was no trace of the spinel. This shows that these oxides were being removed rapidly, since NiO and Cr_2O_3 were not present for sufficient time to form spinel. These results were interpreted by proposing that the erosive stream prevents the formation of a continuous layer of chromia by removing the oxide faster than it can spread laterally. These specimens were described as being in a state of 'erosion-maintained transient oxidation'. This mechanism implies that it would be difficult for protective scales to form and persist on alloys in the presence of erosion. Consequently, the oxidation behaviour of an alloy in the absence of erosion cannot be used as a guide to its behaviour under erosion–oxidation at high temperature.

References

1. I. Finnie, *Wear*, **3** (1960), 87.
2. J. G. Bitter, *Wear*, **6** (1963), 5.
3. J. G. Bitter, *Wear*, **6** (1963), 169.
4. G. P. Tilly, Erosion caused by impact of solid particles. In *Treatise on Materials Science and Technology*, ed., C. M. P. Reese, New York, Academic Press, 1979, p. 287.
5. I. Finnie and S. Vaidyanathan, 'Initiation and propagation of hertzian ring cracks.' In *Fracture Mechanics of Ceramics*, eds., R. C. Bradt, D. P. H. Hasselman, and F. F. Lange, New York, Plenum Press, 1974, p. 231.
6. P. G. Shewmon, *Wear*, **68** (1981), 254.
7. I. M. Hutchings, *Wear*, **35** (1975), 371.
8. G. Sundararajan, *Wear of Materials*, Orlando, FL, ASME, 1991, p. 111.
9. R. Bellman and A. V. Levy, *Wear*, **70** (1981), 1.
10. G. Sundararajan and P. G. Shewmon, *Wear*, **84** (1983), 237.
11. I. M. Hutchings, *Wear*, **70** (1981), 269.
12. A. G. Evans and T. R. Wilshaw, *Acta Met.*, **24** (1976), 939.
13. C. T. Kang, F. S. Pettit, and N. Birks, *Met. Trans.*, **A18** (1987), 1785.
14. S. Hogmark, A. Hammersten, and S. Soderberg, On the combined effects of erosion and corrosion. *Proceedings of the 6th International Conference on Erosion by Liquid and Solid Impact*, J. E. Field and N. S. Corney, eds., Cambridge, 1983, p. 37.
15. D. M. Rishel, N. Birks, and F. S. Pettit, *Mat. Sci. Eng.*, **A143** (1991), 197.
16. G. Sundararajan, *Proceedings of the Conference on Corrosion–Erosion–Wear of Materials at High Temperature*, ed. A. V. Levy, Houstan, TX, NACE, 1991, p. 11–1.
17. S. L. Chang, F. S. Pettit, and N. Birks, *Oxid. Met.*, **34** (1990), 23.
18. V. J. Sethi and I. G. Wright, *Proceedings of the Conference on Corrosion–Erosion–Wear of Materials at High Temperatures*, ed. A. V. Levy, Houstan, TX, NACE, 1991, p. 18-1.
19. S. L. Chang, F. S. Pettit, and N. Birks, *Oxid. Met.*, **34** (1990), 47.
20. C. S. Tedmon, *J. Electrochem. Soc.*, **113** (1966), 766.
21. N. Birks, F. S. Pettit, and B. Peterson, Mat. Sci. Forum, **251–254**, (1997), 475.
22. S. L. Chang, Ph.D. Thesis, University of Pittsburgh, PA, 1987.
23. R. A. Doyle and A. Ball, *Wear*, **151** (1991), 87.
24. I. M. Hutchings and A. V. Levy, *Wear*, **131** (1989), 105.
25. M. M. Stack, F. H. Stott, and G. C. Wood, *Corr. Sci.*, **33** (1992), 965.
26. D. J. Hall and S. R. J. Saunders, *High Temperature Materials for Power Engineering*, Liége, Belgium, Kluwer Academic Publishers, 1990.
27. B. Q. Wang, G. Q. Geng, and A. V. Levy, *Wear*, **159** (1992), 233.
28. A. V. Levy, *Solid Particle Erosion–Corrosion of Materials*, Materials Park, OH, ASM International, 1995.
29. R. J. Link, N. Birks, F. S. Pettit, and F. Dethorey, *Oxid. Met.*, **49** (1998), 213.

10

Protective coatings

Introduction

Coatings have been used for centuries for embellishment or for protection of a substrate that is adequate in all other ways, usually providing shape, stiffness, or strength. The greatest use of coatings has been at room temperature by the jewellry industry for appearance, for ceramics by enamelling to improve appearance and provide impermeability, and by the automobile industry for corrosion protection. In these cases, a very high degree of success has been achieved.

In the use of coatings at high temperatures, the aim is either to protect the surface of an inexpensive, but degradable material, as in the use of coextruded tubing for heat exchangers, or to protect the surface of an alloy that has been developed for strength but does not have sufficient intrinsic resistance to high-temperature corrosion. Typical applications consist of power generation using steam boilers, which are usually fired by fossil fuel; and in gas turbines, which are used for aircraft or marine propulsion, or are land based for power generation. In recent years, the use of ceramic coatings as insulators, thermal-barrier coatings, has also become important for limiting the degradation of alloys, particularly in gas turbines. The various types of coatings are described extensively in ref. 1.

This chapter is organized to give an overview of the fabrication, use, and degradation of the various types of coatings that are available for use in high-temperature systems.

Types of coating systems

Diffusion coatings

In diffusion coating, the substrate surface is enriched in an element that will provide high-temperature corrosion resistance. Typical elements are chromium (chromizing), aluminium (aluminizing), or silicon (siliconizing). The substrate is involved

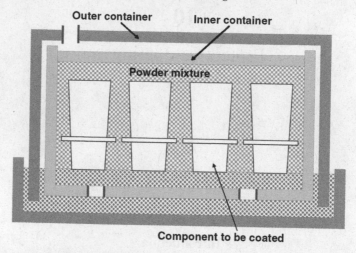

Figure 10.1 Schematic diagram of the apparatus used for aluminizing by pack cementation.

in the formation of the coating and substrate elements are in the coating; a diffusion zone is developed in the substrate beneath the coating. Such enrichment not only allows a protective scale to form by selective oxidation, but also provides a substantial reservoir of the protective element to displace the inevitable breakaway oxidation to substantially longer times. A wide variety of diffusion coatings is used. In this section aluminide coatings on Ni-base alloys will be described in some detail to illustrate key principles, following which the other diffusion coatings will be briefly described.

Aluminizing of nickel-base alloys

The most common method of aluminizing is pack cementation, which has been a commercially viable process for many years.[2] This process, which is shown schematically in Figure 10.1,[3] involves immersing the substrate in a mixture of powders. The mixture contains a powder source of Al, either Al metal or a suitable master alloy, a halide activator, and a filler, which is usually alumina and taken to be inert. A pack will usually contain 2–5% activator, 25% source and the rest filler. The purpose of the filler is to support the component and to provide a porous diffusion path for the gases generated by reaction between the source and activator.

The activator volatilizes within the pack on heating and reacts with the source to produce volatile species of the coating metal. For example, in NaCl-activated packs, the reactions shown in Equations (10.1)–(10.3) take place:

$$3NaCl\,(g) + Al\,(l) = AlCl_3\,(g) + 3Na\,(g), \tag{10.1}$$

$$2AlCl_3\,(g) + Al\,(l) = 3AlCl_2\,(g), \tag{10.2}$$

$$AlCl_2\,(g) + Al\,(l) = 2AlCl\,(g). \tag{10.3}$$

The relative partial pressures of the aluminium chlorides will be determined by the activity of Al in the source, the amount of activator used, and temperature. The pack is usually held under a gas such as argon to minimize oxidation of the source and substrate.

The volatile species diffuse through the pack to the substrate surface where deposition reactions take place. The mixture of volatile gas species reflects the activity of aluminium that exists in the source. The component to be coated represents a low-activity sink for the aluminium so that various deposition reactions, including those given in Equations (10.4)–(10.8), are possible.

$$\textit{Disproportionation} \quad 2\text{AlCl}(g) = \underline{\text{Al}} + \text{AlCl}_2(g) \tag{10.4}$$

$$3\text{AlCl}_2(g) = \underline{\text{Al}} + 2\text{AlCl}_3(g) \tag{10.5}$$

$$\textit{Displacement} \quad \text{AlCl}_2(g) + \underline{\text{Ni}} = \underline{\text{Al}} + \text{NiCl}_2(g) \tag{10.6}$$

$$\textit{Direct reduction} \quad \text{AlCl}_3(g) = \underline{\text{Al}} + \frac{3}{2}\text{Cl}_2(g) \tag{10.7}$$

And, for activators which contain hydrogen, e.g. NH_4Cl, hydrogen reduction occurs, Equation (10.8).

$$\textit{Hydrogen reduction} \quad \text{AlCl}_3(g) + \frac{3}{2}\text{H}_2(g) = \underline{\text{Al}} + 3\text{HCl}(g) \tag{10.8}$$

(The underlined symbols refer to species in the solid substrate.) Displacement reactions are not significant in the aluminizing of Ni-base alloys but often predominate in processes such as chromizing of iron and steels.[4]

The activity of Al established at the substrate surface will be determined by the activity of Al in the source and by the kinetics of transport of the reactive gases through the pack. Both of these factors will be influenced by depletion of Al from the source and activator depletion, and hence by the extent to which the pack vents gases to the surroundings. The transport kinetics during aluminizing have been analyzed by Levine and Caves[5] and Seigle and coworkers.[6–9] The activity of Al will fix the Al concentration and, therefore, the phases, which form in the coating, as can be seen from the phase diagram in Figure 10.2. Figure 10.3[10] shows two typical coatings prepared by pack aluminizing of a Ni-base superalloy. The coating shown in Figure 10.3(a) was formed at low temperature in a high Al activity pack to produce an inward-growing coating. The coating shown in Figure 10.3(b) was formed at high temperature in a low Al activity pack to form an outward-growing coating. The formation mechanisms of these coatings are described in detail in ref. 2. Pack cementation has several advantages over other methods of producing diffusion coatings. One of these is that the pack serves the additional purpose of supporting the substrates. This prevents sagging of large parts and some commercial

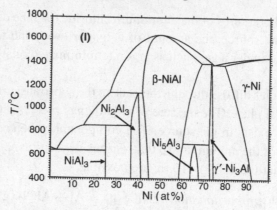

Figure 10.2 Phase diagram for the Ni–Al system.

processes such as Alonizing™ produce Al-rich coatings on tubes, which are many meters in length.

An additional advantage of pack cementation is that the pack and substrate are in contact, which facilitates composition uniformity and gives generally high deposition rates. One disadvantage of this contact, however, is that pack material can be captured in the coating. This is particularly the case for outward-growing, low-activity coatings. In cases where this is undesirable, the substrate can be separated from the pack. In the above-the-pack process,[3] Figure 10.4, the substrates are supported above the pack. The coating gases are generated in the pack and then flow upward over the substrates. Another modification of the process is termed chemical vapour deposition (CVD)[3] although it should be noted that pack cementation and above-the-pack coating, in fact, are CVD processes. In the CVD process, Figure 10.5, the coating gases are generated externally and then fed into an evacuated retort, which contains the substrates. The CVD process has the additional advantages of greater compositional flexibility and the capability to have the coating gases ducted through internal passages, such as cooling holes in turbine blades.

Currently, the improvement in hot corrosion resistance imparted to Ni–Al alloys by Cr additions has resulted in research efforts aimed at codepositing Cr and Al in aluminide coatings. The simultaneous codeposition of Al and Cr via the pack cementation process, using pure elemental powders, is, however, difficult. The large difference in the thermodynamic stabilities of the Al and Cr halides causes aluminium halide species to predominate in the pack atmosphere.[11] However, by employing binary chromium–aluminium (Cr–Al) master alloys, the high relative aluminium halide vapour pressures can be moderated. This is the result of the fact that chromium-rich master alloys exhibit negative deviations from ideality and the activity of Al in the master alloy can be reduced by several orders of magnitude. The reduced thermodynamic activity of Al results in generation of lower vapour

(a)

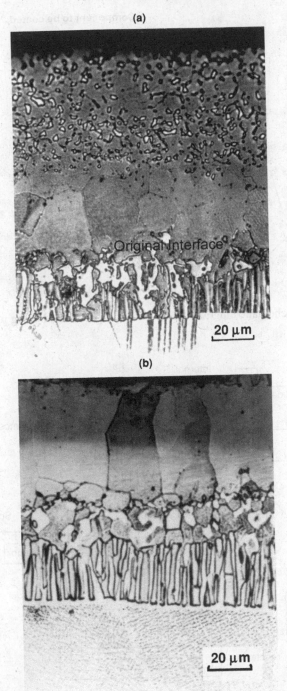

(b)

Figure 10.3 Cross-section micrographs of (a) a 'high-activity' aluminide coating (after annealing) and (b) a 'low-activity' aluminide coating on a nickel-base super-alloy. (These micrographs originally appeared in: G. W. Goward and D. H. Boone, *Oxid. Met.*, **3** (1971), 475, and are reproduced with the kind permission of Springer Science and Business Media.)

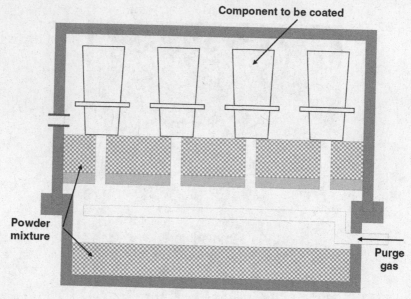

Figure 10.4 Schematic diagram of an above-the-pack aluminizing process.

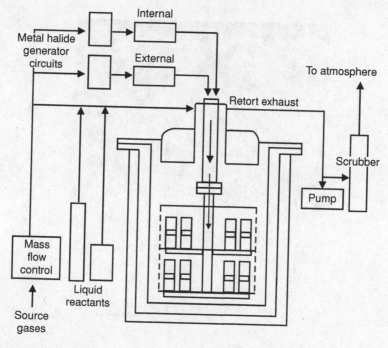

Figure 10.5 Schematic diagram of a CVD aluminizing process.

Figure 10.6 Optical micrograph showing the cross-section of an as-processed modified aluminide coating on a nickel-base superalloy. This particular coating was fabricated under conditions which produced a second phase, PtAl$_2$ (white contrast), in a matrix of β-NiAl.

pressures for the otherwise-favoured aluminium halide species (e.g., AlCl, AlCl$_2$, etc.). Therefore comparable aluminium halide and chromium halide vapour pressures result. Thus, provided a suitable activator and binary master alloy is chosen, the codeposition of Cr and Al into Ni-base materials is possible. Rapp and coworkers[12–14] have used this approach to form coatings containing as much as 13 at% Cr in β-NiAl on Ni-base superalloys and have shown them to have greater hot-corrosion resistance at 900 °C than aluminide coatings without the Cr modification.[14] Da Costa *et al.*[15,16] have achieved Cr concentrations as high as 40 at% using Cr-rich masteralloys and multiple activators (NaCl + NH$_4$Cl).

An important modification of diffusion aluminide coatings is the Pt-modified aluminide coating.[3] These coatings are produced by electroplating a thin layer of Pt or Pt–Ni alloy onto the substrate and annealing it to diffuse the platinum prior to aluminizing. Since aluminium diffuses inward through the platinum layer, the resulting coating consists of β-NiAl with Pt in solution at the coating–gas interface. Under some fabrication conditions a second phase, PtAl$_2$, precipitates in the coating. This phase is evident in Figure 10.6, which shows the as-coated microstructure of a Pt-modified aluminide coating on a Ni-base superalloy substrate. It has been shown, for a given set of annealing and aluminizing conditions, that use of thin Pt layers results in single-phase β-NiAl with Pt in solution; use of thicker Pt layers results in a two-phase (β + PtAl$_2$) coating; use of still thicker layers of Pt results in a complete layer of PtAl$_2$ over the β-NiAl.[17] The presence of the Pt layer has also been clearly shown to increase the Al uptake in the coating, for fixed aluminizing conditions.[18] This is the result of the Pt lowering the activity coefficient of Al in the coating. The Pt-modified aluminide coatings are observed to have markedly improved resistance to cyclic oxidation and hot corrosion,

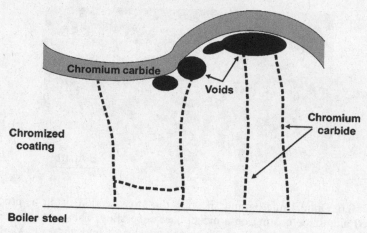

Figure 10.7 Schematic cross-section of a chromized coating formed on a typical boiler steel.

compared to the straight aluminides for a given superalloy substrate.[19] The possible mechanisms for this improvement will be described below in the section on coating degradation.

Chromized coatings

Chromized coatings can be formed by pack cementation in a manner similar to that described for aluminizing.[4,20] Figure 10.7 presents a schematic cross-section of a typical chromized coating on a boiler steel. The coating consists of ferrite and Cr in solution and a Cr concentration gradient running from the surface to the coating–substrate interface. Carbon has diffused out from the steel and formed a chromium-carbide layer on the surface and chromium carbides have precipitated on grain boundaries in the coating. Voids are also evident below the outer carbide layer. When undesirable, the carbide layer and voids can be eliminated by choice of activator.[4,20]

Silicide coatings

The refractory-metal silicides have been used for many years to protect refractory metals from oxidation in very high-temperature, but short-duration applications.[21] These coatings have been highly successful but their use in applications which require long-term stability has been limited by problems with accelerated oxidation and pesting, evaporation of SiO at low oxygen partial pressures, interdiffusion with the substrate, and cracking because of thermal-expansion mismatch between the coating and substrate. These factors have been reviewed in detail by Packer[22] and Kircher and Courtright.[23] Work by Rapp and coworkers[24,25] has been directed at improving the resistance of $MoSi_2$-base coatings for Nb-base alloys, for which there is a good thermal-expansion match, and Mo-base alloys, for which there is a relatively poor thermal-expansion match. The coatings on Nb, formed by pack cementation, consisted of W additions to the $MoSi_2$ to strengthen it, and Ge additions. The latter increases the thermal-expansion coefficient of the protective SiO_2 layer, which improves the cyclic-oxidation resistance and lowers the viscosity, thereby reducing accelerated oxidation at low temperatures. The coatings were reported to provide cyclic-oxidation resistance on Nb for 200 h at 1370 °C.[24] The coatings on Mo included Ge doping and also Na doping,[26] by means of a NaF activator in the coating pack, to successfully limit accelerated oxidation and pesting at low temperatures.[25] Data with regard to coating cracking during thermal cycling from high temperatures, which would be expected to be severe because of the poor match between the thermal-expansion coefficients of Mo and $MoSi_2$, have not been reported.

Other methods of forming diffusion coatings

Diffusion coatings can be formed by a number of methods which have not been described in detail. One such technique is *slurry coating* where the slurry is formed by using a volatile suspension medium that burns off upon heating. This can be accomplished by applying a slurry containing the activator and source metal on the surface of the substrate and heating to form a coating in a manner similar to pack cementation. Another variant involves using an activator-free slurry which will melt when the substrate is heated and then resolidify during interdiffusion with the substrate. Another simple technique is *hot-dip coating* in which the substrate is immersed in a liquid bath. The most common use of this process is for zinc coating (galvanizing) of steels but it is also extensively used to produce Al coatings on steels.

Overlay coatings

Overlay coatings are distinguished from the diffusion coatings in that the coating material is deposited onto the substrate in ways that only give enough interaction with the substrate to provide bonding of the coating. Since the substrate does not enter substantially into the coating formation, in principle, much greater coating-composition flexibility is achievable with overlay coatings. Also, elements that are difficult to deposit into diffusion coatings, can be included in overlay coatings. An import example of this is Cr. As discussed above, Cr is difficult to incorporate into diffusion aluminide coatings, whereas it is readily incorporated into overlay coatings. Overlay coatings based on the Ni–Cr–Al and Co–Cr–Al systems are commonly used to protect superalloys. Also, small amounts of the reactive elements (e.g., Y, Hf) are routinely incorporated into overlay coatings but are difficult (or sometimes impossible) to incorporate into diffusion coatings. The composition flexibility of the overlay coatings also allows mechanical properties to be tailored for a given application.

Overlay coatings are deposited by physical techniques. The most common are physical vapour deposition (PVD), which includes evaporation, sputtering and ion plating, and spray, techniques (plasma spraying, flame spraying, etc.).

Physical vapour deposition (evaporation)

Physical vapour deposition is very versatile. It can be used to deposit metals, alloys, inorganic compounds, or mixtures of these and even some organic materials. The process consists of three principal steps.

- Synthesis of the material to be deposited.
- Transport of vapours from the source to the substrate.
- Condensation, film nucleation, and growth on the substrate.

Physical vapour deposition produces an overlay coating which has several important advantages over some other techniques.

- Composition flexibility.
- The substrate temperature can be varied over a wide range.
- High deposition rates.
- Excellent bonding.
- High-purity deposits.
- Excellent surface finish.

Evaporation is carried out under reduced pressure. The chamber is evacuated and back filled with the required gas at a suitable low pressure. Since evaporation is a line of sight process, the component must be rotated and turned to provide a uniform coating on complex shapes. The adhesion of the coating can be improved by heating the substrate, using quartz lamps or a diffuse electron beam.

The source can be heated by a variety of techniques. Resistance heating can be used, employing W, Mo, or Ta heater wire or tape. These have the advantage that their low vapour pressures do not cause contamination of the deposit. The source can be contained in a crucible made of boron nitride or titanium diboride. Induction heating can also be used to heat a susceptor source contained in a non-susceptor crucible. Electric-arc and laser-beam sources can also be used to heat the evaporant.

The most common heating method for producing PVD coatings for high-temperature systems is electron-beam heating, which uses a target contained on a water-cooled hearth or crucible. This avoids contamination and a high power density

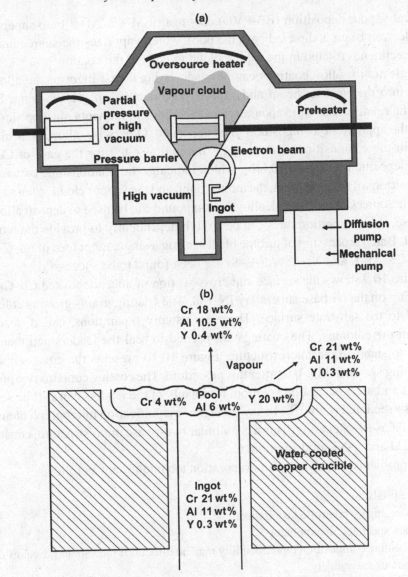

Figure 10.8 (a) Schematic diagram of a PVD apparatus and (b) details of the relationship between the pool and vapour compositions for deposition of a Co–Cr–Al–Y coating.

is available, giving good control over evaporation rates. Electron beams, generated using thermionic guns, are easily manipulated. They do not, therefore, require line of sight to the target but can be bent to impinge around corners. This offers advantages in the spatial arrangement of the equipment. Figure 10.8(a) presents a schematic diagram of the type of equipment that is used to deposit electron-beam physical vapour deposition (EB-PVD) coatings of M–Cr–Al–Y onto superalloys. The electron beam is directed onto the pool, which vaporizes the source material. The specimen is rotated in the vapour cloud to deposit the coating.

Single-source alloy coatings can be made, using a cast ingot of the alloy that is fed into the electron beam and vaporized to deposit on the substrate. In this case, the relative rates of evapouration of the alloy components must be similar to allow the appropriate composition to be deposited. Otherwise, allowances must be made in the composition of the ingot. This is illustrated for the case of Co–Cr–Al–Y deposition in Figure 10.8(b), which indicates the relationships between the compositions of the feed ingot, the molten pool, and the vapour cloud. Alternatively, multiple sources of pure metal, alloy, or compound can be used to deposit alloys. In this case, the evaporation rates can be fixed independently to provide the required deposit. Batch processing of turbine blades, using a single ingot feed of Ni–Cr–Al–Y, Co–Cr–Al–Y, and Ni–Cr–Al–Y–Si has been found to be successful.

Figure 10.9 shows the surface and cross-section of an as-deposited Co–Cr–Al–Y coating on the Ni-base superalloy IN-738. The coating grains grow as columns, normal to the substrate surface. There are usually separations, called 'leaders,' between the columns. The coatings are peened to seal the leaders and then heat treated to stabilize the microstructure. Figure 10.10 presents the cross-section of the coating shown in Fig. 10.9 after this procedure. The coating contains two phases, β-CoAl and a Co-base solid solution. Remnants of the leaders can still be seen in the cross-section. Figure 10.11 shows a higher magnification micrograph of an EB-PVD Ni–Cr–Al–Y coating. This has a similar two-phase microstructure, consisting of β-NiAl and γ-Ni solid solution.

The problems associated with evaporation techniques are as follows.

- High capital cost.
- Complex, highly controlled equipment.
- Requires specimen manipulation.
- Alloy coating composition reproducibility may be difficult if the vapour pressures of the components vary widely.

An alternative approach to deposition of alloys could be to put down layers of pure metals and use a diffusion anneal. Unfortunately, this complicates the process and reduces the overall deposition rate.

Reactive evaporation can be used to deposit compounds. For instance, aluminium could be evaporated into an atmosphere containing oxygen to deposit a layer of alumina. Typically, the gas pressure in a PVD system is so low that the mean free path is greater than the source-to-target distance, which means that the reaction must take place on the substrate surface.

(a)

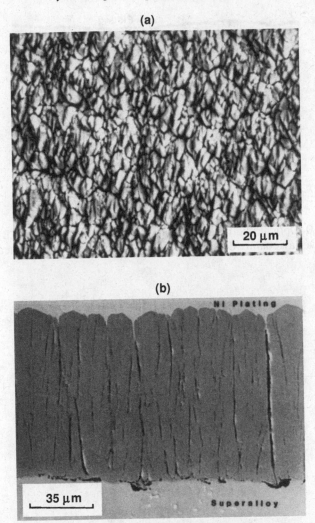

(b)

Figure 10.9 (a) Scanning electron micrographs showing the surface and (b) cross-section of an EB-PVD Co–Cr–Al–Y overlay coating deposited on IN-738.

Plasma spray processes

The plasma spray technique has been used for about 30 years.[27] Figure 10.12 shows a simple schematic diagram of a plasma torch. Basically, a powder is fed into the plasma, which is used as a heater, and the molten product is sprayed onto the substrate to form just about any thickness. The coatings tend to be porous and of poor adhesion. The adhesion can be improved by increased surface roughness of the substrate, which can be produced by grit blasting. An interlayer bond coat of Ni–Al

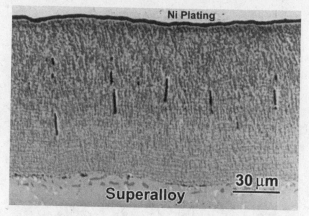

Figure 10.10 Optical micrograph showing the cross-section of an EB-PVD Co–Cr–Al–Y overlay coating deposited on IN-738 after peening and annealing.

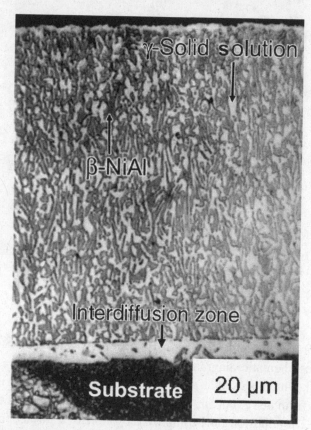

Figure 10.11 Optical micrograph showing the cross-section of an EB-PVD Ni–Cr–Al–Y Coating.

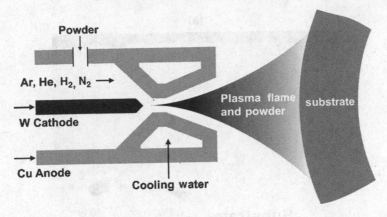

Figure 10.12 Schematic diagram of a plasma torch.

or Mo can also be used to improve adhesion. The porosity can also be virtually eliminated by using appropriate deposition conditions.

Plasma spraying does not cause distortion of the substrate and can be carried out under atmospheric pressure or at reduced pressure. Even the most refractory metals can be melted in the plasma. On the other hand, low-melting-point materials, such as polymers, can also be sprayed. Higher particle velocities improve the coating bond strength and the density. Typically, the velocities of 20 μm diameter particles emerging from the plasma gun are about 275 m s^{-1} at a distance of 5–6 cm from the nozzle. The high rate of heat transfer and the short dwell time reduce the chances of oxidation of the metal particles. Even so there is some entrapment of air in the spray stream and coatings containing more reactive metals (Al, Ti, Y, etc.) are deposited in a vacuum chamber (low-pressure plasma spray, LPPS) or with the plasma shrouded with an inert gas, such as argon. Figure 10.13 shows the microstructure of a Ni–Co–Cr–Al–Y coating, which was deposited by the argon-shrouded technique. The structure is seen to be quite similar to that produced in an EB-PVD coating. The LPPS and argon-shrouded techniques have an important advantage in that virtually any alloy composition that can be produced as a powder can be sprayed to form a coating.

Other spray techniques

In the *detonation gun* technique fixed amounts of powder are accelerated by successive explosions of an acetylene–oxygen mixture. This allows velocities of the order 750 m s^{-1} and temperatures of the order of 4000 °C to be achieved leading to deposits of excellent bond strength but with some porosity.

Flame spraying has been used mainly for oxides and ceramics that will not react with the flame. Heat is provided by an oxy-fuel gas flame, and the material is fed

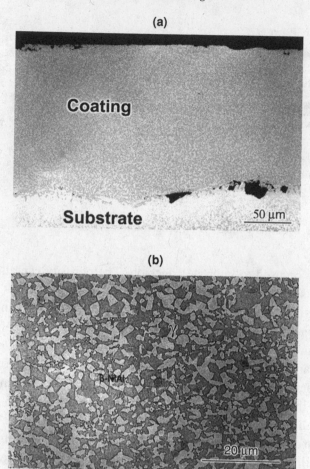

Figure 10.13 Scaning electron micrographs showing (a) the cross-section of a Ni–Co–Cr–Al–Y coating deposited on a single crystal superalloy by argon-shrouded plasma spraying. (b) A higher magnification micrograph showing more details of the microstructure.

in as a wire or powder. Thicknesses of 50 μm or more can be reached, but the coatings are more porous than those obtained using the detonation gun or plasma spray techniques.

Other techniques for depositing overlay coatings

In *sputtering*, which is another form of PVD, the target, or source, is excited under vacuum into the gas phase using momentum transfer from the impact of high-energy ions from an ion gun. The sputtered source material is directed onto the substrate to form the coating.

Cladding is perhaps the oldest form of coating. This consists of simply covering a substrate with a protective surface layer. This is generally done mechanically using roll bonding, coextrusion, or explosive cladding. Coextrusion has been used for applications in the electricity industry where the strength of low-grade, but inexpensive, steel tubes is supplemented with the oxidation resistance of materials such as stainless steel or IN-671 (50 wt% Ni–50 wt% Cr).

Thermal-barrier coatings

Thermal-barrier coatings (TBCs) are ceramic coatings which are applied to components for the purpose of insulation rather than oxidation protection. The use of an insulating coating coupled with internal air cooling of the component, lowers the surface temperature of the component with a corresponding decrease in the creep and oxidation rates of the component. The use of TBCs has resulted in a significant improvement in the efficiency of gas turbines.[28–30]

The earliest TBCs were frit enamels which were applied to aircraft engine components in the 1950s.[28] The first ceramic TBCs were applied by flame spraying and, subsequently, by plasma spraying. The ceramic materials were alumina and zirconia (MgO- or CaO-stabilized), generally applied directly to the component surface. The effectiveness of these coatings was limited by the relatively high thermal conductivity of alumina and problems with destabilization of the zirconia-base materials.[28] Important developments included the introduction of Ni–Cr–Al–Y bond coats and plasma-sprayed Y_2O_3-stabilized zirconia topcoats in the mid 1970s and the development of EB-PVD to deposit the topcoat in the early 1980s.[28] Plasma-sprayed TBCs have been used for many years on combustion liners but, with advanced TBCs, vanes, and even the leading edges of blades, can now be coated. The use of TBCs can achieve temperature differentials across the coating of as much as 175 °C.[31]

Typical systems, shown schematically in Figure 10.14, consist of a nickel-base superalloy substrate coated with M–Cr–Al–Y (M = Ni, Co) or a diffusion aluminide bond coat, which forms an alumina layer (thermally-grown oxide, TGO). Onto this is deposited a yttria-stabilized zirconia (YSZ) TBC. The TBC can be deposited by air plasma spraying (APS), or EB-PVD. The EB-PVD coatings are used for the most demanding applications, such as the leading edges of airfoils. Figure 10.15 presents cross-section micrographs of typical APS and EB-PVD TBCs. The APS coating consists of layers of *splats* with clearly visible porosity and is microcracked. This microcracking is necessary for strain tolerance. The EB-PVD coating consists of columnar grains, which are separated by channels similar to the leaders seen in Figure 10.9 for a metallic overlay coating. These channels are responsible for the EB-PVD TBCs having a high strain tolerance.

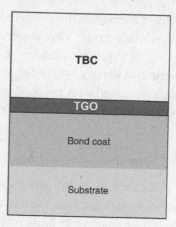

Figure 10.14 Schematic diagram of a typical TBC.

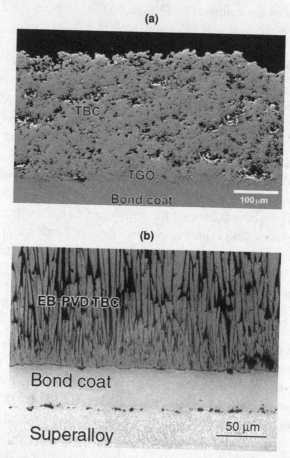

Figure 10.15 Scanning electron micrographs showing cross-sections of (a) an as-processed APS and (b) an as-processed EB-PVD TBC. The APS coating contains an array of fine microcracks, which are not readily visible in the low magnification micrograph shown here.

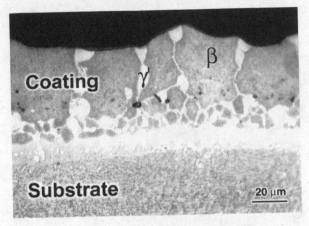

Figure 10.16 Optical micrograph showing the cross-section of a Pt-modified aluminide coating on a nickel-base single-crystal superalloy after oxidation at 1200 °C for 20 h. The original grain structure of the β-phase is evident and γ' has begun to nucleate at β grain boundaries as a consequence of Al depletion.

Coating degradation

Oxidation-resistant coatings

The degradation modes of coatings are essentially the same as those described for cyclic oxidation of alloys in Chapter 5, mixed-oxidant corrosion of alloys in Chapter 7, and hot corrosion of alloys in Chapter 8. However, additional factors arise with coatings; since they are relatively thin, they contain a finite reservoir of the scale-forming elements (Al, Cr, Si), and interdiffusion with the substrate can both deplete the scale-forming element and introduce other elements into the coating. Additional mechanical affects also arise, which can lead to deformation of the coating. Mechanical effects are also critical to the durability of TBCs.

Degradation of diffusion and overlay coatings

The microstructure of a platinum modified aluminide coating after 20 h of exposure at 1200 °C is shown in Figure 10.16. The initial columnar structure of the β-phase is evident. The depletion of Al has resulted in the nucleation of γ' at the β grain boundaries (light areas). Figure 10.17 shows the same type of coating after oxidation at 1200 °C for 200 h. The coating had been converted almost completely to γ', as the result of Al depletion. Continued exposure would result in the γ' transforming to γ. Figure 10.18 shows the presence of Ta-rich phases, which have formed near the coating–scale interface as the result of Ta diffusing up from the superalloy substrate during high-temperature exposure.

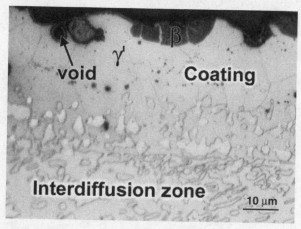

Figure 10.17 Optical micrograph showing the cross-section of a Pt-modified aluminide coating on a nickel-base single crystal superalloy after isothermal oxidation at 1200 °C for 200 h. The coating had been converted almost completely to γ', as the result of Al depletion. Further exposure would result in the γ' transforming to γ.

Figure 10.19 compares the cyclic oxidation behaviour of the uncoated superalloy IN-738 with that of the same alloy with straight aluminide and Pt-modified aluminide coatings at 1200 °C in air. The relative improvement in resistance in going to an aluminde and platinum aluminide coating is clear. If one arbitrarily defines the system life as the time to reach a mass loss of 10 mg cm^{-2}, the use of the straight aluminide increases the life by a factor of six. This is the result of the coating forming an alumina scale whereas the alloy forms a more rapidly growing chromia scale, which is subject to CrO_3 volatilization as well as oxide spallation. The Pt-modified aluminide increases the life by a factor of about twelve, i.e., essentially double the life of the straight aluminide. All of the reasons for the improved behaviour of the Pt-modified aluminide coating are not clear. It has been proposed that the increased Al uptake may contribute to this behaviour.[19] Also, the lower Al activity coefficient in the presence of Pt decreases the diffusive flux into the substrate. Some authors believe the effect of Pt is to improve the adhesion of the alumina to the coating, perhaps by mitigating the effects of sulphur in the alloy and coating.[32]

The surfaces of the Pt-modified aluminides often wrinkle during exposure. The extreme wrinkling (rumpling) of the surface of a bare platinum aluminide coating after exposure at 1100 °C for 1000 cycles is shown in Figure 10.20. Such wrinkling occurs predominantly during cyclic exposures and is minimal during isothermal exposures. This is the result of thermal-expansion mismatch between the coating and the oxide and, particularly, between the coating and the thicker substrate.

(a)

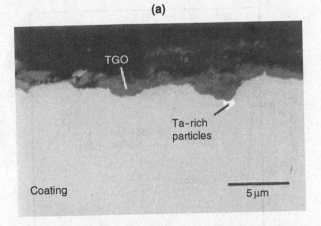

(b)

Figure 10.18 (a) Scanning electron micrographs showing the cross-section and (b) surface from which the TGO has spalled of a Pt-modified aluminide coating on a nickel-base single-crystal superalloy revealing Ta-rich metallic particles at the interface between the coating and the oxide.

Growth stresses in the oxide may also contribute. Wrinkling is affected by the more rapid diffusion of nickel into the coating from the substrate compared to the inward diffusion of aluminium and it is also dependent on the initial surface preparation of the coating with smoother coatings being less susceptible.

The degradation behaviour of overlay coatings is similar to that described for aluminide coatings. Overlay coatings are also susceptible to interdiffusion. This is illustrated in Figure 10.21 which shows cross-sections of a Ni–Co–Cr–Al–Y coating undergoing cyclic oxidation. The Al-rich β-phase begins to dissolve at the coating–scale interface because of Al depletion to form the scale and at the coating–substrate interface because of interdiffusion with the substrate. With longer exposures these depletion zones move toward the center of the coating until the β-phase

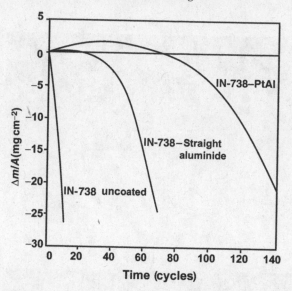

Figure 10.19 Cyclic oxidation data for a straight aluminide and a platinum aluminide on IN-738 at 1200 °C in air.

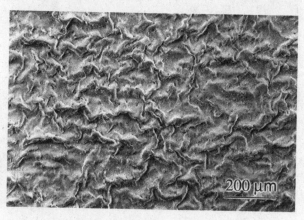

Figure 10.20 Scanning electron micrographs showing the surface of a Pt-modified aluminide coating on a nickel-base single-crystal superalloy after 1000 one-hour cycles at 1100 °C showing severe wrinkling of the coating.

is completely consumed. Overlay coatings can have a wider range of mechanical properties than diffusion coatings. Figure 10.22 is a plot of fracture strain vs. temperature for diffusion aluminide and Co–Cr–Al–Y coatings showing the wide range of properties achievable in the M–Cr–Al–Y coatings relative to those achievable in the aluminides.

(a)

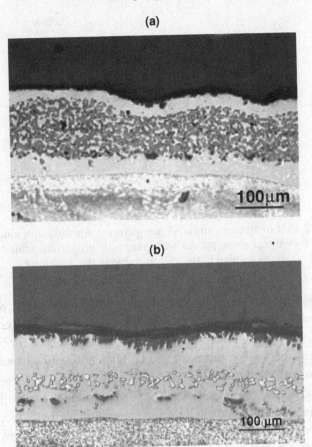

(b)

Figure 10.21 Scanning micrographs showing the cross-sections of a Ni–Co–Cr–Al–Y overlay coating on a single-crystal Ni-base superalloy after (a) 200 and (b) 1000 one-hour cycles at 1100 °C in air.

Coatings on titanium aluminides

The rapid oxidation kinetics of the titanium aluminides (described in Chapter 5) and, particularly, the attendant embrittlement, has created a substantial interest in the possibility of forming a protective coating on these alloys. Coatings of $TiAl_3$ have been formed by pack cementation on α_2-Ti_3Al,[33,34] and γ-TiAl.[35] These coatings form continuous alumina scales but $TiAl_3$ is an extremely brittle compound and tends to crack, particularly for thicker coatings.[34] It is also likely that the presence of a layer of $TiAl_3$ on the surface of α_2 or γ will be as embrittling as a high-temperature oxidation exposure.

The ability of Ti–Cr–Al alloys to form protective alumina scales raises the possibility of applying them as protective coatings.[36] Coatings have been successfully applied to γ substrates by sputtering, low-pressure plasma spraying, high-velocity

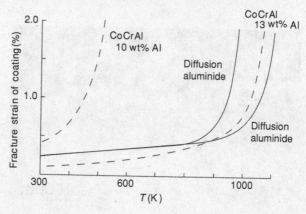

Figure 10.22 Plot of fracture strain vs. temperature for diffusion aluminide and Co–Cr–Al–Y coatings showing the wide range of properties achievable in the M–Cr–Al–Y coatings relative to those achievable in the aluminides.

oxygen fuel spraying and slurry fusion.[36] Figure 10.23 shows the cyclic oxidation kinetics for a sputtered coating and a cross-section of the coating after the oxidation exposure. The oxidation kinetics were similar to those observed for bulk Ti–Cr–Al alloys and there appears to be minimal degradation of the coating by either oxidation or interdiffusion with the substrate.

Cockeram and Rapp[37] have evaluated the oxidation kinetics of silicide coatings on Ti and have used a halide-activated pack-cementation method to form boron- and germanium-doped silicide coatings on orthorhombic Ti–Al–Nb alloy substrates.[38] The coatings greatly decreased the cyclic oxidation kinetics and microhardness measurements did not indicate diffusion of oxygen into the substrate.

The important aspects of coating any of the titanium aluminides are that the coating must prevent interstitial embrittlement and that the coating does not function as an embrittling layer. Additional studies are necessary in this area.

Thermal-barrier coatings

The major challenge for the application of TBCs is coating durability, particularly the resistance of the coating to spalling. The key factors are similar to those described in previous chapters with the additional aspect that the presence of the TBC can affect the performance of the bond coat and thermal stresses in the TBC can contribute to spalling of the TGO. This is important in that first-time spallation removes the TBC, which cannot reform, as is the case with a simple oxidation-resistant coating. There are a number of degradation modes, which can limit the life of a TBC and these must be understood in order to make lifetime predictions for existing systems and to provide the basis for the development of improved TBC systems. These include the following.

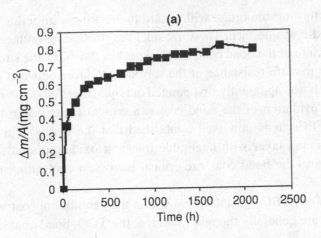

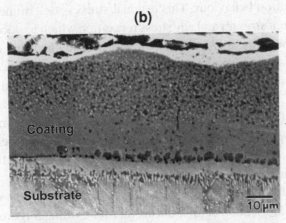

Figure 10.23 (a) Cyclic oxidation kinetics of a sputtered Ti–Cr–Al coating on γ-TiAl at 900 °C and (b) a cross-section of the coating after exposure.

- Cracking within the ceramic topcoat, which leads to spalling of part of the coating.
- Cracking along the interface between the TGO and the bond coat and/or cracking along the interface between the TGO and ceramic topcoat, which result in spalling of the entire TBC.
- Sintering of the TBC at the outer surface, where the temperature is highest. This increases the thermal conductivity of the TBC and can increase the total amount of elastic energy stored in the coating, which provides an additional driving force for cracking and spalling of the coating.
- Particle erosion, which causes a continued wearing-away of the coating and, for large particles, can produce cracks in the coatings and along the interface between the TBC and the bond coat.

Several factors are important with regard to cracking within the TBC or along the TBC–bond-coat interface. These include: the stress state in the zirconia layer, the microstructure of the bond coat, the thickness of the TGO, the stress state in the TGO, and the fracture resistance of the various interfaces between the bond coat and the TGO. It was apparently first pointed out by Miller[39] and it is now generally accepted that oxidation of the bond coat is a critical factor controlling the lives of EB-PVD TBCs. It is now well established that the ability of a bond coat to form an α-alumina layer with negligible transient oxidation, and the adherence of the alumina to the bond coat, are critical factors in controlling the durability of TBCs.

Plasma-sprayed TBCs generally fail within the ceramic topcoat while the EB-PVD coatings are generally thought to spall at the TGO–bond-coat interface or in the alumina layer.[30] This view is somewhat oversimplified, as will be seen below.

The residual stress in the alumina layer on the bond coat also plays a significant role in spallation behaviour. This residual stress is determined by the *growth stresses* and *thermal stresses* and any stress relaxation which has occurred by plastic deformation of the alloy and/or oxide.[40] It is now also clear that residual stresses in the TBC topcoat can also play a role in spalling.

Figure 10.24 presents macroscopic photographs of disc-shaped specimens with an EB-PVD TBC on a platinum aluminide bond coat, which failed after about 1000 cycles at 1100 °C in dry air. One failure initiated at a specimen edge, Figure 10.24(a), and propagated as an elongated buckle, which branched several times. The other, Figure 10.24(b), initiated as a buckle in the centre of the specimen and propagated outward.

The spallation failure of TBCs is generally initiated by defects which are either present following fabrication or arise during thermal cycling. Figure 10.25(a) shows the cross-section of an as-processed EB-PVD TBC on a platinum modified aluminide bond coat indicating defects at the TBC–bond-coat interface which look like 'corn kernels.' These defects, which result from the surface roughness of the bond coat prior to TBC deposition, cause the bond coat to deform during cyclic exposure which results in cracking of the TBC as seen in Figure 10.25(b). The tendency of platinum aluminide coatings to wrinkle, as illustrated in Figure 10.20, involves the same type of deformation. These cracks eventually grow together and a macroscopic TBC failure occurs. Figure 10.26(a) shows the bond-coat surface of a failed coating where the fracture has travelled mainly in the TGO and TBC. Figure 10.26(b) shows a cross-section of a failed coating which shows how the defects evolve. This mode of TBC failure has been termed 'ratcheting'.[41]

Figure 10.27(a) shows a cross-section of an as-processed TBC with a Ni–Co–Cr–Al–Y bond coat. This micrograph illustrates typical defects, which include embedded grit particles, separations between the TBC and bond coat, and

(a)

(b)

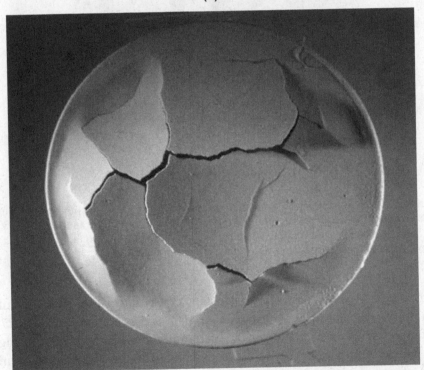

Figure 10.24 Optical micrographs showing the failures of an EB-PVD topcoat with a Pt-modified aluminide bond coat. See text for details.

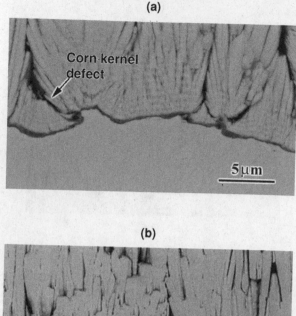

Figure 10.25 Scanning electron micrographs showing cross-sections of (a) an as-processed Pt-modified aluminide bond coat indicating defects in the EB-PVD YSZ TBC; (b) the same coating which has been cycled part way to failure showing how these defects propagate and produce cracks in the TBC.

subsurface oxide inclusions. These defects serve as nucleation sites for failure which propagates along the TGO–bond-coat interface as seen in Figure 10.27(b). This failure contrasts with that for the platinum aluminide bond coat in Figure 10.26 which occurred primarily in the oxide phases. Figure 10.28 illustrates more details of various defects left on the failure surface of a Ni–Co–Cr–Al–Y TBC system.

The failure mechanisms of air-plasma-sprayed TBCs are somewhat different than those described for EB-PVD coatings. Figure 10.29 shows the cross-section of a degraded APS TBC, which is about to spall. Microcracks associated with splat boundaries in the region of the TBC have grown and linked up to form a large crack

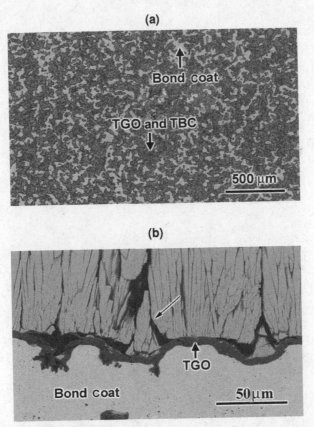

Figure 10.26 Scanning electron micrographs showing the (a) exposed bond-coat surface and (b) cross-section of a failed EB-PVD YSZ TBC system with a Pt-modified bond coat.

which will propagate and cause separation of the TBC. This failure mode is typical of APS TBCs although, depending on the thermal loading, the fracture sometimes occurs farther out into the YSZ topcoat.

The above discussion indicates that TBCs can fail in a number of locations. Wherever a crack nucleates it will propagate if the stored elastic-strain energy exceeds the fracture toughness in that location. The mechanics of TBC failure have been reviewed in detail.[41] The details of the fracture process depend on the type of bond coat and its fabrication, the technique for depositing the topcoat, and even on the nature of the thermal exposure (e.g., cycle frequency).[42]

TBC sintering

Sintering of the TBC has also been observed during high-temperature exposure and can contribute to the elastic-strain energy stored in the system. Figure 10.30

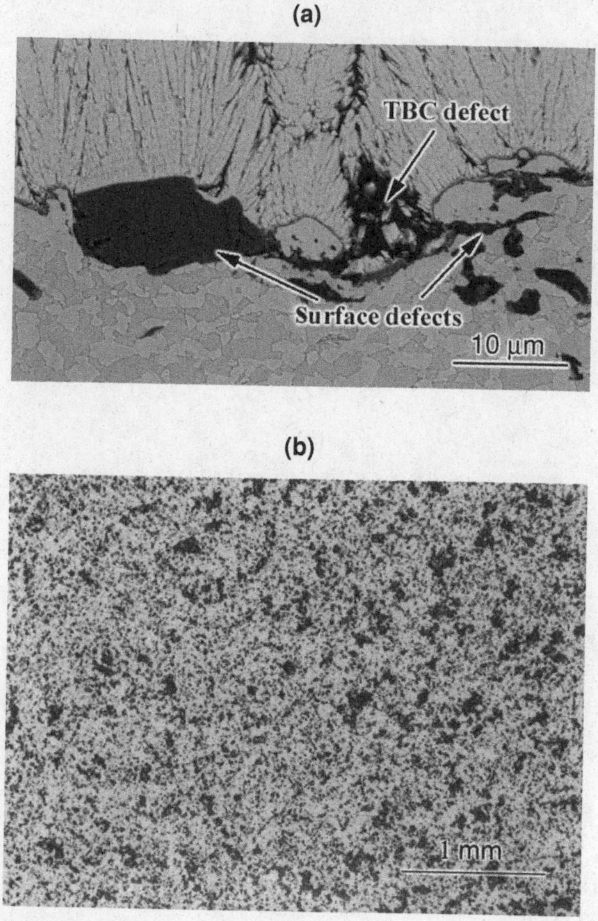

(a)

(b)

Figure 10.27 Scanning electron micrographs showing (a) the cross-section of an as-processed EB-PVD YSZ TBC system with a Ni–Co–Cr–Al–Y bond coat showing typical processing defects (b) low-magnification micrograph of the exposed bond-coat surface of a failed coating.

compares the microstructure of an EB-PVD TBC in the as-processed condition with that of an identical coating, which has been exposed for 10 h at 1200 °C. Sintering between the columns is evident. These observations are consistent with the dilatometry measurements of Zhu and Miller[43] who observed 0.1% shrinkage of cylindrical specimens of plasma-sprayed ZrO_2–8 wt% Y_2O_3 exposed for 15 h at 1200 °C. Figure 10.31 illustrates the sintering behaviour of an APS TBC. The as-processed coating, Fig. 10.31(a), contains a substantial number of microcracks, which have sintered closed after 100 h exposure at 1200 °C, Figure 10.31(b). Sintering has been observed to increase the effective Young's modulus of the TBC to

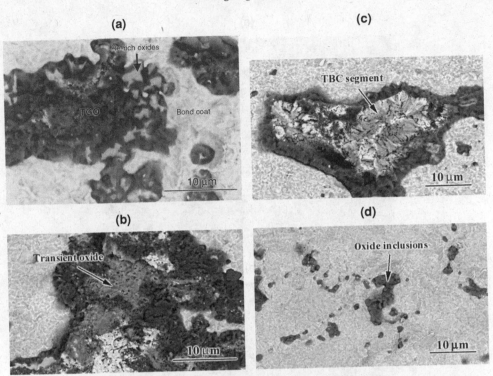

Figure 10.28 Scanning electron micrographs of the exposed bond-coat surface of failed EB-PVD YSZ TBCs with Ni–Co–Cr–Al–Y bond coats which illustrate the influence of various defects.

Figure 10.29 Optical micrograph showing the cross-section of an exposed APS YSZ TBC system with a Ni–Co–Cr–Al–Y bond coat showing a crack propagating in the TBC just above the bond coat.

values approaching that for dense zirconia (200 GPa)[43,44] and to result in residual stresses in the range 200–400 MPa.[44,45] These changes greatly increase the amount of elastic energy stored in the TBC and can enhance the driving force for scale spallation.

(a)

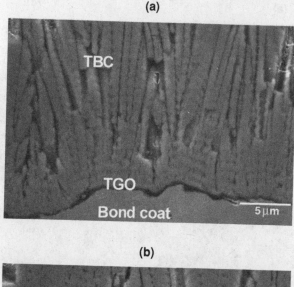

(b)

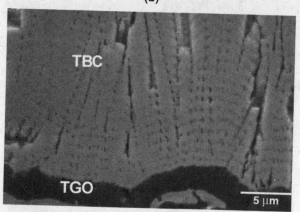

Figure 10.30 Scanning electron micrographs showing the sintering of an EB-PVD YSZ TBC at 1200 °C. Note the decrease in the number of channels after exposure.

Summary

There are a wide variety of coatings used for oxidation and corrosion protection at high temperatures. This chapter has presented a number of these but is by no means encyclopedic. The general purpose of the protective coatings is to provide an extra reservoir of an element (Al, Cr, Si), which can react with oxygen to form a protective surface oxide. The degradation mechanisms of these coatings are similar to those described in previous chapters for bulk alloys. Additional mechanisms arise for coatings because of their small thickness. These include depletion of the scale-forming element, interdiffusion with the substrate, and deformation associated with thermal-expansion mismatch.

(a)

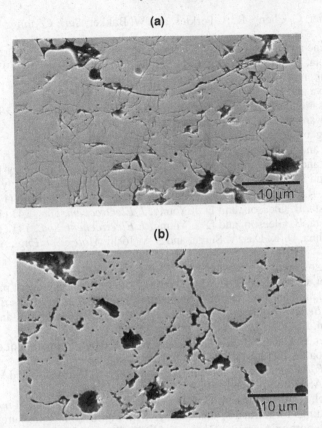

(b)

Figure 10.31 See text for detail. Scanning electron micrographs showing the sintering of an APS YSZ TBC at 1200 °C. Note the decrease in the number of microcracks after exposure.

Thermal-barrier coatings are used as thermal insulators to lower the temperature of the metallic components. Failure of TBCs is generally the result of thermomechanical processes. These processes are complex and depend on many variables associated with the coating and exposure environment.

References

1. K. H. Stern, *Metallurgical and Ceramic Protective Coatings*, London, Chapman and Hall, 1996.
2. G. W. Goward and L. W. Cannon, 'Pack cementation coatings for superalloys: a review of history, theory, and practice', Paper 87-GT-50, Gas Turbine Conference, New York, NY, ASME, 1987.
3. J. S. Smith and D. H. Boone, 'Platinum modified aluminides – present status', Paper 90-GT-319, Gas Turbine and Aeroengine Congress, New York, NY, ASME, 1990.

4. G. H. Meier, C. Cheng, R. A. Perkins, and W. Bakker, *Surf. Coatings Tech.*, **39/40** (1989), 53.
5. S. R. Levine and R. M. Caves, *J. Electrochem Soc.*, **121** (1974), 1051.
6. B. K. Gupta, A. K. Sarkhel, and L. L. Seigle, *Thin Solid Films*, **39** (1976), 313.
7. S. Shankar and L. L. Seigle, *Met. Trans.*, **9A** (1978), 1476.
8. B. K. Gupta and L. L. Seigle, *Thin Solid Films*, **73** (1980), 365.
9. N. Kandasamy, L. L. Seigle, and F. J. Pennisi, *Thin Solid Films*, **84** (1981), 17.
10. G. W. Goward and D. H. Boone, *Oxid. Met.*, **3** (1971), 475.
11. S. C. Kung and R. A. Rapp, *Oxid. Met.*, **32** (1989), 89.
12. R. Bianco and R. A. Rapp, *J. Electrochem. Soc.*, **140** (1993), 1181.
13. R. Bianco and R. A. Rapp, in *High Temperature Materials Chemistry-V*, eds. W. B. Johnson and R. A. Rapp, New York, NY, The Electrochemical Society, 1990, p. 211.
14. R. Bianco, R. A. Rapp, and J. L. Smialek, *J. Electrochem. Soc.*, **140** (1993), 1191.
15. W. Da Costa, B. Gleeson, and D. J. Young, *J. Electrochem. Soc.*, **141** (1994), 1464.
16. W. Da Costa, B. Gleeson, and D. J. Young, *J. Electrochem. Soc.*, **141** (1994), 2690.
17. G. R. Krishna, D. K. Das, V. Singh, and S. V. Joshi, *Mater. Sci. Eng.*, **A251** (1998), 40.
18. M. R. Jackson and J. R. Rairden, *Met. Trans.*, **8A** (1977), 1697.
19. J. Schaeffer, G. M. Kim, G. H. Meier, and F. S. Pettit, 'The effects of precious metals on the oxidation and hot corrosion of coatings'. In *The Role of Active Elements in the Oxidation Behavior of High Temperature Metals and Alloys*, ed. E. Lang, Amsterdam, Elsevier, 1989, p. 231.
20. G. H. Meier, C. Cheng, R. A. Perkins, and W. T. Bakker, Formation of chromium diffusion coatings on low alloy steels for use in coal conversion atmospheres. In *Materials For Coal Gasification*, eds. W. T. Bakker, S. Dapkunas, and V. Hill, Metals Park, OH, ASM International, 1988, p. 159.
21. H. Lavendel, R. A. Perkins, A. G. Elliot, and J. Ong, Investigation of modified silicide coatings for refractory metal alloys with improved low pressure oxidation behavior. Report on Air Force Contract, AFML-TR-65-344, Palo Alto, CA, Lockheed Palo Alto Research Laboratory, 1965.
22. C. M. Packer, in *Oxidation of High-Temperature Intermetallics*, eds. T. Grobstein and J. Doychak, Warrendale, PA, The Mining, Metallurgy, and Materials Society, 1989, p. 235.
23. T. A. Kircher and E. L. Courtright, *Mater. Sci. Eng.*, **A155** (1992), 67.
24. A. Mueller, G. Wang, R. A. Rapp, and E. L. Courtright, *J. Electrochem. Soc.*, **139** (1992), 1266.
25. B. V. Cockeram, G. Wang, and R. A. Rapp, *Mater. Corr.*, **46** (1995), 207.
26. B. V. Cockeram, G. Wang, and R. A. Rapp, *Oxid. Met.*, **45** (1996), 77.
27. R. C., Tucker, Jr., Thermal spray coatings. In *Handbook of Thin Film Process Technology*, eds. S. I. Shah and D. Glocker, Bristol, IOP Publishing Ltd, 1996, p. A4.0:1.
28. R. A. Miller, Thermal barrier coatings for aircraft engines – history and directions. Proceedings of Thermal Barrier Coating Workshop, Cleveland, OH, March 27–29, 1995, p. 17 (NASA CP 3812).
29. A. Maricocchi, A. Bartz, and D. Wortman, PVD TBC experience on GE aircraft engines. Proceedings of Thermal Barrier Coating Workshop, Cleveland, OH, March 27–29, 1995, p. 79 (NASA CP 3312).
30. S. Bose and J. T. DeMasi-Marcin, Thermal barrier coating experience in gas turbine engines at Pratt & Whitney. Proceedings of Thermal Barrier Coating Workshop, Cleveland, OH, March 27–29, 1995, p. 63 (NASA CP 3312).

31. J. T. DeMasi-Marcin and D. K. Gupta, *Surf. Coatings Tech.*, **68–69** (1994), 1.
32. J. A. Haynes, Y. Zhang, W. Y. Lee, B. A. Pint, I. G. Wright, and K. M., Cooley, Effects of platinum additions and sulfur impurities on the microstructure and scale adhesion behavior of single-phase CVD aluminide bond coatings. In *Elevated Temperature Coatings*, eds. J. M. Hampikian and N. B. Dahotre, Warrendale, PA, TMS, 1999, p. 51.
33. J. Subrahmanyam, *J. Mater. Sci.*, **23** (1988), 1906.
34. J. L. Smialek, M. A. Gedwill, and P. K. Brindley, *Scripta met. mater.*, **24** (1990), 1291.
35. H. Mabuchi, T. Asai, and Y. Nakayama, *Scripta met.*, **23** (1989), 685.
36. R. L. McCarron, J. C. Schaeffer, G. H. Meier, D. Berztiss, R. A. Perkins, and J. Cullinan, Protective coatings for titanium aluminide intermetallics. In *Titanium '92*, eds. F. H. Froes and I. L. Caplan, Warrendale, PA, TMS, 1993, p. 1,971.
37. B. V. Cockeram and R. A. Rapp, *Met. Mater. Trans.*, **26A** (1995), 777.
38. B. V. Cockeram and R. A. Rapp, Boron-modified and germanium-doped silicide diffusion coatings for Ti–Al–Nb, Nb–Ti–Al, Nb–Cr and Nb–base alloys. In *Processing and Design Issues in High Temperature Materials*, eds. N. S. Stoloff and R. H. Jones, Warrendale, PA, Mining, Metallurgy and Meterials Society, 1996, p. 391.
39. R. Miller, *J. Amer. Ceram. Soc.*, **67** (1984), 517.
40. M. J. Stiger, N. M. Yanar, M. G. Topping, F. S. Pettit, and G. H. Meier, *Z. Metallkunde*, **90** (1999), 1069.
41. A. G. Evans, D. R. Mumm, J. W. Hutchinson, G. H. Meier, and F. S. Pettit, *Progr. Mater. Sci.*, **46** (2001), 505.
42. G. M. Kim, N. M. Yanar, E. N. Hewitt, F. S. Pettit, and G. H. Meier, *Scripta Mater.*, **46** (2002), 489.
43. D. Zhu and R. A. Miller, *Surf. Coatings Tech.*, **108–109** (1998), 114.
44. C. A. Johnson, J. A. Ruud, R. Bruce, and D. Wortman, *Surf. Coatings Tech.*, **108–109** (1998), 80.
45. J. Thornton, D. Cookson, and E. Pescott, *Surf. Coatings Tech.*, **120–121** (1999), 96.

11

Atmosphere control for the protection of metals during production processes

Introduction

The exposure of metals to gases at high temperatures during production processes may be divided broadly under two headings.

(1) Reheating for subsequent working or shaping.
(2) Reheating for final heat treatment, sometimes of finished components.

When reheating for subsequent working or shaping, the main concern is to heat the components as quickly and economically as possible to the working temperature, within the metallurgical constraint of avoiding element redistribution and thermal cracking. In addition, it is usual to expose the metal directly to a burned-fuel atmosphere containing about 1% excess oxygen necessary to ensure complete combustion and, therefore, the most economic use of the fuel. Under these conditions, as there is little likelihood of operating a controlled-atmosphere policy, surface damage due to oxidation or scaling and, in the case of steels, decarburization, must be removed at a later stage in the process. Indeed, if the surface of the material to be reheated already has some undesirable features, it may be possible to remove, or reduce, them by allowing some scaling to occur.

Alloys which are susceptible to surface deterioration during reheating may be protected either by controlling the atmosphere composition to minimize, or avoid, surface deterioration or by applying a coating to protect the alloy from the atmosphere.

Coatings are not an economic proposition in the treatment of large tonnage production in this way, but are applied to cause selective carburization, etc., during surface-hardening procedures. Alternatively, coating application is common during service (See Chapter 10) and is a fairly successful way to avoid short-term damage to the underlying alloy.

This chapter is concerned with the use of gaseous atmospheres to control surface reactions during reheating for working or heat treatment. The main application of controlled atmospheres is in the area of heat treatment of finished, machined, components or of articles of complex shape, which cannot easily be treated subsequently for the removal of surface damage. In this context, atmosphere control can be considered under two headings: firstly, where the treatment is aimed simply to prevent surface reaction; and secondly, where it is designed to cause a particular surface reaction to proceed such as carburizing or nitriding. We are concerned only with the former.

Prevention or control of oxide-layer formation

This is primarily a question of controlling the oxygen partial pressure of the atmosphere to a value low enough to prevent oxidation, as described in Chapter 2. Very simply, for a metal which undergoes oxidation according to the reaction shown in Equation (11.1),

$$M + \frac{1}{2}O_2 = MO; \quad \Delta G_1^\circ, \tag{11.1}$$

where MO is the lowest oxide of M, the oxygen partial pressure must be controlled so as not to exceed a value, $(p_{O_2})_{M-MO}$, Equation (11.2):

$$(p_{O_2})_{M-MO} = \exp\left(\frac{2\Delta G_1^\circ}{RT}\right). \tag{11.2}$$

Unfortunately $(p_{O_2})_{M-MO}$ is a function of temperature and has lower values at lower temperatures. Thus, if an atmosphere is designed to be effective at high temperatures it may become oxidizing as the temperature is reduced during cooling. A surface oxide layer may, therefore, form as the metal is cooled. Although the metallurgical damage to the surface will be negligible, the surface may be discoloured, i.e., not bright. This condition can be overcome to some extent by rapid cooling or by changing the atmosphere to a lower oxygen partial pressure just before or during cooling.

For alloys, it is clear that the most critical reaction must be considered when deciding on the composition of the atmosphere to be used. For this purpose, the activities of the alloy components must be known since, if the metal M in Equation (11.1) exists at an activity a_M the corresponding equilibrium oxygen partial pressure will be given by Equation (11.3):

$$(p_{O_2})_{M-MO} = \frac{1}{a_M^2} \exp\left(\frac{2\Delta G_1^\circ}{RT}\right). \tag{11.3}$$

If the metal activities in the alloy are not known, then, by assuming the solution to be ideal, mole fractions may be used instead of activities to give a value of the

Table 11.1 *Approximate equilibrium oxygen partial pressures (atm) for oxidation of solid metals*

Metal	a_M	400 °C	600 °C	800 °C	1000 °C	1200 °C
Cu	1	4.8×10^{-19}	3.4×10^{-13}	1.4×10^{-9}	4.4×10^{-7}	Liquid
	0.5	7.6×10^{-18}	5.6×10^{-12}	2.2×10^{-8}	7.0×10^{-6}	Liquid
Ni	1	1.7×10^{-28}	9.1×10^{-20}	2.7×10^{-14}	1.5×10^{-10}	8.4×10^{-8}
	0.5	6.8×10^{-28}	3.6×10^{-19}	1.1×10^{13}	6.2×10^{-10}	3.4×10^{-7}
Fe	1	1.3×10^{-34}	2.4×10^{-25}	1.5×10^{-19}	1.5×10^{-15}	1.2×10^{-12}
	0.5	5.2×10^{-34}	1.8×10^{-24}	6.0×10^{-19}	6.0×10^{-15}	4.8×10^{-12}
	0.1	1.3×10^{-32}	2.4×10^{-23}	1.5×10^{-17}	1.5×10^{-13}	1.2×10^{-10}
Cr	1	7.9×10^{-50}	1.7×10^{-36}	3.8×10^{-28}	2.1×10^{-22}	3.1×10^{-18}
	0.5	3.7×10^{-49}	7.9×10^{-36}	1.8×10^{-27}	9.7×10^{-22}	1.4×10^{-17}
	0.1	1.7×10^{-48}	3.7×10^{-35}	8.2×10^{-27}	4.6×10^{-21}	6.7×10^{-17}

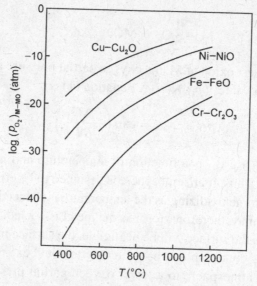

Figure 11.1 Equilibrium oxygen partial pressures for several metal–oxide systems.

oxygen partial pressure around which experiments must be performed to establish the correct atmosphere composition.

Figure 11.1 and Table 11.1 show values of the equilibrium oxygen partial pressure characteristic of copper, iron, nickel, and chromium and their oxides at various activities, derived using thermodynamic data from Kubaschewski et al.[1] It is immediately obvious that the oxygen partial pressures are very low, especially in the case of chromium. The trend to lower oxygen partial pressures as the temperature is reduced is also clear. Finally, the effect of low metal activity, although it leads to higher oxygen partial pressures, is in fact relatively small.

Provision of protective atmospheres in the laboratory

Vacuum

To carry out heat treatment in vacuum is the first, and perhaps the simplest, method of avoiding surface oxide formation to be considered. If the vacuum is against air, as is usual, it is possible to reduce the oxygen partial pressure to values as low as 10^{-10} or 10^{-11} atm. Comparing this with the values in Figure 11.1 it is seen that this should be effective for copper above 670 °C, nickel above 950 °C, and iron above 1200 °C, but not at all for chromium. Due to the very low reaction rates, however, it is generally possible to extend these limits to lower temperatures depending on the time of treatment.

With vacuum treatment there is always the possibility, especially at the higher temperatures, of the loss of volatile metals, such as manganese, from the surface regions of the component. For example, brass treated in this manner will undergo serious, if not total loss of zinc.

Even with modern techniques, vacua of the above levels can only be achieved in relatively small chambers and the technique is largely restricted to laboratory use.

Gaseous atmospheres

Purified inert gases, mainly argon, can be used. However, due to the difficulty of reducing the oxygen partial pressure below 10^{-6} atm, such atmospheres rely for their effect largely on the reduced reaction rate at these reduced oxygen partial pressures. Inert atmospheres are also expensive to produce and are, therefore, largely restricted to the laboratory.

Low oxygen partial pressures can be provided and, more importantly, controlled by using 'redox' gas mixtures. These mixtures consist of an oxidized and a reduced species, which equilibrate with oxygen, e.g., Equation (11.4),

$$CO\,(g) + \frac{1}{2}O_2\,(g) = CO_2\,(g); \quad \Delta G_4^\circ = -282\,200 + 86.7T \text{ J}, \quad (11.4)$$

from which p_{O_2} or, more importantly, p_{CO_2}/p_{CO} can be obtained:

$$p_{O_2} = \left(\frac{p_{CO_2}}{p_{CO}}\right)^2 \exp\left(\frac{2\Delta G_4^\circ}{RT}\right), \quad (11.5a)$$

$$\frac{p_{CO_2}}{p_{CO}} = p_{O_2}^{1/2} \exp\left(\frac{-\Delta G_4^\circ}{RT}\right). \quad (11.5b)$$

Thus from Equation (11.5b) the ratio of carbon dioxide to carbon monoxide may be calculated for any oxygen partial pressure and temperature.

In Figure 11.2 corresponding values of the ratio p_{CO_2}/p_{CO} and p_{O_2} are plotted for the temperatures 400, 600, 800, 1000, and 1200 °C. The lines corresponding

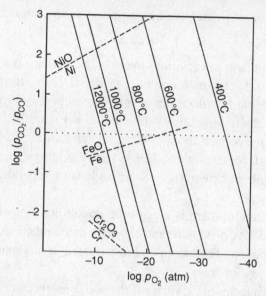

Figure 11.2 Oxidizing/reducing conditions in CO/CO_2 atmospheres.

to the equilibria between metal and oxide are also drawn in for Ni, Fe, and Cr. At $\log(p_{CO_2}/p_{CO})$ values above one of these lines the metal will oxidize, below it the corresponding oxide will be reduced. Since the lines of $\log(p_{CO_2}/p_{CO})$ versus $\log p_{O_2}$ are parallel, by calculating one point for any intermediate temperature the line for that temperature may be drawn in.

Similar, more useful, diagrams have been produced by Darken and Gurry[2] based on the modified Ellingham diagrams by Richardson and Jeffes[3] (see Chapter 2).

Several factors limit the use of CO_2/CO redox atmospheres, however. The usefulness of a redox gas system lies in its buffering capacity. Small concentrations of oxidizing or reducing impurities, or leaks of oxygen into the furnace, are reacted with and removed, thus maintaining the desired oxygen potential, particularly in a flowing gas. This is obviously best achieved when the ratio is close to unity. Thus atmospheres that deviate from unity appreciably will, to some extent, lose their capacity for buffering. For example, in nickel, referring to Figure 11.2, oxide formation will be prevented over the temperature range 400–1200 °C with CO_2/CO ratios varying between about 10^3 and 10^2 depending on temperature. This corresponds to a CO content of only 0.1 to 1%, thus protection of Ni is quite easy.

For iron, the CO_2/CO ratios vary over the same temperature range from about 3 to 0.3. Such atmospheres are easily produced either by mixing the gases or, industrially, by partially burning fuel gas with air to produce the mixture. Furthermore, since the ratio is of the order of unity the mixtures provide good buffering action.

For chromium, the CO_2/CO ratios required are of the order of 2×10^{-4} at 1000 °C decreasing towards lower temperatures. In such atmospheres only about 0.02% of CO_2 may be tolerated before oxidation occurs and such mixtures are expensive to produce and difficult to control. Even if production and control of these atmospheres could readily be achieved, a further problem arises in the form of the carbon activity, or carbon potential, of the gas mixture. Mixtures of CO_2 and CO will have a carbon potential by virtue of the reaction shown in Equation (11.6),

$$2CO\,(g) = CO_2\,(g) + C; \quad \Delta G_6^\circ = -170\,550 + 174.3T\ \text{J}, \tag{11.6}$$

from which a_C may be obtained:

$$a_C = \frac{p_{CO}^2}{p_{CO_2}} \exp\left(\frac{-\Delta G_6^\circ}{RT}\right), \tag{11.7a}$$

$$a_C = \frac{p_{CO}^2}{p_{CO_2}} \exp\left(\frac{20\,606}{T} - 21.06\right). \tag{11.7b}$$

Protective atmospheres based on the CO_2/CO system therefore have a tendency to remove carbon from, or add carbon to, the metal depending on the composition of the metal to be treated. This is particularly important in the case of steels for which carbon is an essential component and the removal of carbon from the surface layers, decarburization, is a deleterious reaction in terms of the strength of the steel surface (see Chapter 5).

In Figure 11.3, carbon activities are shown as a function of p_{CO_2}/p_{CO} ratios for various temperatures. The corresponding metal–metal-oxide equilibria are superimposed. From Figure 11.3 several conclusions may be drawn. At temperatures up to 1200 °C, the Cr–Cr_2O_3 equilibrium corresponds to carbon activities in excess of unity, therefore treatment of Cr in these atmospheres will lead to 'sooting' in the furnace and to severe carburization of the metal. Such atmospheres are, therefore, unsuitable for the complete protection of chromium. For iron, the CO_2/CO ratios required for the prevention of scaling are easily produced and maintained, being within the range 0.1 to 10 for all common temperatures. The problem here is that, during heating of steels, the above CO_2/CO ratios correspond to low carbon activities and therefore lead to decarburization. Consequently, in order to avoid decarburization, it is necessary to use CO_2/CO ratios much lower than those required to protect the iron from scaling. For example, using Figure 11.2 at 1000 °C, the value of $\log(p_{CO_2}/p_{CO})$ required to protect iron is -0.3, corresponding to a CO_2/CO ratio of 0.50, but this corresponds to a carbon activity of 0.01 or a carbon content of 0.021 wt% according to the relationship of Ellis *et al.*[4] between carbon activity,

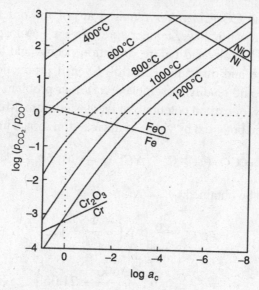

Figure 11.3 Carbon activities in CO/CO_2 atmospheres.

a_C, carbon mole fraction X_C, and temperature, T, for plain carbon steels in the austenite phase field, Equation (11.8):

$$\log a_C = \log \left[\frac{X_C}{(1 - 5X_C)} \right] + \frac{2080}{T} - 0.64. \tag{11.8}$$

Thus, if the steel has a carbon content of 0.8%, which corresponds to a mole fraction of 0.036 and, from Equation (11.8), to a carbon activity 0.433, the atmosphere composition required to prevent decarburizaton is obtained by locating log (0.433) = −0.364 and reading the corresponding $\log(p_{CO_2}/p_{CO}) = -1.75$ on the line for 1000 °C. This corresponds to an atmosphere of p_{CO_2}/p_{CO} ratio of 0.0178 or, since $p_{CO} + p_{CO_2} = 1$, to a composition of 98.25% CO and 1.75% CO_2. Thus, in order to protect a plain carbon steel containing 0.8% C, the above atmosphere should be used. If the atmosphere contains more CO_2 then some decarburization will occur, although it may contain up to 50% CO_2 before the iron begins to oxidize.

Similar calculations may be carried out for nickel although, from Figure 11.3, it is obvious that this metal will be protected by atmospheres of CO_2 containing only 1% CO or less, depending on temperature.

It is clear that, in order to protect alloys containing chromium from oxidation, atmospheres based on the CO_2–CO system cannot be used if carburization is to be avoided. Here, and in other cases where strong carbide formers are involved, protection may be achieved by using the H_2-H_2O system for which the relevant

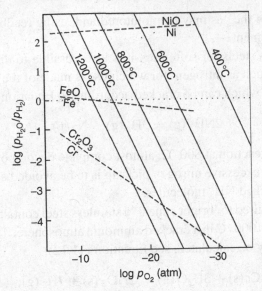

Figure 11.4 Oxidizing/reducing conditions in H_2/H_2O atmospheres.

equilibrium is shown in Equation (11.9),

$$H_2\,(g) + \frac{1}{2}O_2\,(g) = H_2O(g); \quad \Delta G_9^\circ = -247\,000 + 55T \text{ J}, \tag{11.9}$$

for which we obtain Equations (11.10a) or (11.10b):

$$\frac{p_{H_2O}}{p_{H_2}} = p_{O_2}^{1/2} \exp\left(\frac{-\Delta G_9^\circ}{RT}\right), \tag{11.10a}$$

$$\frac{p_{H_2O}}{p_{H_2}} = p_{O_2}^{1/2} \exp\left(\frac{29\,844}{T} - 6.65\right). \tag{11.10b}$$

From Equation (11.10b), the ratio of water vapour to hydrogen in the atmosphere may be calculated for any oxygen partial pressure and temperature.

Protective atmospheres based solely on the H_2O–H_2 gas system are hardly used outside the laboratory. As may be seen from Figure 11.4, however, both nickel and iron can be protected at all temperatures in hydrogen containing 10% H_2O and so, for laboratory use, cylinder hydrogen, usually containing a few parts per million of water vapour, can be used without further purification.

The system is of most use in the protection of alloys containing high chromium concentrations, such as stainless steels for which the required water content is as low as, or less than, 1000 ppm, depending on temperature. The advantage of this system is that the water-vapour content can be controlled by cooling the gas to a low temperature, thereby removing water from the gas by condensation. Similarly the water content of the gas may be monitored and, using feedback, controlled by dew-point measurement.

To heat stainless steels for softening, it is also possible to dilute the redox gas with an inert gas, such as nitrogen, for which use is made of the relatively cheaply available ammonia which can be cracked according to Equation (11.11):

$$2NH_3 (g) = 3H_2 (g) + N_2 (g). \tag{11.11}$$

By catalyzing this reaction at 800 °C, almost complete decomposition is achieved. This is important if excessive nitrogen pick-up is to be avoided as this occurs most readily from uncracked NH_3 molecules.

Suppose it is required to 'bright anneal' a stainless steel containing 20% Cr, 10% Ni, and 0.05% C at 1000 °C in a cracked ammonia atmosphere. The maximum dew point of the gas may be calculated, see Equation (11.12):

$$2Cr (s) + 3H_2O (g) = Cr_2O_3 (s) + 3H_2 (g);$$
$$\Delta G^{\circ}_{12} = -378\,300 + 94.5T\ J. \tag{11.12}$$

Thus the expressions in Equation (11.13a) and, for unit activity of Cr_2O_3, Equation (11.13b), are derived:

$$\frac{a_{Cr_2O_3}}{a_{Cr}^2} \left(\frac{p_{H_2}}{p_{H_2O}} \right)^3 = \exp \left(\frac{378\,300 - 94.5T}{RT} \right)$$

$$= 4.31 \times 10^{10} \text{ at } 1000\,°C; \tag{11.13a}$$

$$\frac{p_{H_2}}{p_{H_2O}} = 3.51 \times 10^3 a_{Cr}^{2/3}. \tag{11.13b}$$

In these alloys the activity coefficient of chromium is almost 2, also the mole fraction of Cr corresponding to 20% in iron is $X_{Cr} = 0.19$. Thus $a_{Cr} = 0.35$. The required ratio is therefore that shown in Equation (11.14a):

$$\frac{p_{H_2}}{p_{H_2O}} = 3.51 \times 10^3 (0.35)^{2/3} = 1.84 \times 10^3. \tag{11.14a}$$

In ammonia, cracked according to Equation (11.11), there are three parts hydrogen and one part nitrogen, thus the hydrogen partial pressure in this atmosphere is 0.75 atm. Therefore, from Equation (11.14a) Equation (11.14b) is obtained:

$$p_{H_2O} = \frac{0.75}{1.84 \times 10^3} = 4.08 \times 10^{-4} \text{ atm.} \tag{11.14b}$$

The dew point of the gas mixture must therefore be controlled to produce a water-vapour partial pressure no greater than 4.08×10^{-4} atm. The relevant dew point may be obtained from steam tables ($-29\,°C$) or may be estimated as follows.

The Clausius–Clapeyron equation relates the saturated vapour pressure, p, to temperature and the latent heat of evaporation, L, as shown in Equation (11.15):

$$\frac{\partial \log p}{\partial \left(\frac{1}{T}\right)} = -\frac{L}{2.303R}. \tag{11.15}$$

Integrating this equation between T_1 and T_2 yields Equation (11.16a):

$$\log \left(\frac{p_1}{p_2}\right) = -\frac{L}{2.303R} \left(\frac{1}{T_1} - \frac{1}{T_2}\right). \tag{11.16a}$$

For water $p_2 = 1$ atm at 373 K and $L = 41000 \, \text{J mol}^{-1}$. Thus

$$\log p_1 = \frac{-41000}{2.30R} \left(\frac{1}{T_1} - \frac{1}{373}\right). \tag{11.16b}$$

The required dew point is the temperature at which $p_1 = 4.08 \times 10^{-4}$ atm, i.e., 234.3 K (or $-38.7\,°C$) according to Equation (11.16b). Correspondingly the dew points of such atmospheres, used for bright annealing of stainless steels, would be controlled to about $-40\,°C$.

Even with such atmospheres, however, there may be complications due to hydrogen uptake, decarburization, and nitriding. Hydrogen uptake and nitriding in cracked-ammonia atmospheres are both avoided largely by ensuring that the time of treatment is the minimum for the metallurgical requirements. In addition, by restricting the treatment to items of thin section, the heating time is reduced and any hydrogen dissolved is able to diffuse out quickly. Decarburization may occur by the reaction given in Equation (11.17):

$$C + H_2O\,(g) = CO\,(g) + H_2\,(g);$$

$$\Delta G^{\circ}_{17} = 135\,00 - 142.6T \, \text{J}. \tag{11.17}$$

which proceeds rapidly and is, therefore, dangerous. Fortunately, the most usual application of the cracked-ammonia atmosphere is to stainless steels where a low carbon content is an advantage.

Provision of controlled atmospheres in industry

When a process is considered for use in industry it must satisfy certain criteria.

(1) It must be effective, i.e., achieve the technical objective.
(2) It must be reliable, i.e., capable of monitoring and control.
(3) It must be economical, i.e., basically inexpensive to install and use and capable of accepting full production throughput.

For most applications the above requirements rule out systems based on high vacuum and the general practice is to use atmospheres derived from fuels. The gases used are therefore mixtures of N_2, CO, CO_2, H_2, H_2O, and CH_4 which make up the products of combustion of fuels. Recently, atmospheres based on nitrogen have been used at an increasing rate.

In the early days of atmosphere control, town gas was used for the preparation of atmospheres; however, being derived from coal, this inevitably involved the need for a plant for sulphur removal, and variation in analysis made consistent control of the atmosphere composition difficult.

Recently, the availability of liquid petroleum gas (LPG), e.g., propane, with reproducible analysis and low sulphur content has improved the situation together with the use of natural gas for the same purpose.

It will also be seen that the large quantities of high-purity nitrogen, produced when producing liquid oxygen for steelmaking, represent a convenient feedstuff for controlled atmospheres.

Types of atmospheres

Starting from the fuel and air, various types of atmospheres can be produced. The main, or common, differentiation is between 'exothermic' and 'endothermic' atmospheres. The nomenclature is ambiguous and it is as well to be clear about its meaning.

An exothermic atmosphere is produced exothermically by burning the fuel with measured amounts of air. This type of atmosphere has the highest oxygen potential.

An endothermic atmosphere is produced by heating, by external means, a mixture of fuel gas with air over a catalyst to provide a gas containing reducing species. This atmosphere has a low oxygen potential and heat is absorbed during its preparation, hence the atmosphere is described as endothermic.

Exothermic atmospheres

The first stage of production is a combustion chamber where gas and air are reacted, the mixture being adjusted between limits, capable of supporting combustion, described as 'rich' or 'lean' depending on whether the gas has low or high concentrations of oxidized species, respectively.

From the combustion chamber, the gas produced is cooled by water spray which removes sulphurous gases and reduces the water content usually to a dew point of about 35 °C, equivalent to a water-vapour content of around 6.5% by volume.

Typical rich exothermic and lean exothermic compositions are given, in volume percentages as, *rich:* 5% CO_2, 9% CO, 9% H_2, 0.2% CH_4, 7% H_2O, 70% N_2; *lean:*

10% CO_2, 0.5% CO, 0.5% H_2, 7% H_2O, 82% N_2. The rich atmosphere can be used for the bright annealing of low carbon steels.

The atmosphere can be refined further by reducing the water-vapour content. This can be accomplished by refrigeration causing the water to condense. The temperature of this treatment is normally restricted to 5 °C. Otherwise ice may form and block the process whereas water can easily be drained away. More effective drying down to dew points in the region of −40 °C can be achieved using activated alumina or silica-gel towers. Used in pairs, normally one absorber is operational while the other is regenerated.

These treatments lead to dew points of about 4 °C (0.8% H_2O) and −40 °C (0.04% H_2O), and, apart from the reduced water-vapour content, these atmospheres are identical with those given above and could be referred to as 'rich-dried' or 'lean-dried' atmospheres. Such atmospheres simply present a way of slightly reducing the oxygen potential, since, at temperature, the reaction shown in Equation (11.18),

$$CO_2\,(g) + H_2\,(g) = H_2O\,(g) + CO\,(g), \tag{11.18}$$

will tend to equilibrate. Thus, the dew point of the cold, dried gas cannot be maintained but the total oxygen content and the oxygen potential are reduced.

A really protective gas, or even reducing gas, can be obtained, if, after drying, the gas is 'stripped' of CO_2. This can be achieved in several ways. High-pressure water scrubbing will remove some CO_2 but the process is not widely used. More usual are methods of absorbing CO_2 either chemically, using monoethanolamine (MEA) aqueous solution, or physically, using a molecular sieve.

MEA method
In this technique the gas is passed counter-current to 15% solution of MEA in water and CO_2 is absorbed down to about 0.1%. Sulphurous gases are also absorbed. The used MEA solution is regenerated by heating when the CO_2 and other absorbed gases are driven off. The heat required for this is frequently obtained by combining the regenerator and combustor in one unit.

After stripping the CO_2, the gas must be dried, usually using silica gel or activated alumina.

Molecular-sieve method
The molecular sieves used are artificial zeolites, which absorb both carbon dioxide and water vapour on the surface of structural cavities of molecular size. The zeolite can be regenerated and the method has the advantage of supplying stripped and dried gas in one operation.

These synthetic zeolites are also used in twin towers; both CO_2 and H_2O are removed, by absorption, down to 0.05% by volume in both cases. Alternative adsorption and regeneration cycles are used, as reactivation on modern zeolites can be carried out at room temperature using a CO_2- and H_2O-free purge gas.

Typical dried and stripped exothermic gases have analyses such as 0.05% CO_2, 0.5% CO, 0.5% H_2, 0% CH_4, 0.04% H_2O, and 99% N_2, which is almost pure nitrogen and may be applied to the bright annealing of carbon steels.

Hydrogen-enriched gases

In order to reduce the oxygen potential of the above gas still further it is necessary to increase the hydrogen content. This can be done by producing an initially richer gas by reducing the air-to-fuel ratio in the combustor and so, after drying and stripping, the gas contains extra H_2 and CO. Alternatively, steam may be added to the stripped gas when, on reheating, the reaction $H_2O + CO = H_2 + CO_2$ occurs over a catalyst. The H_2O and CO_2 are once more removed as described above.

Hydrogen may also be added in the form of ammonia, which will crack over a catalyst. A typical dried, stripped, hydrogen-enriched atmosphere would have the following composition: 0.05% CO_2, 0.05% CO, 3–10% H_2, 0% CH_4, 0% H_2O, and 90–97% N_2.

Endothermic atmospheres

Endothermic atmospheres are produced by reacting mixtures of fuel and air that are not capable of supporting combustion. Usually the aim is to crack the hydrocarbons over a catalyst at about 1050 °C and convert them mainly to CO and H_2. If the mixture is too rich, deposition of carbon could occur in the catalysis chamber and so the mixture is adjusted to avoid this.

Steam can also be used as the oxidant as partial, or total, replacement for air, in which case a hydrogen-enriched gas is produced of composition typically 5% CO_2, 17% CO, 71% H_2, 0.4% CH_4, 0.5% H_2O, and 0% N_2. Further addition of steam will, on passing once more over a catalyst, push the equilibrium ($H_2O + CO = H_2 + CO_2$) to the right and, after drying and stripping of CO_2 produce compositions such as 0–0.5% CO_2, 0–5% CO, 75–100% H_2, 0–0.5% CH_4, 0% H_2O, and 0–25% N_2, depending on the air–steam mixture. Such gases are quite suitable for the bright annealing of stainless steel due to the oxygen and carbon potentials. The generation of endothermic atmospheres has been described by Banerjee.[5]

Nitrogen-based atmospheres

A glance at the analysis of dried, stripped, exothermic atmospheres will confirm that they are predominantly pure nitrogen. Basically the fuel has been used to remove oxygen from the air.

An alternative method would be to liquefy the air and fractionate it to produce pure nitrogen. This is, already, a common process carried out to produce oxygen for bulk steelmaking purposes. The nitrogen that is a convenient by-product will be available in considerable volume at steelmaking sites that have on-plant oxygen facilities. Using vacuum-insulated transporters, liquid nitrogen can be delivered over long distances and is being used in increasing quantities in bright treatment facilities.

The typical analysis of fractionated nitrogen is 99.99% N_2, 8 ppm O_2, a dew point of 76 °C. Due to the method of production the analysis can be guaranteed. Since there is no reducing gas species present, there is no fire or explosion hazard. Neither is the gas toxic although, of course, it will not support life.

During inert-gas heat treatment of alloys for solution treatment, for instance, argon is the gas of choice. Recent trends have been to substitute nitrogen for argon in order to reduce costs. The danger in this is that some nitrogen may be absorbed, particularly if the alloy contains elements that form stable nitrides. Tests using argon, nitrogen, and their mixtures indicated that alloys such as 304 stainless steel, 410 stainless steel, and 4140 steel with significant chromium content would pick up nitrogen from all atmospheres that contained nitrogen.[6]

Even where reducing gases must be used for carburizing, for example, a good precaution is to have a standby nitrogen connection to the furnace for emergency flushing. This may be necessary during hazardous conditions following failure of the gas-power generator or water supply. Not only will a nitrogen flush under such conditions avoid a possible catastrophe but the workpiece will be protected during cooling. In addition, there is no pollution.

It is also possible to mix nitrogen with controlled amounts of hydrogen, hydrocarbons, and/or ammonia to produce atmospheres capable of most, if not all, controlled-atmosphere treatments. The only exceptions arise when metals containing strong nitride formers are to be treated, in which case a similar atmosphere based on argon is preferred. The actual choice will depend on the duration of the treatment compared with the rate of reaction with nitrogen gas. In fact, the rate of uptake of nitrogen from N_2 gas is much lower than when ammonia is present as the uncracked molecular species.

Where nitrogen is used as the base of heat-treating atmospheres, as opposed to inert atmospheres, the furnace must be heated either electrically or indirectly, for instance by burning gas in incandescent, radiant tubes. The furnace atmosphere is primarily nitrogen containing the active methane–hydrogen, or other, system as minor species. This results in improved control and much reduced explosion hazard, compared with alternative systems. A good overview of nitrogen-based atmospheres and other heat-treating atmospheres can be found in ref. 7.

The economics of using nitrogen as a controlled-atmosphere source become more attractive when such factors as improved safety, reliability, and therefore, productivity are considered.[8] Furthermore, the present tendency is to move away from oil towards electricity, in which case nitrogen atmospheres will be particularly attractive.

Modern techniques are currently available using carburizing and nitriding systems under vacuum. In these processes of vacuum carburizing and plasma carburizing, the components are heated under vacuum to around 950 °C. Methane is leaked into the chamber to a pressure of between 3 and 30 mbar to add carbon to the system. In the absence of a plasma, the methane will only decompose to the extent of about 3%, probably on the surface of the components according to a sequence such as that shown in Equation (11.19):

$$CH_4 \rightarrow CH_3 + H \rightarrow CH_2 + 2H \rightarrow CH + 3H \rightarrow C + 4H. \quad (11.19)$$

These reactions may be stimulated to provide 80% decomposition by using a plasma process to excite the methane molecule. In this case, the molecular breakdown may occur in the plasma to produce charged species. Hydrocarbons other than methane may be used as the feedstock. The usual operating sequence involves flushing and evacuation, heating to temperature under the inert atmosphere, carburizing for a pre-determined time followed by a diffusion anneal in a carbon-free atmosphere. This cycle is designed to provide the optimum surface carbon content and carburized depth.[9–11]

Monitoring and control

For most applications, the analysis can be defined in terms of CO_2 and H_2O content and these can be measured conveniently, but not continuously, using chemical absorption techniques. Once the correct CO_2 and H_2O levels have been established it is usually simply a matter of controlling the feeds of fuel and air to the appropriate levels.

For CO_2 monitoring on a more sophisticated level, infra-red absorption can be used; H_2O can also be monitored using this technique.

A convenient method of continuously monitoring the gas for H_2O content is to measure the dew point using a commercially available dew-point meter. Certain meters, working on a conductance or capacitance principle, can be set to control the input, or drying, parameters, or to trigger alarms if a certain limiting value is exceeded.

Whereas the above methods allow the oxygen potential of an atmosphere to be deduced from measurements of the CO_2 and H_2O contents, a more recent development allows the oxygen potential, or partial pressure, to be measured directly and

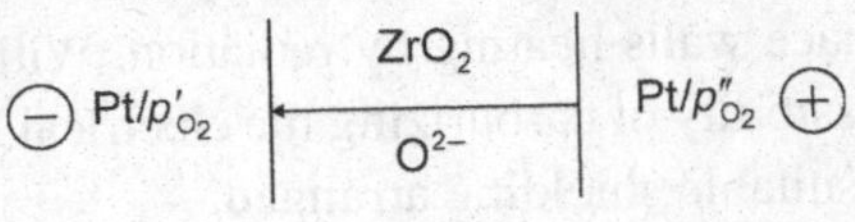

$$\ominus \ Pt/p'_{O_2} \quad \boxed{\begin{array}{c} ZrO_2 \\ \hline O^{2-} \end{array}} \quad Pt/p''_{O_2} \ \oplus$$

Figure 11.5 Schematic diagram of an electrochemical cell for monitoring oxygen partial pressures.

continuously. This technique is based on the use of a high-temperature galvanic cell. The instrument is simply a tube of zirconia stabilized with CaO or Y_2O_3. Within the tube a known (reference) oxygen partial pressure is imposed either by flushing with a gas of known oxygen potential or by using a metal–metal-oxide mixture. A platinum wire establishes contact with the inner surface of the tube. The tube is then placed in the furnace atmosphere at temperature. Since the stabilized zirconia has conductivity for oxygen ions only, the galvanic cell illustrated in Figure 11.5 is established. The free-energy change for the cell reaction is given by Equation (11.20):

$$\Delta G = RT \ln \left(\frac{p'_{O_2}}{p''_{O_2}} \right). \tag{11.20}$$

Consequently the emf of the cell is given by Equation (11.21),

$$E = \frac{RT}{4F} \ln \left(\frac{p''_{O_2}}{p'_{O_2}} \right), \tag{11.21}$$

where F is Faraday's constant. In practice, the reference electrode could be controlled by flushing air or by establishing a reference oxygen partial pressure by a mixture of a suitable metal and its oxide, such as Ni and NiO. Clearly the emf of such a cell can be indicated continuously and also used for feedback to correct the feed settings of fuel and air.

Difficulties encountered include thermal shock requiring relatively slow heating- and cooling-rates, although if mounted permanently in the furnace these should be achieved. It is important to mount the cell in a position that is exposed to the same temperature and gas composition as the working volume of the furnace to avoid unrepresentative readings. In addition, the cell will only indicate the equilibrium oxygen partial pressure, regardless of whether or not the atmosphere has equilibrated. This is because the gas will equilibrate locally on the platinum contact of the cell.

Heating methods

It is clear that an independent source of heating is essential. Electrical heating is clean, controllable, and simple. Alternatively, gas can be burned within incandescent tubes built into the furnace walls heating by radiation. With atmospheres of high carbon potential, the possibility of carburizing the electrical heating elements must also be considered and suitable shielding arranged.

References

1. O. Kubaschewski, E. H. Evans, and C. B. Alcock, *Metallurgical Thermochemistry*, Oxford, Pergamon Press, 1967.
2. L. S. Darken and R. W. Gurry, *Physical Chemistry of Metals*, New York, McGraw-Hill, 1953.
3. F. D. Richardson and J. H. E. Jeffes, *J. Iron Steel Inst.*, **160** (1948), 261.
4. T. Ellis, S. Davidson, and C. Bodworth, *J. Iron Steel Inst.*, **201** (1963), 582.
5. S. N. Banerjee, *Adv. Mater. Proc.*, **153** (June) (1998), 84SS.
6. J. Conybear, *Adv. Mater. Proc.*, **153** (June) (1998), 53.
7. P. Johnson, 'Furnace Atmospheres'. In *ASM Handbook*, 10th edn, Metals Park, OH, ASM International, 1991, vol. 4, p. 542.
8. R. G. Bowes, *Heat Treatment Met.*, **4**, (1975), 117.
9. F. Preisser, K. Loser, G. Schmitt, and R. Seeman, ALD Vacuum Technologies Information, East Windsor, CT.
10. F. Schnatbaum and A. Medber, ALD Vacuum Technologies Information, East Windsor, CT.
11. M. H. Jacobs, T. J. Law, and F. Ribet, *Surf. Eng.*, **1** (1985), 105.

Further reading

L. H. Fairbank and L. G. W. Palethorpe, 'Heat treatment of Metals,' ISI Special Report. No. 95, London, ISD 1966, p. 57.
R. V. Cutts and H. Dan, Proceedings of the Institute of Iron and Steel Wire Manufacturers Conference, Harrogate, March 1972, paper 7.

Appendix A

Solution to Fick's second law for a semi-infinite solid

Consider an alloy A–B which has B being removed from the surface, e.g., by evaporation. Initially the concentration of B (expressed as mole fraction) will be uniform at N_B^o, as indicated in Figure A1(a). If the surface concentration is fixed at a constant value $N_B^{(S)}$, the concentration profile of B will appear as that in Figure A1(b). For a constant interdiffusion coefficient in the alloy, Fick's second law may be written for this case as in Equation (A1), where $\tilde{D}$ is the interdiffusion coefficient:

$$\frac{\partial N_B}{\partial t} = \tilde{D}\frac{\partial^2 N_B}{\partial x^2}. \tag{A1}$$

The solution to this equation, following Gaskell,[1] can be accomplished by letting $Z = \dfrac{x}{2\sqrt{\tilde{D}t}}$ so that we obtain Equations (A2) and (A3):

$$\frac{\partial Z}{\partial x} = \frac{1}{2\sqrt{\tilde{D}t}}, \tag{A2}$$

$$\frac{\partial Z}{\partial t} = \frac{-x}{4\sqrt{\tilde{D}t^3}}. \tag{A3}$$

Thus, the expressions in Equations (A4) and (A5) are derived:

$$\frac{\partial N_B}{\partial t} = \frac{dN_B}{dZ}\frac{\partial Z}{\partial t} = \frac{-x}{4\sqrt{\tilde{D}t^3}}\frac{dN_B}{dZ}, \tag{A4}$$

$$\frac{\partial^2 N_B}{\partial x^2} = \frac{\partial}{\partial x}\left[\frac{dN_B}{dZ}\frac{\partial Z}{\partial x}\right] = \frac{d^2N_B}{dZ^2}\left(\frac{\partial Z}{\partial x}\right)^2 = \frac{1}{4\tilde{D}t}\frac{d^2N_B}{dZ^2}. \tag{A5}$$

Substitution of Equations (A4) and (A5) into Equation (A1) yields Equation (A6):

$$\frac{-x}{4\sqrt{\tilde{D}t^3}}\frac{dN_B}{dZ} = \tilde{D}\frac{1}{4\tilde{D}t}\frac{d^2N_B}{dZ^2}. \tag{A6}$$

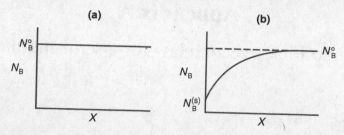

Figure A1 (a) Initial uniform concentration of solute B in an alloy A–B; (b) the concentration profile at some time t during which B has been removed from the alloy surface.

Defining a new variable $y \equiv \dfrac{dN_B}{dZ}$ and substituting in Equation (A6) yields Equation (A7),

$$y = -\frac{\sqrt{\tilde{D}t}}{x}\frac{d^2 N_B}{dZ^2} = -\frac{1}{2Z}\frac{d^2 N_B}{dZ^2}, \tag{A7}$$

which is a simple differential equation,

$$y = -\frac{1}{2Z}\frac{dy}{dZ}. \tag{A8}$$

Separation of variables yields Equation (A9),

$$-2ZdZ = \frac{dy}{y}, \tag{A9}$$

which, after indefinite integration, yields Equation (A10),

$$-Z^2 = \ln y - \ln A, \tag{A10}$$

where $-\ln A$ is the integration constant. Equation (A10) may be rewritten as Equation (A11):

$$\frac{dN_B}{dZ} = A\exp(-Z^2) \tag{A11}$$

This general solution may now be applied to the specific case described by Figure A1(b) and integrated with the following conditions: (a) initial condition, $N_B = N_B^o$ at $t = 0$ (i.e., $Z = \infty$) for $x > 0$; (b) boundary condition, $N_B = N_B^{(S)}$ at $x = 0$ (i.e., $Z = 0$) for $t > 0$. This results in Equation (A12):

$$\int_{N_B^{(S)}}^{N_B^o} dN_B = A\int_0^\infty \exp(-Z^2)dZ. \tag{A12}$$

The integral on the right-hand side of Equation (A12) is a standard integral with the value $\frac{\sqrt{\pi}}{2}$ which identifies the constant A as expressed in Equation (A13):

$$A = \frac{2}{\sqrt{\pi}}\left(N_B^o - N_B^{(S)}\right). \tag{A13}$$

Substitution of Equation (A13) into Equation (A11) allows integration of Equation (A11) to solve for the concentration profile at any time $t > 0$:

$$\int_{N_B^o}^{N_B} dN_B = \frac{2\left(N_B^o - N_B^{(S)}\right)}{\sqrt{\pi}} \int_{\infty}^{Z} \exp(-Z^2) dZ. \tag{A14}$$

Integrating the left-hand side and reversing the limits on the right results in Equation (A15),

$$N_B - N_B^o = -\frac{2\left(N_B^o - N_B^{(S)}\right)}{\sqrt{\pi}} \int_{Z}^{\infty} \exp(-Z^2) dZ$$

$$= -\left(N_B^o - N_B^{(S)}\right) \mathrm{erfc}(Z), \tag{A15}$$

where $\mathrm{erfc}(Z)$ is the complementary error function of Z. Reversing the order of terms to remove the negative sign yields Equation (A16):

$$\frac{N_B - N_B^o}{N_B^S - N_B^o} = \mathrm{erfc}(Z). \tag{A16}$$

This may be rearranged to Equations (A17) or (A18),

$$\frac{N_B - N_B^{(S)}}{N_B^o - N_B^{(S)}} = 1 - \frac{N_B - N_B^o}{N_B^{(S)} - N_B^o} = 1 - \mathrm{erfc}(Z); \tag{A17}$$

$$\frac{N_B - N_B^{(S)}}{N_B^o - N_B^{(S)}} = \mathrm{erf}(Z). \tag{A18}$$

where $\mathrm{erf}(Z)$ is the error function of Z which may be found in standard mathematical tables. Thus, the particular solution for the conditions describing Figure A1 is given in Equation (A19):

$$N_B = N_B^{(S)} + \left(N_B^o - N_B^{(S)}\right)\mathrm{erf}\left(\frac{x}{2\sqrt{\tilde{D}t}}\right). \tag{A19}$$

This is a specific case of the general solution to Fick's second law for diffusion into or out of a semi-infinite solid, which may be written as in Equation (A20),

$$N_B = A_1 + B_1 \,\mathrm{erf}\left(\frac{x}{2\sqrt{Dt}}\right) \tag{A20}$$

where A_1 and B_1 are constants which are determined by the initial and boundary conditions on the particular problem.

Reference

1. D. R. Gaskell, *An Introduction to Transport Phenomena in Materials Engineering*, New York, Macmillan Publishing Co., 1992, ch. 10.

Appendix B

Rigorous derivation of the kinetics of internal oxidation

The following is a treatment of the kinetics of internal oxidation of planar specimens based on an analysis by Wagner.[1] This treatment is more general than that presented in the text in that it considers counter-diffusion of solute.

Consider the case of atomic oxygen diffusing into the specimen from the surface ($x = 0$) in the positive x-direction and combining with the outward-diffusing solute at $x = X$ to form BO_ν precipitates. The corresponding concentration profiles are presented in Figure B1. It is assumed that the oxide is so stable that its solubility product in the alloy is negligible, i.e., both N_B and N_O go to zero at the internal oxidation front. Since the internal-oxidation-front penetration will be controlled by oxygen diffusion into the alloy let X be expressed as in Equation (B1),

$$X = 2\gamma\sqrt{D_O t} \tag{B1}$$

where γ is an undetermined factor giving the proportionality between X and the characteristic diffusion length $\sqrt{D_O t}$. Fick's second law for the oxygen diffusion is given in Equation (B2),

$$\frac{\partial N_O}{\partial t} = D_O \frac{\partial^2 N_O}{\partial x^2} \tag{B2}$$

which will be subject to following conditions. (a) Initial conditions:

$$t = 0; N_O = N_O^{(S)}, \text{ for } x = 0; N_O = 0, \text{ for } x > 0.$$

(b) Boundary conditions:

$$t = t, N_O = N_O^{(S)}, \text{ for } x = 0; N_O = 0, \text{ for } x = X.$$

The general solution to Equation (B2), from Equation (A20), is shown in Equation (B3).

$$N_O = A_1 + B_1 \text{erf}\left(\frac{x}{2\sqrt{D_O t}}\right) \tag{B3}$$

327

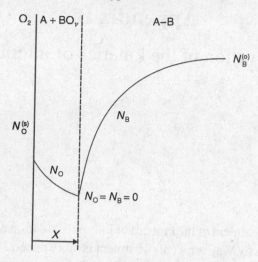

Figure B1 Concentration profiles for internal oxidation of A–B.

Solution for A_1 and B_1 using the boundary conditions yields Equation (B4):

$$N_O = N_O^{(S)} \left[1 - \frac{\text{erf}\left(\dfrac{x}{2\sqrt{D_O t}}\right)}{\text{erf}(\gamma)} \right] \tag{B4}$$

Fick's second law for the solute diffusion is given in Equation (B5),

$$\frac{\partial N_B}{\partial t} = D_B \frac{\partial^2 N_B}{\partial x^2} \tag{B5}$$

subject to the following conditions. (a) Initial conditions:

$$t = 0, \ N_B = 0, \text{ for } x < 0; \ N_B = N_B^{(o)}, \text{ for } x > 0.$$

(b) Boundary conditions:

$$t = t, \ N_B = 0, \text{ for } x = X; \ N_B = N_B^{(o)}, \text{ for } x = \infty.$$

The general solution to Equation (B5) is given in Equation (B6):

$$N_B = A_2 + B_2 \text{erf}\left(\frac{x}{2\sqrt{D_B t}}\right). \tag{B6}$$

Solution for A_2 and B_2 using the boundary conditions yields Equation (B7),

$$N_B = N_B^{(o)} - \frac{N_B^{(o)}\left[1 - \text{erf}\left(\dfrac{x}{2\sqrt{D_B t}}\right)\right]}{\text{erfc}(\Theta^{1/2}\gamma)}, \tag{B7}$$

where $\Theta \equiv D_O/D_B$. The internal oxidation kinetics (i.e., γ) are obtained from writing a flux balance at $x = X$:

$$\lim_{\varepsilon \to 0}\left(-D_O \frac{\partial N_O}{\partial x}\right)_{x=X-\varepsilon} = v\left(D_B \frac{\partial N_B}{\partial x}\right)_{x=X+\varepsilon}. \tag{B8}$$

Differentiating Equation (B4) yields Equation (B9a) or (B9b):

$$\left(\frac{\partial N_O}{\partial x}\right)_{x=X} = \frac{N_O^{(S)}\cancel{2}}{\mathrm{erf}(\gamma)\sqrt{\pi}}\frac{1}{\cancel{2}\sqrt{D_O t}}\exp\left(\frac{-X^2}{4D_O t}\right); \tag{B9a}$$

$$\left(\frac{\partial N_O}{\partial x}\right)_{x=X} = \frac{N_O^{(S)}}{\mathrm{erf}(\gamma)}\frac{1}{\sqrt{\pi}}\frac{1}{\sqrt{D_O t}}\exp\left(-\gamma^2\right). \tag{B9b}$$

Differentiating Equation (B7) yields Equations (B10a) or (B10b):

$$\left(\frac{\partial N_B}{\partial x}\right)_{x=X} = \frac{N_B^{(o)}}{\mathrm{erfc}(\Theta^{1/2}\gamma)}\frac{\cancel{2}}{\sqrt{\pi}}\frac{1}{\cancel{2}\sqrt{D_B t}}\exp\left(\frac{-X^2}{4D_B t}\right); \tag{B10a}$$

$$\left(\frac{\partial N_B}{\partial x}\right)_{x=X} = \frac{N_B^{(o)}}{\mathrm{erfc}(\Theta^{1/2}\gamma)}\frac{1}{\sqrt{\pi}}\frac{1}{\sqrt{D_B t}}\exp(-\Theta\gamma^2). \tag{B10b}$$

Substitution of Equations (B9b) and (B10b) into the flux balance, Equation (B8), yields Equation (B11a),

$$-D_O \frac{N_O^{(S)}}{\mathrm{erf}(\gamma)\sqrt{\cancel{\pi}}\sqrt{D_O t}}\exp(-\gamma^2)$$

$$= vD_B \frac{N_B^{(o)}}{\mathrm{erfc}(\Theta^{1/2}\gamma\sqrt{\cancel{\pi}})\sqrt{D_B t}}\exp(-\Theta\gamma^2). \tag{B11a}$$

which may be rearranged to Equation (B11b) or (B11c):

$$\frac{N_O^{(S)}}{N_B^{(o)}} = \frac{vD_B}{D_O}\frac{\mathrm{erf}(\gamma)}{\mathrm{erfc}(\Theta^{1/2}\gamma)}\frac{\exp(-\Theta\gamma^2)}{\exp(-\gamma^2)}\left(\frac{D_O}{D_B}\right)^{1/2}, \tag{B11b}$$

$$\frac{N_O^{(S)}}{N_B^{(o)}} = \frac{v}{\Theta^{1/2}}\frac{\mathrm{erf}(\gamma)}{\mathrm{erfc}(\Theta^{1/2}\gamma)}\frac{\exp(\gamma^2)}{\exp(\Theta\gamma^2)}. \tag{B11c}$$

Equation (B11c) may be solved numerically for any condition but can be solved analytically for two important limiting cases.

Case 1: *Negligible counterdiffusion of solute*

A situation is often encountered in which solute diffusion is very slow relative to that of oxygen, and the oxygen solubility is small relative to the solute content of the alloy, i.e. $\dfrac{D_B}{D_O} \ll \dfrac{N_O^{(S)}}{N_B^o} \ll 1$, for which $\gamma \ll 1$ and $\gamma \Theta^{1/2} \gg 1$. This allows the simplifications of terms in Equation (B11c), shown in Equations (B12)–(B14):

$$\mathrm{erf}(\gamma) = \frac{2}{\sqrt{\pi}}\gamma, \tag{B12}$$

$$\exp(\gamma^2) = 1 + \gamma^2 \approx 1, \tag{B13}$$

$$\mathrm{erfc}\left(\gamma \Theta^{1/2}\right) = \frac{2}{\sqrt{\pi}} \frac{\exp\left(-\gamma^2 \Theta\right)}{2\gamma \Theta^{1/2}}. \tag{B14}$$

Insertion of Equations (B12–B14) in Equation (B11) yields Equation (B15a),

$$\frac{N_O^{(S)}}{N_B^{(0)}} = \frac{\nu}{\Theta^{1/2}} \frac{\left(\frac{2}{\sqrt{\pi}}\gamma\right)(1 + \gamma^2)}{\frac{2}{\sqrt{\pi}} \frac{\exp(-\gamma^2\Theta)}{2\gamma\,\Theta^{1/2}} \exp(\Theta\gamma^2)}, \tag{B15a}$$

which simplifies to Equation (B15b) or (B15c):

$$\frac{N_O^{(S)}}{N_B^{(0)}} = 2\nu\gamma^2; \tag{B15b}$$

$$\gamma = \left(\frac{N_O^{(S)}}{2\nu\,N_B^{(0)}}\right)^{1/2}. \tag{B15c}$$

Substitution of Equation (B15c) into Equation (B1) gives the front penetration at a given time as shown in Equation (B16):

$$X = \left[\frac{2N_O^{(S)}D_O}{\nu\,N_B^{(0)}}t\right]^{1/2}. \tag{B16}$$

This is the same result obtained using the quasi-steady-state approximation Equation (5.9).

Case 2:   *Significant counterdiffusion of solute*

For cases where the solute counterdiffusion becomes important, the approximations $\gamma \ll 1$ and $\gamma \Theta^{1/2} \ll 1$ become valid, and we obtain the expressions given in Equations (B17)–(B19):

$$\mathrm{erf}(\gamma) = \frac{2}{\sqrt{\pi}}\gamma, \tag{B17}$$

$$\mathrm{erfc}(\gamma \Theta^{1/2}) = 1 - \frac{2}{\sqrt{\pi}}\gamma \Theta^{1/2} \approx 1, \tag{B18}$$

$$\exp(\gamma^2) \approx \exp(\gamma^2\Theta) \approx 1. \tag{B19}$$

Substitution of these equations into Equation (B11c) yields Equations (B20a) and (B20b):

$$\frac{N_O^{(S)}}{N_B^{(0)}} = \frac{\nu}{\Theta^{1/2}} \frac{\frac{2}{\sqrt{\pi}}\gamma}{1} \left(\frac{1}{1}\right),$$ (B20a)

$$\gamma = \frac{\sqrt{\pi}\Theta^{1/2}N_O^{(S)}}{2\nu N_B^{(0)}}.$$ (B20b)

A more general treatment of the internal oxidation kinetics has been presented by Böhm and Kahlweit[2] who modified Wagner's analysis to include a finite solubility product of the oxide in the alloy matrix. The key differences in this case are that N_B and N_O do not go to zero at the reaction front and the concentration of oxygen maintains finite values into the alloy ahead of the front. The case of oxides with large solubility products has been treated by Laflamme and Morral.[3]

Morral[4] has criticized the Wagner analysis and argued that concentration profiles such as that shown in Figure B1 violate principles of local equilibrium and that solute enrichment in the zone of internal oxidation is not possible for the case of zero solubility product. Analysis of this issue is beyond the scope of this book but it should be remarked that many experimental observations are consistent with conclusions based on Wagner's model.

References

1. C. Wagner, *Z. Elektrochem.*, **63** (1959), 772.
2. G. Böhm and M. Kahlweit, *Acta met.*, **12** (1964), 641.
3. G. R. Laflamme and J. E. Morral, *Acta met.*, **26** (1978), 1791.
4. J. E. Morral, *Mater. High Temp.*, **20** (2003), 275.

Appendix C

Effects of impurities on oxide defect structures

This aspect of the theory of defect structures of non-stoichiometric compounds is usually covered in the main text of books on high-temperature oxidation. The subject of doping is interesting for its own sake, and it is vitally important for the study of the physical chemistry and electrochemistry of ionic compounds. In the case of an introduction to high-temperature oxidation our opinion is that, since the control of oxidation rates by controlling the ionic and electronic transport properties of oxides by impurity solution is not generally used as a technique for the development of oxidation-resistant alloys, this subject should be dealt with in an appendix. This allows it to be covered adequately without over-emphasizing its importance.

In the following discussion ZnO will be used as a typical n-type oxide and NiO as a typical p-type oxide.

Negative (n-type) oxides

The native defect structure of an oxide involving excess cations on interstitial sites with electrons in the conduction band may be represented as in Equations (C1) and (C2):

$$ZnO = Zn_i + e' + \frac{1}{2}O_2\,(g), \tag{C1}$$

$$ZnO = Zn_i^{\cdot} + 2e' + \frac{1}{2}O_2\,(g). \tag{C2}$$

To represent the addition of a more positive cation to the ZnO lattice, consider Al_2O_3 as the dopant oxide. The solution of Al_2O_3 in ZnO may occur in two ways.

(a) The Al^{3+} ions occupy normal Zn^{2+}-ion lattice sites. Since only two corresponding anion lattice sites are available for the three oxide ions, one must be discharged releasing

332

oxygen to the atmosphere and putting two electrons in the conduction band according to Equation (C3):

$$Al_2O_3 = 2Al^{\cdot}_{Zn} + 2e' + 2O^X_O + \frac{1}{2}O_2 \,(g). \tag{C3}$$

(b) The introduction of extra electrons to the conduction band upsets the equilibrium between them and interstitial zinc ions according to Equations (C1) and (C2). Thus, Al_2O_3 also dissolves in a manner allowing some interstitial zinc ions to be eliminated according to Equations (C4) or (C5):

$$Al_2O_3 + Zn^{\cdot\cdot}_i = 2Al^{\cdot}_{Zn} + 3O^X_O + Zn^X_{Zn}, \tag{C4}$$

$$Al_2O_3 + Zn^{\cdot}_i = 2Al^{\cdot}_{Zn} + e' + 3O^X_O + Zn^X_{Zn}. \tag{C5}$$

The overall result of doping ZnO with Al_2O_3, i.e., the oxide of a higher cationic charge, is to decrease the concentration of cation interstitials and to increase the concentration of conduction-band electrons, thus reducing the cationic conductivity and increasing the electronic conductivity. Such an effect would decrease the oxide growth rate on an alloy on which such an oxide could form.

To represent the addition of a less positive ion onto the ZnO lattice consider the solution of lithium oxide, Li_2O. This may also occur in two ways.

(a) The two Li^+ ions occupy normal Zn^{2+} cationic sites but only one anion site is occupied. The second anion site is filled by taking O_2 from the atmosphere and withdrawing electrons from the conduction band in order to ionize it:

$$Li_2O + 2e' + \frac{1}{2}O_2 \,(g) = 2Li'_{Zn} + 2O^X_O. \tag{C6}$$

(b) Since the removal of conduction-band electrons upsets the equilibrium conditions of Equations (C1) and (C2), an accompanying mechanism would be for two Li^+ ions to displace a Zn^{2+} ion from an existing cation site and for oxygen to be evolved according to Equations (C7a) or (C7b):

$$Li_2O + 2Zn^X_{Zn} = 2Li'_{Zn} + 2Zn^{\cdot\cdot}_i + \frac{1}{2}O_2 \,(g), \tag{C7a}$$

$$Li_2O + 2Zn^X_{Zn} = 2Li'_{Zn} + O^X_O + 2Zn^{\cdot\cdot}_i. \tag{C7b}$$

This occurs together with the mechanism of Equation (C6) so as to maintain the equilibrium between conduction-band electrons and interstitial zinc ions.

The net result is to increase the concentration of interstitial zinc ions and reduce the concentration of conduction-band electrons, thus increasing the cationic conductivity and reducing the electronic conductivity. Such doping should, therefore, lead to increased oxidation rates.

Similar doping reactions can also occur when the n-type conduction behaviour arises through the existence of anion vacancies such as Equation (C8):

$$MO = V^{\cdot\cdot}_O + 2e' + \frac{1}{2}O_2 \,(g) + M^X_M. \tag{C8}$$

The corresponding doping reactions with Al_2O_3 would be those shown in Equations (C9) and (C10):

$$Al_2O_3 + V_{\ddot{O}} = 2Al_M^{\cdot} + 3O_O^X, \tag{C9}$$

$$Al_2O_3 = 2Al_M^{\cdot} + 2e' + 2O_O^X + \frac{1}{2}O_2\,(g). \tag{C10}$$

The doping reactions with Li_2O would be those shown in Equations (C11) and (C12):

$$Li_2O + 2e' + \frac{1}{2}O_2\,(g) = 2Li_M' + 2O_O^X, \tag{C11}$$

$$Li_2O = 2Li_M' + V_{\ddot{O}} + O_O^X. \tag{C12}$$

Positive (p-type) oxides

The intrinsic defect structure involving cation vacancies and electron holes can be expressed, using NiO as an example, as in Equation (C13):

$$\frac{1}{2}O_2\,(g) = O_O^X + V_{Ni}'' + 2h^{\cdot}. \tag{C13}$$

The consequences of dissolving cations of higher (e.g., Al^{3+}) and lower (e.g., Li^+) valence than nickel may be considered similarly to the cases of n-type oxides as follows.

Dissolution of Al_2O_3 may be achieved in two ways which occur together to preserve the equilibrium of Equation (C13).

(a) Two Al^{3+} ions occupy normal nickel sites and extra oxygen from Al_2O_3 is evolved as oxygen gas and contributes two electrons, which neutralize two electron holes:

$$Al_2O_3 + 2h^{\cdot} = 2Al_{Ni}^{\cdot} + 2O_O^X + \frac{1}{2}O_2\,(g). \tag{C14}$$

(b) The two Al^{3+} ions occupy normal nickel-ion sites and the three oxide-ions can occupy normal oxide-ion sites, thus creating a nickel-ion vacancy, Equation (C15):

$$Al_2O_3 = 2Al_{Ni}^{\cdot} + V_{Ni}'' + 3O_O^X. \tag{C15}$$

Thus the dissolution of higher-valence cations into a cation-deficit p-type oxide, such as NiO, leads to creation of more cation vacancies and reduction of the concentration of electron holes, thus increasing the cation conductivity and decreasing the electronic conductivity. The net effect would be to increase the oxidation rate, at least in the case where lattice diffusion controls the rate.

The effects of dissolving a cation of lower valence follow similar lines, e.g., using Li_2O one has the following possibilities.

(a) The two Li$^+$ ions occupy normal nickel sites, an oxygen atom is ionized from the gas phase to fill the second anion site, and two electron holes are created, Equation (C16):

$$\text{Li}_2\text{O} + \frac{1}{2}\text{O}_2\,(\text{g}) = 2\text{Li}'_{\text{Ni}} + 2\text{h}^\cdot + 2\text{O}_\text{O}^\text{X}. \tag{C16}$$

(b) One of the two Li$^+$ ions occupies a nickel vacancy, Equation (C17):

$$\text{Li}_2\text{O} + \text{V}''_{\text{Ni}} = 2\text{Li}'_{\text{Ni}} + \text{O}_\text{O}^\text{X}. \tag{C17}$$

The net result is that nickel vacancies are consumed and electron holes are produced to maintain equilibrium in Equation (C13). Thus the cation conductivity is reduced and electronic conductivity is increased. A corresponding reduction in oxidation rate would be expected.

It is also possible to produce p-type behaviour by having interstitial excess anions according to Equation (C18),

$$\frac{1}{2}\text{O}_2(\text{g}) = \text{O}''_\text{i} + 2\text{h}, \tag{C18}$$

although the number of examples are limited because of the large strain energy associated with moving a large oxide ion into an interstitial position.

The corresponding dissolution mechanism for Al$_2$O$_3$ and Li$_2$O in a p-type oxide MO with excess anions would be as follows, in order to maintain equilibrium in Equation (C18).

(a) The Al$_2$O$_3$ dissolution could proceed as shown in Equations (C19) and (C20):

$$\text{Al}_2\text{O}_3 = 2\text{Al}'_\text{M} + 2\text{O}_\text{O}^\text{X} + \text{O}''_\text{i}, \tag{C19}$$

$$\text{Al}_2\text{O}_3 + 2\text{h}^\cdot = 2\text{Al}'_\text{M} + 2\text{O}_\text{O}^\text{X} + \frac{1}{2}\text{O}_2\,(\text{g}). \tag{C20}$$

This will result in increased oxide-ion interstitials and higher ionic conductivity, together with lower concentrations of electron holes and lower electronic conductivity.

(b) The Li$_2$O dissolution could proceed as in Equations (C21) and (C22):

$$\text{Li}_2\text{O} + \text{O}''_\text{i} = 2\text{Li}'_\text{M} + 2\text{O}_\text{O}^\text{X}, \tag{C21}$$

$$\text{Li}_2\text{O} + \frac{1}{2}\text{O}_2\,(\text{g}) = 2\text{Li}'_\text{M} + 2\text{h}^\cdot + 2\text{O}_\text{O}^\text{X}. \tag{C22}$$

This will produce lower interstitial anion concentrations with correspondingly reduced ionic conductivity, together with increased electron-hole concentration and associated increased electronic conductivity.

Index